30

DATE			

Statistical mechanics is one of the crucial fundamental theories of physics. Lawrence Sklar, one of the preeminent philosophers of physics, offers a comprehensive, non-technical introduction to the theory and to historical attempts to understand its foundational elements.

Among the topics covered in depth are probability and statistical explanation, basic issues in equilibrium and non-equilibrium statistical mechanics, the role of cosmology, the reduction of thermodynamics to statistical mechanics, and the alleged foundation of the notion of time-asymmetry in the entropic asymmetry of systems in time.

The book emphasizes the interaction of scientific and philosophical modes of reasoning. In this way, it will interest philosophers of science and physicists and chemists concerned with philosophical questions. It will also be read by general readers interested in the foundations of modern science.

Physics and chance

Physics and chance

Philosophical issues in the foundations of statistical mechanics

LAWRENCE SKLAR

University of Michigan, Ann Arbor

CAMBRIDGE
UNIVERSITY PRESS

Published by the Press Syndicate of the University of Cambridge
The Pitt Building, Trumpington Street, Cambridge CB2 1RP
40 West 20th Street, New York, NY 10011–4211, USA
10 Stamford Road, Oakleigh, Victoria 3166, Australia

First published 1993

Printed in the United States of America

Library of Congress Cataloging-in-Publication Data
Sklar, Lawrence.
Physics and chance : philosophical issues in the foundations of
statistical mechanics / Lawrence Sklar.
p. cm.
Includes bibliographical references and index.
ISBN 0–521–44055–6
1. Statistical mechanics.   2. Physics – Philosophy.   I. Title.
QC174.8.S55   1993   92–46215
530.1′3 – dc20   CIP

A catalog record for this book is available from the British Library.

ISBN 0–521–44055–6 hardback

To the memory of
my father

Contents

Preface

The aim of this work is to continue the exploration into the foundational questions on the physical theory that underpins our general theory of thermodynamics and that, for the first time, introduced probabilistic considerations into the fundamentals of our physical description of the world. That physical theory is statistical mechanics.

The history of this foundational quest is a long one. It begins with an intense examination of the premises of the theory at the hands of James Clerk Maxwell, Ludwig Boltzmann, and their brilliant critics. It continues in the work of Hans Reichenbach, who introduced their deep questions to the philosophical community. And the quest has persisted as a set of difficult conceptual challenges, in a study made ever richer by the development of ever more sophisticated technical resources with which to treat the problems. I hope that this book will encourage others in the philosophical community to join with those in physics who continue to look for the elusive resolutions of the puzzles.

I have benefited enormously from discussions with John Earman, who has often guided me to the crucial questions to be addressed. James Joyce and Robert Batterman, as students and as colleagues, have been enormously helpful to me in my thinking about these issues. I have learned much also from David Malament, Sandy Zabell, Paul Horwich, Frank Artzenius, and Abner Shimony.

Lou Case and Michele Vaidic were of great help in preparing the manuscript of this book. I am also grateful to Terence Moore of Cambridge University Press, to the two referees for the Press for their help in bringing the book to publication, and to Ronald Cohen for his editorial work on the manuscript.

The research contained in this book has been supported by a number of grants from the National Science Foundation, whose support is gratefully acknowledged here.

1

Introduction

I. Philosophy and the foundations of physics

There are four fundamental theories that constitute, at present, the foundational pillars of our physical theory of the world: general relativity, quantum mechanics, the theory of elementary particles, and statistical mechanics. Physics is of course far from a finished discipline, and each of these fundamental theories presents its own budget of scientific and philosophical problems. But the kinds of problems faced by those who would examine the so-called foundational issues in these areas vary in a marked and interesting way from theory to theory.

General relativity – at present the most plausible theory of the structure of space-time, a domain of inquiry that, since Einstein, is usually taken to include the theory of gravitation – is in many ways the most fully worked out of the theories. Many fascinating scientific questions remain: Should we accept general relativity, or some alternative to it like the Dicke–Brans scalar-tensor theory? Are there generalizations of the theory that might encompass other forms of interaction over and above gravitation? Should the allowable worlds be restricted by some conditions of causal "niceness," for example? But we are, at least, clear about what the theory itself amounts to. Scientifically, the most dubious aspect of the theory is its totally classical nature, and all expect that some day we will have a new quantized theory to take its place. But at least we know what theory we are talking about when we begin to explore the foundations of our space-time theory.

Philosophically, too, the situation is particularly clean in the case of space-time theories. Although dispute rages among philosophers and philosophically inclined physicists over what the right answers are to foundational questions in the metaphysics and epistemology of space-time theories, there is at least general agreement about what the issues are. Subjects such as the epistemology of geometric theories, conventionalism vs. empiricism, substantivalism vs. relationism, the status of the "definitions" in special relativity, the causal theory of space-time topology, and so on have been discussed long enough and hard enough by physicists and philosophers so that even if we don't agree that we have the answers, at least there is a general consensus about the right questions to ask.

Quantum mechanics is, again, a theory whose *scientific* claims are at least fairly clear. The basic structure of the fundamental theory is available to us in the refined and elegant versions of von Neumann, et al. Once again, of course, there does come a point where we aren't yet sure what the theory should be. Just as general relativity runs into the fundamental problem that it is not a quantized theory, leaving us to speculate about just what our ultimate quantized space-time theory will be, so quantum mechanics runs into difficulties when one tries to extend it to cover relativistic aspects of systems with an infinite number of degrees of freedom. We still await the final formulation of a quantum theory of fields, whether in the vein of Wightmannian axiomatic theories, theories of scattering of the LSZ type, S-matrix theories, or something else.

Philosophically, quantum mechanics, like general relativity, presents us with a now well-defined set of foundational questions to be answered. In particular we must understand how to reconcile the measurement process with dynamic evolution, given that it is treated so very differently in the orthodox theory. This understanding will require the resolution of such subsidiary problems as wave-particle duality, the non-causal interdependence of separated systems, the nature of uncertainty relations, the existence or non-existence of hidden variables, and so on.

Here the situation is interestingly different from that facing us in the philosophy of space-time theories. In both cases we have a good idea what the questions are that need to be answered. In the latter case we at least know quite definitively what some possible answers are, even if we aren't sure which are right and which are wrong. In the quantum case I think many will agree that most (perhaps all) of the answers themselves are insufficiently clear in their formulation to begin to decide which answer is correct. Before we are to decide who is right, we first must understand much more clearly what they are saying.

The theory of elementary particles again has its own special flavor as an area for foundational research. Here we are concerned with the age-old question of the fundamental constitution of matter. This theory presents a comparatively phenomenological aspect compared with the two mentioned earlier. What we have are enormous quantities of data to be assimilated. Of course, quantum theory provides severe constraints upon what our theory of matter must look like. But we also need much additional structure. This is provided in part by a hierarchical theory of composition: Matter is made up of molecules, which are composed of atoms, which are composed of "elementary" particles, which are, perhaps, composed of quarks. Atoms are bound together by electromagnetic forces, their nuclei by strong and weak interactions, and some of the "elementary" particles by, perhaps, gluon forces. But just *why* there are the structures and forces there are, and just *what* fundamental laws govern the composition of macro-objects out of their micro-constituents remains something of a mystery to us, despite the recent progress of theories such as gauge field theory and supersymmetry.

Not surprisingly, the philosophical issues at the foundations of this theory also remain less well characterized and far less well explored than those concerned with space-time theories and quantum mechanics. Although some attention has been paid to such questions as why explanation so frequently consists in reducing macro-objects to their microscopic constituents, and as to whether this process should be viewed as proceeding to ever more microscopic levels, or, instead coming to a halt in some final "self-sustaining" level, little attention of a systematic and rigorous sort has been paid by the philosophical community to the foundational issues to which even our present, only partially formulated, theory of the constitution of matter gives rise.

Finally, we come to the fascinating pair of theories to which this book is devoted: thermodynamics and statistical mechanics. Originating as theories devoted solely to accounting for the thermal behavior of macroscopic matter, these branches of physics now stand beside the other three as fundamental components of our scientific world-view. Independent of the other theories, they are an indispensable supplement to them. Just how independent is a matter of some controversy that we will discuss in detail. Without invoking these theories we cannot begin to move from the fundamental theory governing the behavior of the microscopic constituents of matter and the theory of the constitution of macroscopic objects out of their microscopic parts to a full explanation of the macroscopic behavior in terms of the microscopic constitution and its laws. Thermal phenomena, magnetic phenomena, the structure of matter in all its phases of solid, liquid, gas, and plasma, the spectral distribution of radiation and the laws of particle scattering are among the various aspects of the world whose description and explanation requires the application of thermodynamics and statistical mechanics.

Originally formulated to provide the general laws governing the thermal properties of matter and the interplay between heat and mechanical work, thermodynamics evolved to become a far more general theory of extraordinarily wide scope. The flow of heat in matter, driven by temperature gradients; the interchange of heat and mechanical work in heat engines; the distribution of energy between matter and radiation; the response of the internal magnetization of matter to imposed magnetic fields all may be systematized under the same general concepts of equilibrium, temperature, and entropy, concepts originally invented to describe the thermal phenomena in the narrower sense. Aspects of the world ranging from the distribution of elementary particles upon interaction to the distribution of stars interacting by gravity in massive clusters can all be subjected to the thermodynamic viewpoint.

 Thermodynamics has a peculiar place when viewed in the context of the panoply of fundamental theories that physics uses to describe the world. Special relativity provides the space-time framework in which all physical phenomena are to be described. This framework is modified and generalized by general relativity in the curious revolutionary shift due to Einstein which makes the spatio-temporal arena of phenomena a dynamic and causal participant in events and assimilates the gravitational interaction to the space-time structure itself. Quantum theory provides a radical and novel kinematics, relative to which all phenomena must be described. It provides the basic language in which states of systems are to be described, the rules for the dynamical evolution of those states when subject to causal influence, and the rule, which, if philosophically mysterious and infuriatingly hard to truly understand, provides the definitive principle for moving from states described in a quantum mechanical way to probabilities of outcomes described in the older, non-quantum terms. In addition to these two fundamental theories (theories not yet properly reconciled to each other) we need the field-theoretic principles that describe the specific nature of material existents in their most primitive and elementary form – that is, the field theory of the elementary particles.

What is surprising is that there is any place at all in this picture for a discipline such as thermodynamics. If we consider the dynamic evolution of the world as it would be characterized by the theories noted, there seems little place for additional fundamental lawlike structure. We can describe any system at a time by its quantum state, and project its evolution into the future using the general picture of the space-time arena of dynamics, the overall kinematic dynamical principles provided by the general quantum theory, and the specific rules of dynamical interaction provided by the field theory of the elementary particles. What else is there to do? That there is another, fundamental, level of description that is vital and fruitful – a level involving such concepts as equilibrium, irreversibility, entropy, and temperature – and that this mode of description and its laws are essential and useful over a wide range of phenomena of apparently radically different fundamental natures, is the surprising truth of thermodynamics.

Many no longer think of thermodynamics as an autonomous science, however. Since the middle of the nineteenth century there has been a continual program in physics designed to show us that thermodynamics itself can be reduced to, and explained in terms of, a deeper understanding of the world. Beginning with early clever guesses that heat was a form of internal motion of microscopic parts of matter, progressing through the development of the kinetic theory of gases in the nineteenth century, and evolving into the complex and multi-faceted discipline of statistical mechanics, there has been a continuous program designed to show us that the familiar thermodynamic results are consequences of a broader and more profound structure of the world.

Part of the motivation of at least some of the discoverers and developers of statistical mechanics has been the hope that the deeper understanding of thermodynamic phenomena provided by this new approach would eliminate the need to invoke thermodynamic-like principles of the world as independent laws. Strenuous efforts have been made to try to show us that, properly understood, thermodynamic features of the world are reducible to, and intelligible in terms of, the features posited by the other fundamental theories of the world and the lawlike nature of those features as these other fundamental theories describe them. This hope for the elimination of thermodynamic principles as autonomous elements in our picture of the world has, however, been continuously thwarted by the seeming necessity, at ever deeper levels, of invoking some hypotheses that remain underivable from the fundamental laws of kinematics or dynamics. These are now taken to be of a probabilistic or statistical nature.

Just what these seemingly autonomous principles of statistical mechanics are, how they might be used to allow us to derive from them the basic thermodynamical laws, and the degree to which they can or cannot themselves be derived from the fundamental laws of general kinematical and dynamical theories is largely the subject matter of this book.

Unlike the relatively closed disciplines of general relativity and quantum mechanics, statistical mechanics presents us not with a single, well-formulated and easily presented theory, but, rather, with something of a hodgepodge of approaches, formulations, and schools. Although some aspects of the theory are well understood and universally accepted, such crucial areas as the correct approach to the introduction into the theory of irreversibility and the approach to equilibrium and the proper statistical mechanical definition of entropy are the subject of intense and seemingly interminable controversy. One hundred years after the explosive birth of the theory at the hands, primarily, of Maxwell and Boltzmann, we find the same fundamental issues at the foundational level arousing controversy among contemporary theorists as aroused controversy in the early years of the theory. Not that no progress has been made nor that the issues are as unclear now as they were then. But it is still true that there exists no single, coherent, and clear understanding as to just what the theory of statistical mechanics *is* or ought to be. The theory of the approach to the state of equilibrium has itself not settled down to that placid, if somewhat boring, state!

One would expect that this reign of controversy and confusion over fundamentals would attract the vigorous attention of those philosophers concerned with an application of the conceptually clarifying techniques of philosophy to fundamental issues in the physical sciences. Surprisingly, however, statistical mechanics has received far less attention from philosophers than either our theory of space-time or quantum mechanics. Not that foundational issues in statistical mechanics are ignored, of course, for there exists a large group of physicists who devote substantial effort to the task of clarifying this theory at its most fundamental level. But aside from occasional forays into particular problem areas and a very few major attempts at overall clarification in the context of a philosophical attack on the so-called problem of the direction of time, philosophers have generally kept their distance from the conceptual perplexities of statistical mechanics.

I believe that some of this reticence can be accounted for by the very disarray in the science itself. In the theory of space-time one can start immediately with a clear understanding of what the theory says. The conceptual problems then appear in a neat and orderly array. Indeed, it takes little time before the outlines of the possible philosophical options open to one become clear. Philosophical dispute is not instantly resolved, but at least we usually know what the dispute is. In quantum mechanics, again, the fundamental problems are fairly clear. Here, of course, it isn't obvious that any fully internally coherent, much less universally acceptable, solutions are forthcoming. But in the case of statistical mechanics the diversity of approaches and the lack of a single universally accepted account of what the theory is has, I think, discouraged many from even beginning to explore the philosophical ramifications of the fundamental science. Even finding one's way through the essential literature is, as I have discovered to my dismay, an arduous task.

This context fixes to a large degree the basic aim of this book. I hope to bring together in one place a sufficiently elementary, comprehensive, and organized survey of the fundamental physical and philosophical issues in this area to provide an introduction to this under-explored portion of the philosophy of physics for philosophers and scientists who would like to understand what the crucial debates are all about. Further, I hope that this work will be sufficiently comprehensive, and yet sufficiently accessible, that it will encourage some to proceed to the more detailed, technical works of the physicists that are more difficult of access, and will lead to a more extensive application of the talents of conceptual clarification available to philosophers to this confused field, a field desperately in need of all the understanding that can be thrown upon it.

I doubt if I shall here resolve the major conceptual questions to everyone's satisfaction. On some questions I will offer my own appraisal of the answers and my own conclusions about the right direction in which to go. On other questions I will be happy enough to at least get straight what the important questions are and how they depend upon one another. I will consider my ambitions amply fulfilled if this work can stimulate the same application of energy and talent in attacking the fundamental problems I will survey and expound as is presently devoted to the fundamental philosophical questions in space-time theories and quantum mechanics. On these questions, my "answers" will amount to no more than a few tentative suggestions about the kinds of approaches that I can think of and that might bear fruit under a kind of extensive investigation that is not possible for them here.

II. The structure of this book

Let me begin by surveying the range of interconnected problems with which we will be concerned. As is not uncommon in philosophy, these form a rather dense network of issues, so that resolving one question seems to presuppose prior resolution of all the others. To some extent, then, the sorting out of issues and the order in which they will be treated has a degree of arbitrariness to it. Nonetheless, I hope the scheme I am following will present the questions in at least one reasonable order. The exploration of the substantive issues will follow a preliminary outline of the history of thermodynamics and statistical mechanics. This survey of the high points in the development of these fields is essential to introduce to the reader the main problems and their overall context.

1. Probability

Both kinetic theory and statistical mechanics are frequent users of probabilistic or statistical statements. Molecular distributions in gases are sometimes asserted to be "most probable" distributions. Probability distributions over collections of all possible micro-states compatible with given macro-constraints are posited. What is the *meaning* of these probabilistic and statistical assertions?

As we shall see, philosophers have offered us a wide variety of accounts of just what probabilistic or statistical assertions mean in general. These accounts range from simple identifications of probability with relative frequencies of phenomena in the world to accounts that deny any objective or physical content to the assertions at all, reserving probability as

a measure of subjective degree of belief. Some of the conceptual issues
concerning the role of probability in statistical mechanics, we shall see,
hinge on a prior understanding of probability from a general philosophical
point of view. That is, our understanding of just what is being asserted
in the particular theories we are examining will vary with our conception
of the meaning of probabilistic assertions in the general context.

But, we shall see, there are other conceptual problems regarding the
role of probability in the particular context of statistical mechanics that
are rather more specific to the very special function played by probabilities in this particular realm of physics. In other words, even having adopted
some general philosophical stance, there will be additional conceptual
problems in attempting to apply this philosophical interpretation to the
context of statistical mechanics.

I do not expect to resolve to general philosophical satisfaction the
issue of the correct interpretation of probabilistic assertions. I will lay out
briefly what some of the major alternatives are and what some difficulties with them are taken to be. I will opt for one interpretation as most
plausible in understanding the role of probability in statistical mechanics,
allowing that making this choice stick requires some answers to philosophical questions I certainly don't have available. Attention will be paid,
however, to the problems more specific to the role of probability in
statistical mechanics, because these have been rather neglected in the
literature in comparison with the broader, and admittedly more fundamental, general questions of interpretation.

2. Statistical explanation

Scientific theories are supposed to explain what happens in the world.
The features of the world that statistical mechanics sets out to explain
are, as we shall see, curious and varied. The existence of equilibrium
states, the "inevitable" approach of non-equilibrium systems to them, the
existence of a small set of macroscopic parameters sufficient to characterize equilibrium and a somewhat larger set sufficient to characterize
some approaches to equilibrium, the lawlike inter-relationships among
these macroscopic parameters – these are the sorts of things to be accounted for in our theory.

The explanations proffered by the theory make essential use of the
statistical and probabilistic assertions postulated by it. What is the nature
of an explanation that rests upon a statistical or probabilistic assumption?
Again, we have available to us a rather wide variety of accounts in
methodological philosophy of science as to just what statistical explanation amounts to. Here again I will lay out the fundamentals of these
varying philosophical accounts and spend at least some time examining

their virtues and deficiencies in the general methodological context. But once again it will be seen that statistical explanation as it functions in statistical mechanics has its own idiosyncratic features that deserve special attention in their own right. We shall see that many of the perennial debates in the foundations of statistical mechanics rest upon assumptions about just what explanation in this context ought to be. Bringing those assumptions to the surface will, I believe, at least clarify the foundational debates even if not resolving them.

3. The equilibrium problem

Accounting for the equilibrium properties of systems, we invoke, in statistical mechanics, certain probabilistic assumptions. Using them we calculate certain quantities. Associating these quantities with macroscopically measured parameters we offer an explanation of the equilibrium properties of the system.

But what justified our choice of fundamental statistical assumptions? Further, what *explains* their correctness? As we shall see, these may be two quite different questions. Further, why is the property we calculate the appropriate one to choose to associate with the macroscopic quantity?

There is an attempt to offer at least partial answers to some of these questions that relies heavily on certain features characteristic of the laws governing the micro-constituents of the macroscopic system. This is so-called ergodic theory. We shall see that there is a very subtle interplay in the theory between certain features of the system engendered by the laws governing its micro-constituents and other features that are the result, rather, of the vast number of micro-constituents that make up a macrosystem. I shall try, by disentangling a number of distinct questions that tend to get muddled together in the literature, to make clearer just how each result following from a study of the laws governing the micro-constituents can be used in resolving a particular quandary. Basically the line will be that there are a number of quite distinct questions one can ask and that some results are answers to some questions but not to others. Putting this simply, the response is likely to be, "of course *that* is true," but, as we shall see, failure to make the fine distinctions here has led to substantial confusion on foundational issues.

4. The non-equilibrium problem

The study of non-equilibrium in statistical mechanics presents quite a different set of scientific problems than those that arise in the equilibrium case. In fact there is at present no truly unified, coherent theory of non-equilibrium. We shall briefly explore the reasons for this situation and outline some of the varied, related, but quite distinct, proposals that have been made for treating the general non-equilibrium problem.

As in the equilibrium case, predicting and explaining in the case of non-equilibrium calls for the postulating of statistical assertions fundamental to the derivations used. But, as we shall see, the kinds of statistical assertions invoked, the rationale for adopting them, and the proposed explanations offered as to why they are correct differ, sometimes radically, in this case from the equilibrium case. In particular, the relationship of the generalizations to the underlying dynamical laws governing the micro-constituents is quite different in the two cases.

The roles of initial conditions and laws in the explanations, the modes of putting irreversibility into the theory, and the accounts offered to rationalize and to explain the fundamental postulates will be examined.

5. Cosmology and statistical mechanics

Since Boltzmann, it has been frequently alleged that a full understanding of the place statistical mechanics plays in our theoretical structure requires reference to the global, cosmological structure of the universe. Recent work on observational and theoretical cosmology has made the interrelationship of cosmological and statistical mechanical features of the world a vital area of scientific exploration.

The aim of this section will be to concisely present the present state of scientific knowledge and speculation regarding cosmology and its relation to thermal phenomena, and then to examine the arguments offered philosophically. Once again, what I hope to show is that there are a number of quite distinct questions to be answered, questions that can easily be confused with one another. When these are disentangled from one another, the earlier exposition of statistical explanation will allow us to clarify the extent to which cosmology does and does not resolve the deep perplexities still remaining in statistical mechanics, such as those concerning the origin of irreversibility.

6. The reduction of thermodynamics to statistical mechanics

The original goal of statistical mechanics was to account for that range of macroscopically observable phenomena whose behavior is accounted for in the theory of thermodynamics. The alleged relationship of thermodynamics to statistical mechanics, then, is that of one theory that has been reduced to another. We will examine a number of standard philosophical accounts of the reductive relationship and seek to place the reduction of thermodynamics to statistical mechanics in these philosophically developed frameworks. As in the consideration of probability and statistical explanation, emphasis will not be placed on resolving still outstanding general methodological issues. Indeed, in some cases there will have to be a simple dogmatic adoption of one plausible position among others. Rather, I want to focus on the problems that are special to this particular reduction, maintaining, as I will, that once again the general has been emphasized to the detriment of an adequate treatment of the particular. We shall see that even in its own right thermodynamics has very special features vis-à-vis other physical theories. Its descriptive concepts have peculiarities not shared in general by theoretical concepts of physics. And certainly this will be true of the concepts of statistical mechanics as well. Naturally, then, the interrelation of these two theories will require kinds of interconnections unfamiliar in other, more commonly examined, cases of reduction.

7. The direction of time

Most of the attention that has been paid by philosophers to the foundational questions in statistical mechanics has not been directed primarily to the conceptual problems in this area for their own intrinsic interest. The issue has been explored rather as a means of providing resources useful in discussing the so-called problem of the direction of time. Here, the resources of statistical mechanics have been invoked in order to offer a reductionistic account of the very notion of past-future asymmetry itself. The idea has been that in some sense or another, it is the asymmetry of the world in entropic aspects, described by phenomenological thermodynamics and, allegedly, explained by statistical mechanics, that grounds our very notion of the distinction between past and future. The idea here is present in Boltzmann's work, and it has received its most detailed exegesis by Reichenbach.

I shall examine in some detail in just what sense it is plausible to even suggest that the notion of temporal asymmetry is grounded in or reducible to the entropic asymmetry of the world in time. Failure to clearly understand just what *kind* of reduction one ought to have in mind here has led, I believe, to some confusion in the literature. I will then explore, to some degree, the plausibility of the reductionist program. It is to be hoped that the earlier clarification of statistical mechanics we have obtained will allow at least some deeper penetration into this extremely difficult problem than has previously been gained.

III. Notes to the reader

Finally, some notes about the comprehensiveness and technical level of this book. The number of conceptual problems explored here is rather large. Each of these problems requires for an adequate treatment reference to, and exposition of, a substantial amount of physics, the history of science, and philosophy. In order to keep the work within reasonable bounds, many fascinating conceptual problems in the foundations of statistical mechanics will simply have to be ignored altogether or, at best, noted in passing. I will have little to say, for example, about the kinds of explanation utilized in recent extremely elegant work of the structure of phase changes.

The scientifically knowledgeable reader will be especially struck by the overall focus in this book on classical statistical mechanics and the apparent neglect of quantum statistical mechanics. Because we know that the correct micro-theory underlying the macroscopic and thermodynamic features of the world is quantum mechanical and not classical, what is the point in directing our philosophical attention to a version of the theory we know to be incorrect and outmoded? The reason is clear. It is that the particular conceptual problems on which we focus – the origin and rationale of probability distribution assumptions over initial states, the justification of irreversible kinetic equations on the basis of reversible underlying dynamical equations, and so on – appear, for the most part, in similar guise in the development of both the classical and quantum versions of the theory. The hope is that by exploring these issues in the technically simpler classical case, insights will be gained that will carry over to the understanding of the corrected version of the theory.

There are indeed places where reference to the quantum version of the theory is essential. For example, quantum ensembles of finite systems cannot escape recurrence results as can classical ensembles of finite systems. And, some allege, only by reference to quantum limitations can the limits on initial ensembles be correctly delineated. Where one simply must pay attention to the quantum nature of the underlying microtheory, I will do so. But the bulk of the development will focus on the varieties of classical statistical mechanics as the objects for conceptual exploration. This way of doing things is not idiosyncratic, but common in the physics literature devoted to foundational issues. Whether things can be carried off profitably in this way can only be determined by continuing with the project in this vein and judging the results.

My intention is to make this work accessible to many who will be interested in the problems treated but who lack an extensive background either in physics, philosophy of science, or both. For this reason, resort to formalism and technical apparatus has been kept to an absolute minimum. Results will frequently simply appear with the note "it can be shown," with proofs not presented at all. I believe that the conceptual problems treated here are too fundamental and too interesting to keep them reserved only to those equipped with the resources of modern mathematical physics.

Naturally this approach will frequently make the treatment of both physics and philosophy skeletal. Some of the difficulty incurred by this approach will be remedied, I hope, by the bibliographical information appended to each chapter. The references cited are not intended to be comprehensive or exhaustive. Rather I have chosen a sampling of those works that I have found most useful in laying out the essential philosophical and physical issues in a clear way. The bibliographies are annotated in the hope that they will provide a road map into the literature for those who would like to pursue the material covered in this work further, either by mastering the formal scientific material touched on, seeing for themselves proofs stated to exist but not presented, or following out in greater depth the exposition of philosophical positions expounded and criticized here. Given the lack of coherence in the physics, noted earlier, this explicit guide to the literature is, I believe, a particularly essential part of this work.

Reference material that demands a modicum of familiarity with intermediate mathematics or a fair familiarity with the more basic parts of physical theory is marked with an asterisk (*). Material demanding comprehension of more advanced mathematics and physics is marked with two asterisks (**).

2

Historical sketch

The aim of this chapter is to present an extremely abbreviated and selective historical survey of the development of thermodynamics, kinetic theory, and statistical mechanics. No attempt will be made to be comprehensive. Nor is the material presented in a manner that would suit historians of science. Neither issues of chronology and attribution nor the far more important questions of placing the specific scientific results in their broader scientific and cultural context are our primary concern. The goal is merely to present some important developments in the history of these theories in such a way as to provide a background or context that will facilitate our later conceptual exploration.

Many of the scientific results noted in this chapter will only be mentioned here and not referred to again. They are being noted simply to place other, more relevant, results in their historical context. Other aspects of the historical development – those dealing with the conceptual problems at the foundational level – will be treated in much greater detail, especially in Chapters 5 through 7. The reader ought not to be discouraged, then, if certain conceptual and foundational issues seem to be too sketchily treated in this chapter to be grasped with full clarity, nor upset that many questions posed in the historical context are left in abeyance. These conceptual and foundational questions and issues will be taken up at the appropriate point in far greater detail.

I have chosen to organize the material here in topical vein. Hence, thermodynamical questions are taken up before those of kinetic theory and statistical mechanics, even if chronologically the development of the two areas was, of course, interwined. I first take up the evolution of thermodynamics in its long progress from the first phenomenological constitutive laws to its fully general and abstract form and its later extensions into novel areas. I then take up the development of the kinetic approach to matter, the development of kinetic theory at the hands of Maxwell and Boltzmann toward a genuinely statistical or probabilistic theory, the development of formal statistical mechanics, and some aspects of the evolution of foundational questions through the probing critiques of the theory that were early responses to it. Finally, some more recent developments are noted. Once again these developments are treated here only to put things in historical perspective. Where they bear on foundational issues they will be taken up in detail in Chapter 7.

I. Thermodynamics

1. Phenomenological properties and laws

The discovery of the right concepts with which to describe nature and the discovery of the lawlike relations among these features are processes that cannot be disentangled from one another. The appropriate concepts are the ones that allow the formulation of lawlike regularities among the phenomena so described.

There is a familiar progression of concepts, at least in the earlier stages of development of a science, a progression that requires great philosophical subtlety in order to be fully characterized, but whose superficial nature is clear. In the beginning we characterize the world in terms of concepts directly connected to our sensory awareness of things. Gradually we move to concepts whose applicability is fixed by the use of standardized physical instruments, a step frequently allowing the transition from qualitative and comparative to fully metric concepts. Finally, as the concepts are utilized to construct laws describing the regularities we discover in nature, we move to a stage where the very meaning of the relevant conceptual terms is more closely associated with the role of the terms in the theory than with any specific class of instrumental devices for detection or quantification of the relevant magnitude.

One point of origin of the full thermodynamic theory is the discovery of the first so-called equation of state for gases. The relevant law is formulated using concepts whose scientific role is already well-established in kinematics (the description of motion) and dynamics (the theory of the causal sources of change of motion). A gas confined to a box is found to obey the simple law that the product of its volume and the pressure imposed upon it, and that it exerts on the walls of its container, is constant.

But this law neglects the additional relevant parameter of temperature, and it is in the introduction of the concept of temperature, and the associated concept of heat, that we get the beginnings of a full thermodynamic theory. Temperature is first associated, in the manner described, with felt hotness and coldness. The introduction of thermometers detaches the concept from felt sensory qualities and allows us to invoke a metric scale founded upon the degree of expansion of a substance with its changing temperature. The very existence of the physical feature of thermal expansion that allows for the construction of the first thermometers immediately suggests the search that leads to the generalization of the equation of state for gases to the famous ideal gas law: $PV = kT$, where k is a constant.

Even the consideration of simple experiments on the resulting temperature when amounts of a given substance at two temperatures are mixed suggests the necessity of an extensive quantity – heat – to parallel the intensive notion of temperature. If we have double the amount of a substance at a given temperature, we have another thing with the same temperature but twice as much heat content. Moving to the results obtained on the mixing of different substances, there develops a slow awareness of the need for such notions as heat capacity – how much heat needs to be added to a given substance to raise its temperature to a specified amount.

Although the equations of state and the calorimetric (mixing) results deal with static properties of matter, it is soon realized that a theory of the laws of change of the features introduced is needed as well. Temperature difference is invoked as the driving force for heat *flow*, and, invoking the notion of the distribution of temperature over matter, soon the heat flow equation is derived and its solutions studied.

Additional work shows that there is a hierarchy of laws that can be formulated governing the combined flow of matter and heat, although the derivation of these equations requires the concomitant progress in the development of the full abstract thermodynamic theory. Here, progress is obtained by imposing stringent assumptions upon the flow of matter and heat, deriving the correct equations of motion, and then generalizing upon the result by relaxing the stringent constraints a little at a time to move to more complex equations for which those first derived prove only approximations. Euler's equations, for example, describe the motion of an incompressible, non-viscous fluid. Retaining incompressibility but allowing viscosity (momentum transfer by diffusion rather than by transfer of bulk matter), one gets the Navier–Stokes equations. Continuing the process to allow such things as compressibility, shock wave discontinuity, and so on leads to the ability to characterize ever more complex situations in which matter and heat are both in a changing state drive by their interactive forces.

2. Conservation and irreversibility

Conservation. Concurrent with the gradual development of ever more complex and general constitutive equations – the various equations relating the lawlike interrelations among the descriptive parameters of particular kinds of matter such as ideal gases in their macroscopically static states, incompressible viscous fluids in flow, and so on – another development of more immediate interest to our foundational project took place. This was the development of the basic principles of thermodynamics, the general principles that are taken to hold for all particular material systems, and that together with the more particularized constitutive equations, provide the resources needed for the full theory. There are two parallel paths in this progress, and insight into both strands develops simultaneously. One path is that in which heat is identified as a kind of internal energy of the system, and the assimilation of this insight into the overarching principle of the conservation of energy. The other is the growing realization of the fundamental role played by irreversibility in the theory, assimilated after the conceptual innovations of absolute temperature and entropy into the famous Second Law of Thermodynamics.

Let us take up the development of the energetic theory of heat first. At least as early as Francis Bacon, the connection between the ability of frictional motion to raise the temperature of an object suggested the speculation that heat, the gain of which accounted for a rise in temperature, might be nothing more than some kind of motion of internal constituents of matter. This internal motion was not directly evident because, presumably, of the microscopic nature of the components. In inelastic collision, momentum is strictly conserved, but kinetic energy as macroscopically determined is not. Could it be that the overtly observable motion has not vanished entirely from the world but, instead, merely hidden itself in the inner motion of microscopically invisible components of matter?

But for a long time, this energetic theory was mostly overshadowed in popularity by the idea of heat as a quasi-substantial entity. Simple calorimetric experiments on mixing can be accounted for by the idea that heat is some *thing* that can inhabit matter and be transferred from one piece of it to another. Something like the density of this so-called caloric substance in the matter is taken to account for temperature, with subtle modification in the theory being needed to account for the varying specific heats of different materials. Although energetic theory is never fully forgotten, the caloric-substantival theory remains in the forefront of speculation for a long period of time. Authors can be found – D. Bernoulli, for example – who mention the energetic account, even going beyond it to farsighted speculations about the particular form of internal motion heat amounts to, but who, in the bulk of their work, opt for the caloric approach. It must be understood that the preference for caloric theories as opposed to energetic theories of heat is no mere blindered dogmatism. As caloric theorists pointed out, the energetic theory did suffer from observational difficulties of some magnitude. How, for example, if heat is nothing but energy of motion of microscopic constituents of matter, can it be propagated from the sun to the earth over the intervening matterless vacuum?

The victory of the energetic theory over the caloric theory is founded on a multiplicity of factors. First, there are the changes in background science that offer the answers to the objections to the energetic theory like that mentioned. One example is the growing treatment of radiation as a quasi-substance that can carry energy across matterless space. Next, there is the growing importance of the conversion of heat to overtly observable mechanical motion in the heat engines so vitally important to the Industrial Revolution. Most important of all, though, is the growing ability to carefully quantify amounts of heat and amounts of mechanical motion, and the ability to show that the disappearance of a given quantity of overt motion always results in a corresponding increase in a specific amount of generated heat. Further, the destruction of a given quantity of heat always results in the generation of a specific quantity of overtly observable motion. When this is combined with the growing prominence of conservation principles in the foundation of physics, the suggestion is unavoidable that there is one kind of quality of things – energy – that is never created or destroyed. The apparent disappearance of quantity of motion in inelastic collision or friction, with a corresponding generation of heat, and the apparent generation of motion *ex nihilo* in engines, at the cost of a corresponding disappearance from the world of heat, is then really nothing more than the *conversion* of motion from the realm of the overtly observable to that of the invisible microscopic and vice versa. By the midpoint of the nineteenth century, Clausius is writing confidently of "the kind of motion we call heat."

Irreversibility. The facts about the interconvertibility of mechanical energy and heat outlined in the last section led, by means of the principles of the conservation of energy and the identification of heat as non-overt internal energy, to a degree of assimilation of the physics of thermal phenomena into the general theory of motion and energy. The observations we focus on here, however, although also arising out of the study of the interconversion of heat and motion, especially in heat-driven engines designed to perform mechanical work, served instead to emphasize the necessity within a general theory of thermal phenomena of principles radically unlike those familiar from the general kinematic and dynamical theory of motion. These are the principles of irreversibility.

The theory was inspired originally by S. Carnot's brilliant study of the efficiency of heat engines. The question is what is the maximum amount of mechanical work that can be obtained from an engine, like a steam engine, that is driven by the expansion and contraction of a heated gas working medium? First, there is the realization that engines can deviate from ideal efficiency: Heat can be allowed to flow from high to low temperatures without doing work in the process, fast motion of the working fluid can result in energy being uselessly dissipated in turbulent motion of the working fluid, and so on.

Far more important, though, is the realization that even an ideally efficient engine is limited in its ability to convert input heat into output usable mechanical motion. The crucial observation is that in order for heat to be converted into mechanical work, it must be delivered to the machine at a higher temperature than that at which residual heat in the working fluid is discharged at the end of the mechanical cycle. For example, input steam from the boiler must be at a higher temperature than the temperature of the condenser that turns the output steam into water to eliminate back pressure in the steam engine. Heat can only do work by "falling downhill' from high to low temperature. Insofar as this fall bypasses mechanical conversion, it is wasted heat. But no more work can be extracted from the heat than is allowed by the temperature gap between input and output reservoirs. Crudely, heat at high temperature is valuable, because work can be extracted from it by using an extraction reservoir, at the end of the cycle, of lower temperature. Heat available at a temperature no higher than that of an available sink for discharged heat at the end of the cycle is useless for the extraction of mechanical work.

Of course we can, by mechanical work, raise the temperature of a reservoir of heat – say, by compressing a gas without heating it. But an ideal temperature raiser operating off mechanical work as input will consume the same amount of mechanical work as the output of an ideal engine that operates between the same extreme temperatures. Our ability to get useful mechanical work out of heat energy is limited not simply by the fact that heat is energy and that the conservation law discussed earlier prohibits our extracting mechanical energy greater than the heat energy consumed. It is limited also by the necessity to operate our engines between sources of heat at a high temperature and a sink at a lower temperature.

The results are soon generalized into a principle of the irreversibility of processes in the world. Energy must be available in a "high quality" state to do work. Performing the work degrades the energy into "low quality." If the conversion of heat into work in the process is ideal, only as much work is created as would be sufficient to restore the energy left at the end to its high quality state. But engines are not ideal, and the processes that reduce them from ideality are all those that allow the quality of energy to degrade without producing work in the process. The net result is that there is a steady degradation of the quality of energy throughout the world. Very soon, speculation arises about the ultimate fate of the entire universe. Is it one in which all energy is conserved, so never vanishing, yet becomes so degraded (into a uniform temperature) as to be unavailable for work? The famous "heat death" of the universe becomes a cliche of popular science.

3. Formal thermostatics

At the hands of R. Clausius and later developers, Carnot's brilliant insights are converted into a formal theory of surpassing elegance, usually called thermodynamics. Throughout, the trick is to borrow as much conceptualization as one can from the familiar mathematical resources of dynamics, adding such intrinsically thermal concepts as are necessary, and modifying the dynamical principles to account for the special features of the new situation. In particular, the essential role played by the irreversibility in the new context must be taken into account.

One begins with the notion of a system. The concepts of kinematics and dynamics are appropriated as allowing for descriptions of systems and of their changes over time. Thus we can talk about the volume of a system, the pressure applied upon it, and so on. The crucial distinction is made between those systems energetically isolated from the remainder of the world (adiabatic systems) and those that can exchange energy either with an indeterminate environment or with one another. The total energy flow into or out of a system is divided into those parts of the flow that are the result of overtly observable mechanical work, and the remainder, called the flow of heat. This is in fact merely a definition of heat flow, but the definition is rationalized by the recoverability of total energy from an isolated system summarized in the First Law of Thermodynamics. This is just the applicability to the thermodynamic context of the general principles that energy is neither created nor destroyed, the so-called Principle of the Conservation of Energy.

Perhaps the most important conceptual discovery of all is that of equilibrium. If we allow an isolated system enough time, it settles to an overtly unchanging state called the equilibrium state. Two systems, each individually in equilibrium when energetically isolated, may or may not leave their respective equilibrium states when they are allowed to exchange energy with one another. When the condition of the systems is such that they remain in their initial equilibrium state when so energetically joined, they are said to be in mutual equilibrium with one another. An observation (originally covert and only later made fully explicit as the Zeroth Law of Thermodynamics) shows us that two systems each in equilibrium with a third system will be in equilibrium with one another (the transitivity of mutual equilibrium). This, together with the observation that it is the possession of a common thermometer temperature that characterizes systems in mutual equilibrium, allows the introduction of the notion of empirical temperature as the numerical indicator of an equivalence class of systems in mutual equilibrium.

At this point, the work of Carnot on heat engines is applied. The temperature of the reservoir into which heat is discharged in an ideal engine able to convert all of its input heat into overt work is taken as the zero point of an absolute temperature scale. The ratio metric for this scale is further defined by the ratios of available work obtainable from heat in differing ideal engines. The net result is the definition of the full absolute temperature scale.

One then examines the routes by which a system in one equilibrium state can be converted into a system in a distinct equilibrium state. One cannot characterize these states by any definite heat content, because even by ideal reversible processes, systems can be brought from one state to another by routes that differ in their use of mechanical work compared with their use of the flow of heat. But, Clausius shows, a new quantity, entropy, can be defined, such that the change of entropy from one equilibrium state to another will be the same independently of the reversible path traversed in getting from the initial state to the final state. This allows the assignment (relative to an arbitrary zero point) of a definite entropy value to each equilibrium state of a system. Entropy is so defined that a system whose energy is less degraded (and so more available for transformation into mechanical work) has a lower entropy than its more degraded counterpart. The crucial fact needed to justify the introduction of such a definite entropy value is the irreversibility of physical processes. It is the fact that heat engines cannot not only not generate mechanical work without the consumption of heat, but that they cannot be run in such a way as to produce work without degrading the quality of heat in the world, that is crucial to the proof of the existence of entropy.

The fundamental fact of irreversibility is summarized in the Second Law of Thermodynamics, which is given many different formulations. In the Clausius version, "it is impossible to construct a device that, operating in a cycle, will produce no effect other than the transference of heat from a cooler to a warmer body." In the Kelvin–Planck version, "it is impossible to construct a device which, operating in a cycle, will produce no effect other than the extraction of heat from a reservoir and the performance of an equivalent amount of work." Or, given the entropy concept, "the entropy of the final state of an energetically isolated system undergoing a transition is never less than its initial state."

Refined and elegant versions of the formal theory of thermodynamics are available. Perhaps the high point is the 1909 formulation of the Second Law by C. Carathéodory. His basic postulate, "In the neighborhood of the initial state of a system there exist states not accessible from the original state along any adiabatic path," provides an elegant mathematical constraint upon thermodynamic transitions from which the formulations of the Second Law in terms of engines and cycles can be derived.

4. Extending thermodynamics

Constitutive equations. While the fundamental theory of thermo-dynamics, with its universal laws, was developing, there was continued growth in the ability to construct constitutive equations that are both able to cover situations outside the range of the original gas laws, and able to characterize the phenomenology of matter with greater accuracy than those first equations.

There was, for example, the discovery by J. van der Waals of a form of equation of state for gases more accurate than the original Ideal Gas Law. This discovery was motivated in large part by his knowledge of the atomic-kinetic underpinnings of the macroscopic theory. More pro-found in its implications was the systematic study of so-called phase changes and their regularities. The familiar radical change of water upon heating from solid ice to liquid to gas was generalized. Such important phenomenological observations as the existence of critical points (above which the gas-liquid distinction vanishes) and formal restrictions on possible phase diagrams, were developed. The observation that systems apparently radically different in their constitution can be characterized in terms of clear similar regularities in their behavior under phase changes was extremely important.

The extension of thermodynamics concepts beyond their original range of characterizing kinematic-dynamic-thermal features of matter in the narrow sense was also of profound importance. J. Gibbs showed how energetically thermally driven processes of chemical combination and dissociation can be assimilated into the thermodynamic scheme by treat-ing chemical concentrations in a manner parallel to volume and so-called chemical potentials in a manner parallel to pressure. Another important example of such an extension is the thermodynamic treatment of the magnetization of matter, where the imposed magnetic field is the driving force and the resulting magnetization of the matter is the result. Overall, the theory is extended into a universally comprehensive scheme in which the exchange of energy between that which is macroscopically overt and that which is transmuted into internal energy of microscopic constituents becomes formalizable in the thermodynamic scheme. In each case a dual role is played by the constituent equations particular to the system in question and the universal principles of conservation and irreversibility. Both are needed for an adequate account.

Statistical equilibrium thermodynamics. In the orthodox thermo-dynamic theory, equilibrium is taken to be a state characterized by unchanging values of the macroscopic parameters. An isolated system settles to an equilibrium state of fixed pressure, volume, and so on. A system in perfect energetic contact with an "infinite" reservoir at a given temperature settles to an equilibrium state in which its temperature and energy content are fixed. As we shall see in Sections II,6; III; and IV this chapter, the development of the kinetic theory, and especially the responses offered by its proponents to important critical assaults on the theory, lead to a subtler notion of equilibrium. Here the idea is that equilibrium itself is not to be taken as characterizing a single macroscopic state of system, but as descriptive of a class of states each with its intrinsic "probability." Precisely what that means, of course, remains to be explored.

This revised notion of equilibrium originally arises in a context in which a particular theory of the nature of the energy that is heat is being considered. It its original formulation, this new notion of equilibrium is tied to the specification of the atomic structure. Yet, as we shall explore in detail, the degree to which the probabilistic or statistical assumptions of this revised kinetic theory are derivative from the laws governing the behavior of the micro-constituents of the system is problematic. In at least some formulations of the theory, the probabilistic assumptions must be "put in" as additions to the other components of the theory. This suggests that it might be reasonable to return to the original thermodynamic theory of equilibrium, and see if these probabilistic assumptions could be built into that theory at its own level, instead of bringing in these probabilistic considerations only at the level where atomicity and micro-mechanics are also introduced.

Even if we ultimately decide that the probabilistic assumptions are not independent of the micro-mechanics, but, in some sense, derivable from them, there is still a good motivation for seeking to construct, on independent postulates, a thermodynamics that allows for fluctuations. There will be a thermodynamics whose equilibrium is characterized not by a single macroscopic state but by a probability distribution over a class of such states. We will discuss the so-called reduction of thermodynamics to statistical mechanics in Chapter 9. It is a truism of the philosophical theory of reduction that in the reducing process the reduced theory is often modified from the form it had prior to the recognition of the possibility of reduction. If thermodynamics is reducible to an underlying probabilistic theory, we would not be surprised to find ourselves under pressure to reformulate the thermodynamic theory so that the existence of fluctuations, even in the equilibrium state, is already accounted for in this now modified, reduced theory.

There is an elegant postulational version of thermodynamics that does just what we have described. In purely macroscopic-phenomenological terms, and on a postulational basis, a novel thermodynamics is derived from the orthodox theory and that is designed to characterize equilibrium in this new probabilistic vein. The idea descends from early work of A. Einstein. In this new theory, for example, a small system kept in perfect energetic contact with an infinite reservoir at a constant temperature suffers a range of temperatures and internal energy contents, centered, of course, around the values of those quantities predicted by the older thermodynamic theory. An energetically isolated system, in this new account, can be conceived as broken up into a large number of energetically connected component systems. Instead of the static uniformity throughout all these pieces that one would expect in the older account, in this new account a distribution of numbers of components in a diversity of macroscopic states is now predicted by the theory.

The theory is of surpassing elegance. Only a few postulates of great simplicity regarding the nature of the probability distribution functions, and the application of the Zeroth Law of Thermodynamics, suffice to generate the specific form of the fluctuational nature of equilibrium identical to the forms familiar from the probabilistic theory built on the underlying micro-structure and micro-mechanics.

Non-equilibrium thermodynamics of systems near equilibrium. In the orthodox presentation of thermodynamics, we attribute thermodynamic features (temperature, entropy) only to systems that are in equilibrium. In the formal version of the theory, the very meaningfulness of such attributions is restricted to equilibrium states by the way the concepts are introduced into the theory. For this reason it is often suggested that the theory be called thermostatics, and not thermodynamics. Of course this does not mean that the theory tells us nothing about change. The very problems the theory was originally designed to solve are those that arise when we ask, for example, what states can be obtained from others by processes that leave the system energetically isolated from its environment, or when we ask how much overt mechanical work can be obtained by a process that takes a system in an original equilibrium state and ends up with it in another. But we have no resources within this theory to adequately describe systems not in equilibrium, nor to deal thermodynamically with such details of the transition from one equilibrium to another as the rates of flow of quantities.

But of course we can deal with systems not in equilibrium in a thermodynamic way. Long before the expansion of thermodynamics into non-equilibrium thermodynamics, one dealt with temperature distributions over a system and with the flows of heat driven by temperature differences. Other such laws, connecting fluxes or flows with their driving forces, exist as well, and provide the starting point from which a fuller non-equilibrium theory can be developed.

The key to such a theory is to allow for *fields* of quantities to describe a system. In the manner familiar from continuum mechanics, we can introduce such notions as the density of mass at a point, a field of velocities of mass components, a field of energy density, and so on. The trick is to extend this notion of a field of quantities to such purely thermodynamic quantities as temperature and entropy. But how can it be legitimate to talk of varying temperature from point to point in a system and, even worse, of a varying density of entropy, when temperature and entropy are only defined, within orthodox thermostatics, for systems in equilibrium? The solution is to invoke the notion of local equilibrium. If thermodynamic features of things change slowly enough from point to point and from moment to moment, we can hope to think of a non-equilibrium system as having small enough regions over small enough times that are describable "as if" they constituted equilibrium systems with appropriate temperatures and entropies. The full understanding of just what constitutes "slow and small enough change" would take us into an analysis that goes beyond the phenomenological level to which we are now restricted, but the theory can be developed on a postulational basis that assumes the legitimacy of our new field-type thermodynamic attributions.

Once the appropriate field-like concepts have been introduced, one invokes balance equations, again familiar from continuum mechanics. These are the local versions of the various principles of the conservation of mass, the conservation of energy and momentum, and so on. The most crucial additional move is to understand how to formulate the Second Law of Thermodynamics for this new non-equilibrium situation. Entropy is, of course, not generally conserved. In a small region it can even decrease by "flowing out" through the boundaries of the region. What is crucial is that within a small region the entropy generated is never smaller than the entropy that "flows out," so that we have a local version of the principle of the non-decrease of entropy over time. This can be captured in various differential or integral inequalities (such as the so-called Clausius–Duhem inequality).

In non-equilibrium theory the notion of a steady-state process takes a role analogous to that taken by the equilibrium state in thermostatics. Even in a system that is prevented from reaching equilibrium by inter-vention from the outside, there may be stationarity of processes. As a typical example, consider a fluid between two plates externally main-tained at a slight temperature difference. The steady flow of heat from the higher to the lower temperature plate through the fluid is a steady-state diffusion process describable by non-equilibrium thermodynamics.

In simple cases we can think of a steady-state process as being the steady change of a single extensive quantity driven by a single conjugate force – for example, heat flow driven by temperature difference or chemical diffusion of a single species driven by the relevant density variation in the species. More interesting is the interaction of forces and fluxes when more than one quantity and driving force is involved. Heat and matter flow, driven by temperature and density variation, can intricately interact with one another. In the case of a system that is globally close to equilibrium, a suggestion of Lord Kelvin, developed in detail by L. Onsager and others, becomes fruitful. This is to postulate a reciprocity relation among flows, so that the effect of the driving force of one flow on the flow of another quantity is matched by a reciprocal effect. The very meaningfulness of the formal statement of the principle first of all requires that the forces and flows be characterizable in a linear way, a restriction that is grounded in the closeness to equilibrium of the system, just as linearity of restoring force is often a justifiable assumption mechanically when the system is near its "neutral" position. The reciprocity relations can then be derived from certain very general postulates of invariance. For example, a principle of "detailed balance" among states leads to these reciprocity relations.

The situation of global nearness to equilibrium, linearity, and reciprocity leads to a number of elegant additional results. In traditional thermostatics it is possible to show that the equilibrium state is stable against small perturbations. In this extended theory it is possible to show that the steady-state processes also possess a stability property. Any small deviation of the system from its steady-state flow will die out, restoring the original steady-state condition. Again, whereas the equilibrium states of thermostatics are the states consistent with external constraints that maximize the entropy of the system in question, in near equilibrium non-equilibrium theory one can show (assuming the linearity and reciprocity conditions) that the stable steady-state flows will be just those processes compatible with the restraints that *minimize* the production of entropy.

Non-equilibrium thermodynamics of systems far from equilibrium. When the system with which we deal is in a state that is far from equilibrium, many of the tractable features of systems near-equilibrium no longer apply. We cannot hope that in generic situations, forces and fluxes will be related in a simple linear way; that there will be simple reciprocity relations among the cross effects that forces of one kind generate on non-conjugate fluxes; or that the nice stability relations for steady-state systems will hold. Nor can we expect an elegant principle of minimal entropy production to hold generally.

If a system is deviant enough from equilibrium, so that even locally we cannot view it as at least momentarily in something like an equilibrium state, then we have no hope of capturing its behavior in anything like thermodynamic terms. But if the system, as before, can be thought of as being in a kind of momentary equilibrium over small enough spatial subregions, then we can, once again, apply the field-like generalizations of the thermodynamic notion. As before we introduce field-like mechanical features such as velocity and momentum fields, and field-like thermo-dynamic features such as a field of temperatures and of entropy density. Just as in the case of systems near equilibrium, we introduce balance equations that summarize the various conservation principles of mass, momentum, and energy. Again, as before, a field-like surrogate for the Second Law is introduced. Although entropy in a region may decrease due to flow of entropy across the region's boundaries, the net result of entropy flow and entropy production in the region is always positive.

It is steady-state situations that attract our attention. Even if the system is not in equilibrium, and is maintained out of equilibrium by imposed boundary conditions, it may maintain an unchanging state of flow. Some-times these steady-state situations will be stable and easily reproducible, so that a far-from-equilibrium system, no matter how often prepared, will always fall into the same steady-state situation. In other cases, though, the steady-state situations will give the appearance of spontaneously being generated or suddenly decaying. In other cases, steady-state flows will apparently spontaneously change into different steady-state flows. One situation of particular interest is where we manipulate a far-from-equilibrium system by slowly varying an externally controllable parameter. In some cases this results in the generation of a unique, always repro-duced, steady-state flow. In other cases, at various critical values of the parameters, the system may, unpredictably, jump into one of a variety of available options for steady-state flow. Under other appropriate cir-cumstances for particular values of parameters, the system will show a fascinating oscillatory behavior, switching back and forth with clock-like regularity between a number of distinguishable steady-state flow modes. This varied behavior is accounted for by the possibility of small fluctuations not dying out, as they do in the equilibrium case or in the case of simple stable steady-state behavior of near equilibrium systems, but instead becoming amplified by various kinds of positive feedback mechanisms. As a parameter changes, the system may, as a matter of small fluctuation, pick a particular steady-state mode, but once having moved into that mode by a random choice, become locked into it in preference to the other modes it could have initially chosen.

Some elegant experimental instances of steady-state behavior of sys-tems far from equilibrium have been constructed. If one heats a fluid between plates of differing temperature, one gets steady, stable diffusion of heat from the hotter to the cooler plate when the temperature dif-ference between the plates is small. As this temperature difference is increased, however, one sees, at a critical difference of temperature, a sudden change of the motion into one in which macroscopic *convection* of the fluid sets in. If one carefully controls the experiment, the convection takes the form of the creation of stable hexagonal convection cells that transmit heat from the hotter to the cooler plate by a stable, steady-state, flow of heated matter. More complex situations can be generated by using solutions of a variety of chemicals that can chemically combine with one another and dissociate chemically from one another. Here, by varying chemical concentrations, temperatures, and so on, one can generate cases of steady-state flow, of oscillatory behavior with repetitive order both in space and in time, or of bifurcation in which the system jumps into one or another of a variety of possible self-sustaining flows, and so on.

There is at least some speculative possibility that the existence of such "self-organizing" phenomena as those described may play an important role in biological phenomena (biological clocks as generated by oscillatory flows, spatial organization of an initially spatially homogeneous mass by random change into a then self-stabilizing spatially inhomogeneous organization, and so on).

II. Kinetic theory

1. Early kinetic theory

Just as the theory of heat as internal energy continued to be speculated on and espoused, even during the period in which the substantival-caloric theory dominated the scientific consensus, so throughout the caloric period there appeared numerous speculations about just what *kind* of internal motion constituted that energy that took the form of heat. Here, the particular theory of heat offered was plainly dependent upon one's conception of the micro-constitution of matter. Someone who held to a continuum account, taking matter as continuous even at the micro-level, might think of heat as a kind of oscillation or vibration of the matter. Even an advocate of discreteness – of the constitution of matter out of discrete atoms – would have a wide variety of choices, especially because the defenders of atomism were frequently enamored of complex models in which the atoms of matter were held in place relative to one another surrounding clouds of aether or something of that sort.

As early as 1738, D. Bernoulli, in his *Hydrodynamics*, proposed the model of a gas as constituted of microscopic particles in rapid motion. Assuming their uniform velocity, he was able to derive the inverse relationship of pressure and volume at constant temperature. Furthermore, he reflected on the ability of increasing temperature to increase pressure at a constant volume (or density) of the gas, and indicated that the square of the velocity of the moving particles, all taken as having a common identical velocity, would be proportional to temperature. Yet the caloric theory remained dominant throughout the eighteenth century.

The unfortunate indifference of the scientific community to Bernoulli's work was compounded by the dismaying tragi-comedy of Herepath and Waterston in the nineteenth century. In 1820, W. Herepath submitted a paper to the Royal Society, again deriving the ideal gas laws from the model of independently moving particles of a fixed velocity. He identified heat with internal motion, but apparently took temperature as proportional to particle velocity instead of particle energy. He was able to offer qualitative accounts of numerous familiar phenomena by means of this model (such as change of state, diffusion, and the existence of sound waves). The Royal Society rejected the paper for publication, and although it appeared elsewhere, it had little influence. (J. Joule later read Herepath's work and in fact published a piece explaining and defending it in 1848, a piece that did succeed, to a degree, in stimulating interest in Herepath's work.)

J. Waterston published a number of his ideas in a similar vein in a book in 1843. The contents of the kinetic ideas were communicated to the Royal Society in 1845. The paper was judged "nothing but nonsense" by one referee, but it was read to the Society in 1846 (although not by Waterston, who was a civil servant in India), and an abstract was published in that year. Waterston gets the proportionality of temperature to square of velocity right, understands that in a gas that is a mixture of particles of different masses, the energy of each particle will still be the same, and even (although with mistakes) calculates on the model the ratio of specific heat at constant pressure to that at constant volume. The work was once again ignored by the scientific community.

Finally, in 1856, A. Krönig's paper stimulated interest in the kinetic theory, although the paper adds nothing to the previous work of Bernoulli, Herepath, Waterston, and Joule. Of major importance was the fact that Krönig's paper may have been the stimulus for the important papers of Clausius in 1857 and 1858. Clausius generalized from Krönig, who had idealized the motion of particles as all being along the axes of a box, by allowing any direction of motion for a particle. He also allowed, as Krönig and the others did not, for energy to be in the form of rotation of the molecular particles or in vibrational states of them, as well as in energy of translational motion. Even more important was his resolution of a puzzle with the kinetic theory. If one calculates the velocity to be expected of a particle, it is sufficiently high that one would expect particles released at one end of a room to be quickly found at the other. Yet the diffusion of one gas through another is much slower than one would expect on this basis. Clausius pointed out that the key to the solution was

in molecular collisions, and introduced the notion of mean free path –
the average distance a molecule could be expected to travel between
one collision and another.

The growing receptiveness of the scientific community to the kinetic
theory was founded in large part, of course, on both the convincing
quantitative evidence of the interconvertibility of heat and overt mechanical energy with the conservation of their sum, and on the large
body of evidence for the atomic constitution of matter coming from other
areas of science (chemistry, electro-chemistry, and so on).

2. Maxwell

In 1860, J. Maxwell made several major contributions to kinetic theory.
In this paper we find the first language of a sort that could be interpreted
in a probabilistic or statistical vein. Here for the first time the nature of
possible collisions between molecules is studied, and the notion of the
probabilities of outcomes treated. (What such reference to probabilities
might mean is something we will leave for Section II,5 of this chapter.)
Although earlier theories generally operated on some assumption of
uniformity with regard to the velocities of molecules, Maxwell for the
first time takes up the question of just what kind of distribution of velocities of the molecules we ought to expect at equilibrium, and answers it
by invoking assumptions of probabilistic nature.

Maxwell realizes that even if the speeds of all molecules were the
same at one instant, this distribution would soon end, because in collision
the molecules would "on average" not end up with identical speeds. He
then asks what the distribution of speeds ought to be taken to be. The
basic assumptions he needs to derive his result are that in collisions, all
directions of impact are equally likely, and the additional posit that for
any three directions at right angles to one another, the distribution law
for the components of velocity will be identical. This is equivalent to the
claim that the components in the y and z directions are "probabilistically
independent" of the component in the x direction. From these assumptions he is able to show that "after a great number of collisions among
a great number of identical particles," the "average number of particles
whose velocities lie among given limits" will be given by the famous
Maxwell law:

Number of molecules with velocities between v and $v + dv$ =
$Av^2 \exp(-v^2/b)dv$

It is of historical interest that Maxwell may very well have been influenced by the recently published theory of errors of R. Adrian, K. Gauss,

and A. Quetelet in being inspired to derive the law. Maxwell is also aware that his second assumption, needed to derive the law, is, as he puts it in an 1867 paper, "precarious," and that a more convincing derivation of the equilibrium velocity distribution would be welcome.

But the derivation of the equilibrium velocity distribution law is not Maxwell's only accomplishment in the 1860 paper. He also takes up the problem of transport. If numbers of molecules change their density from place to place, we will have transport of mass. But even if density stays constant, we can have transfer of energy from place to place by molecular collision, which is heat conduction, or transfer of momentum from place to place, which is viscosity. Making a number of "randomness" assumptions, Maxwell derives an expression for viscosity. His derivation contained flaws, however, and was later criticized by Clausius.

An improved theory of transport was presented by Maxwell in an 1866 paper. Here he offered a general theory of transport, a theory that once again relied upon "randomness" assumptions regarding the initial conditions of the interaction of molecules on one another. And he provided a detailed study of the result of any such molecular interaction. The resulting formula depends upon the nature of the potential governing molecular interaction, and on the relative velocities of the molecules, which, given non-equilibrium, have an unknown distribution. But for a particular choice of that potential – the so-called Maxwell potential, which is of inverse fifth power in the molecular separation – the relative velocities drop out and the resulting integrals are exactly solvable. Maxwell was able to show that the Maxwell distribution is one that will be stationary – that is, unchanging with time, and that this is so independently of the details of the force law among the molecules. A symmetry postulate on cycles of transfers of molecules from one velocity range to another allows him to argue that this distribution is the *unique* such stationary distribution. Here, then, we have a new rationale for the standard equilibrium distribution, less "precarious" than that offered in the 1860 paper.

The paper then applies the fundamental results just obtained to a variety of transport problems: heat conduction, viscosity, diffusion of one gas into another, and so on. The new theory allows one to calculate from basic micro-quantities the values of the "transport coefficients," numbers introduced into the familiar macroscopic equations of viscous flow, heat conduction, and so on by experimental determination. This allows for a comparison of the results of the new theory with observational data, although the difficulties encountered in calculating exact values in the theory, both mathematical and due to the need to make dubious assumptions about micro-features, and the difficulties in exact experimental determination of the relevant constants, make the comparison less definitive than one would wish.

3. Boltzmann

In 1868, L. Boltzmann published the first of his seminal contributions to kinetic theory. In this piece he generalizes the equilibrium distribution for velocity found by Maxwell to the case where the gas is subjected to an external potential, such as the gravitational field, and justifies the distribution by arguments paralleling those of Maxwell's 1866 paper.

In the second section of this paper he presents an alternative derivation of the equilibrium distribution, which, ignoring collisions and kinetics, resorts to a method reminiscent of Maxwell's first derivation. By assuming that the "probability" that a molecule is to be found in a region of space and that momentum is proportional to the "size" of that region, the usual results of equilibrium can once again be reconstructed.

In a crucially important paper of 1872, Boltzmann takes up the problem of non-equilibrium, the approach to equilibrium, and the "explanation" of the irreversible behavior described by the thermodynamic Second Law. The core of Boltzmann's approach lies in the notion of the distribution function $f(x,t)$ that specifies the density of particles in a given energy range. That is, $f(x,t)$ is taken as the number of particles between some specified value of the energy x and $x + dx$. He seeks a differential equation that will specify, given the structure of this function at any time, its rate of change at that time.

The distribution function will change because the molecules collide and exchange energy with one another. So the equation should have a term telling us how collisions effect the distribution of energy. To derive this, some assumptions are made that essentially restrict the equation to a particular constitution of the gas and situations of it. For example, the original equation deals with a gas that is, initially, spatially homogeneous. One can generalize out of this situation by letting f be a function of position as well as of energy and time. If one does so, one will need to supplement the collision term on the right-hand side of the equation by a "streaming" term that takes account of the fact that even without collisions the gas will have its distribution in space changed by the motion of the molecules unimpeded aside from reflection at the container walls. The original Boltzmann equation also assumes that the gas is sufficiently dilute that only interactions of two particles at a time with one another need be considered. Three and more particle collisions/interactions need not be taken into account. In Section III,6,1 I will note attempts at generalizing beyond this constraint.

In order to know how the energy distribution will change with time, we need to know how many molecules of one velocity will meet how many molecules of some other specified velocity (and at what angles) in any unit of time. The fundamental assumption Boltzmann makes here is the famous *Stosszahlansatz*, or Postulate with Regard to Numbers of Collisions. One assumes the absence of any "correlation" among molecules of given velocities, or, in other words, that collisions will be "totally random." At any time, then, the number of collisions of molecules of velocity v_1 and v_2 that meet will depend only on the proportion of molecules in the gas that have the respective velocities, the density of the gas, and the proportion of volume swept out by one of the molecules. This – along with an additional postulate that any collision is matched by a time-reverse collision in which the output molecules of the first collision would, if their directions were reversed, meet and generate as output molecules that have the same speed and reverse direction of the input molecules of collisions of the first kind (a postulate that can be somewhat weakened to allow "cycles" of collisions) – gives Boltzmann his famous kinetic equation:

$$\left(\frac{\partial f_1}{\partial t}\right)_{coll} = \int d^3v \int d\Omega \sigma(\Omega)|v_1 - v_2|(f_2'f_1' - f_2 f_1)$$

Here the equation is written in terms of velocity, rather than in terms of energy, as it was expressed in Boltzmann's paper. What this equation describes is the fraction of molecules with velocity v_1, f_1 changing over time. A molecule of velocity v_1 might meet a molecule of velocity v_2 and be knocked into some new velocity. On the other hand, molecules of velocities v_1' and v_2' will be such that in some collisions of them there will be the output of a molecule of velocity v_1. The number f_2 gives the fraction of molecules of velocity v_2, and the numbers f_1', and f_2' give the respective fractions for molecules of velocities v_1' and v_2'. The term $\sigma(\Omega)$ is determined by the nature of the molecular collisions, and rest of the apparatus on the right-hand side is designed to take account of all the possible ways in which the collisions can occur (because of the fact that molecules can collide with their velocities at various angles from one another).

The crucial assumption is that the rate of collisions in which a molecule of velocity v_1 meets one of velocity v_2, $f(v_1, v_2)$, is proportional to the product of the fraction of molecules having the respective velocities, so that it can be written as $f(v_1)f(v_2)$. The two terms, $f_2' f_1'$ and $-f_2 f_1$, on the right-hand side of the equation can then be shown to characterize the number of collisions that drive molecules into a particular velocity range from another velocity range, and the number of those that delete molecules from that range into the other, the difference being the net positive change of numbers of molecules in the given range.

Introducing the Maxwell–Boltzmann equilibrium velocity distribution function into the equation immediately produces the result that it is stationary in time, duplicating Maxwell's earlier rationalization of this distribution by means of his transfer equations.

But how can we know if this standard equilibrium distribution is the *only* stationary solution to the equation? Knowing this is essential to justifying the claim that the discovery of the kinetic equation finally provides the micro-mechanical explanation for the fundamental fact of thermodynamics: the existence of a unique equilibrium state that will be ceaselessly and monotonically approached from any non-equilibrium state. It is to justifying the claim that the Maxwell–Boltzmann distribution is the unique stationary solution of the kinetic equation that Boltzmann turns.

To carry out the proof, Boltzmann introduces a quantity he calls *E*. The notation later changes to *H*, the standard notation, so we will call it that. The definition of *H* is arrived at by writing $f(x,t)$ as a function of velocity:

$$H = \int d^3 v f(v,t) \, \log f(v,t)$$

Intuitively, *H* is a measure of how "spread out" the distribution in velocities of the molecules is. The logarithmic aspect of it has the virtue that the total spread-outness of two independent samples proves to be the sum of their individual spreads. Boltzmann is able to show this as long as $f(v,t)$ obeys the kinetic equation,

$$dH/dt \leq 0$$

and that $dH/dt = 0$ only when the distribution function has its equilibrium form. Here, then, is the needed proof that the equilibrium distribution is the uniquely stationary solution to the kinetic equation.

4. Objections to kinetic theory

The atomistic-mechanistic account of thermal phenomena posited by the kinetic theory received a hostile reception from a segment of the scientific community whose two most prominent members were E. Mach and P. Duhem. Their objection to the theory was the result of two programmatic themes, distinct themes whose difference was not always clearly recognized by their exponents.

One theme was a general phenomenalistic-instrumentalistic approach to science. From this point of view, the purpose of science is the production of simple, compact, lawlike generalizations that summarize the fundamental regularities among items of observable experience. This view of theories was skeptical of the postulation of unobservable "hidden" entities in general, and so, not surprisingly, was skeptical of the postulation of molecules and their motion as the hidden ground of the familiar phenomenological laws of thermodynamics.

The other theme was a rejection of the demand, common especially among English Newtonians, that all phenomena ultimately receive their explanation within the framework of the *mechanical* picture of the world. Here the argument was that the discovery of optical, thermal, electric, and magnetic phenomena showed us that mechanics was the appropriate scientific treatment for only a portion of the world's phenomena. From this point of view, kinetic theory was a misguided attempt to assimilate the distinctive theory of heat to a universal mechanical model.

There was certainly confusion in the view that a phenomenalistic-instrumentalistic approach to theories required in any way the rejection of atomism, which is, after all, a theory that can be given a phenomenalistic-instrumentalistic philosophical reading if one is so inclined. Furthermore, from the standpoint of hindsight we can see that the anti-mechanistic stance was an impediment to scientific progress where the theory of heat was concerned. It is only fair to note, however, that the anti-mechanist rejection of any attempt to found electromagnetic theory upon some mechanical model of motion in the aether did indeed turn out to be the route justified by later scientific developments.

More important, from our point of view, than these philosophical-methodological objections to the kinetic theory were specific technical objections to the consistency of the theory's basic postulates with the mechanical theory of atomic motion that underlay the theory. The first difficulty for kinetic theory, a difficulty in particular for its account of the irreversibility of thermal phenomena, seems to have been initially noted by Maxwell himself in correspondence and by W. Thomson in publication in 1874. The problem came to Boltzmann's attention through a point made by J. Loschmidt in 1876–77 both in publication and in discussion with Boltzmann. This is the so-called *Umkehreinwand*, or Reversibility Objection.

Boltzmann's *H*-Theorem seems to say that a gas started in any non-equilibrium velocity distribution must monotonically move closer and closer to equilibrium. Once in equilibrium, a gas must stay there. But imagine the micro-state of a gas that has reached equilibrium from some non-equilibrium state, the gas energetically isolated from the surrounding environment during the entire process. The laws of mechanics guarantee to us that a gas whose micro-state consists of one just like the equilibrium gas – except that the direction of motion of each constituent molecule is reversed – will trace a path through micro-states that are each the "reverse" of those traced by the first gas in its motion toward equilibrium. But because *H* is indifferent to the direction of motion of the molecules and depends only upon the distribution of their speeds, this

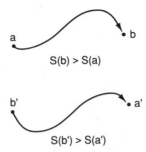

Figure 2-1. Loschmidt's reversibility argument. Let a system be started in micro-state a and evolve to micro-state b. Suppose, as is expected, the entropy of state b, S(b) is higher than that of state a, S(a). Then, given the time-reversal invariance of the underlying dynamical laws that govern the evolution of the system, there must be a micro-state b', that evolves to a micro-state a' and such that the entropy of b', S(b'), equals that of b and the entropy of a' equals that of a, S(a'), (as Boltzmann defines statistical entropy). So for each "thermo-dynamic" evolution in which entropy increases, there must be a corresponding "anti-thermodynamic" evolution possible in which entropy decreases.

means that the second gas will evolve, monotonically, *away* from its equilibrium state. Therefore, Boltzmann's *H*-theorem is incompatible with the laws of the underlying micro-mechanics. (See Figure 2-1.)

A second fundamental objection to Boltzmann's alleged demonstration of irreversibility only arose some time after Maxwell and Boltzmann had both offered their "reinterpretation" of the kinetic theory to overcome the Reversibility Objection. In 1889, H. Poincaré proved a fundamental theorem on the stability of motion that is governed by the laws of Newtonian mechanics. The theorem only applied to a system whose energy is constant and the motion of whose constituents is spatially bounded. But a system of molecules in a box that is energetically isolated from its environment fits Poincaré's conditions. Let the system be started at a given time in a particular mechanical state. Then, except for a "vanishingly small" number of initial states (we shall say what this means in Section 3,I,3), the system will eventually evolve in such a way as to return to states as close to the initial state as one specifies. Indeed, it will return to an arbitrary degree of closeness an unbounded number of times. (See Figure 2-2.)

In 1896, E. Zermelo applied the theorem to generate the *Wiederkehreinwand*, or Recurrence Objection, to Boltzmann's mechani-cally derived *H*-Theorem. The *H*-Theorem seems to say that a system started in non-equilibrium state must monotonically approach equilib-rium. But, according to Poincaré's Recurrence Theorem, such a system, started in non-equilibrium, if it does get closer to equilibrium, must at

Figure 2-2. Poincaré recurrence. We work in phase-space where a single point represents the exact microscopic state of a system at a given time – say the position and velocity of every molecule in a gas. Poincaré shows for certain systems, such as a gas confined in a box and energetically isolated from the outside world, that if the system starts in a certain microscopic state o, then, except for a "vanishingly small" number of such initial states, when the system's evolution is followed out along a curve p, the system will be found, for any small region E of micro-states around the original micro-state o to return to a micro-state in that small region E. Thus, "almost all" such systems started in a given state will eventually return to a microscopic state "very close" to that initial state.

some point get back to a state mechanically as close to its initial state as one likes. But such a state would have a value of H as close to the initial value as one likes as well. Hence Boltzmann's demonstration of necessary monotonic approach to equilibrium is incompatible with the fundamental mechanical laws of molecular motion.

5. The probabilistic interpretation of the theory

The result of the criticisms launched against the theory, as well as of Maxwell's own critical examination of it, was the development by Maxwell, Boltzmann, and others of the probabilistic version of the theory. Was this a revision of the original theory or merely an explication of what Clausius, Maxwell, and Boltzmann had meant all along? It isn't clear that the question has any definitive answer. Suffice it to say that the discovery of the Reversibility and Recurrence Objections prompted the discoverers of the theory to present their results in an enlightening way that revealed more clearly what was going on than did the original presentation of the theory.

As we relate what Maxwell, Boltzmann, and others said, the reader will find himself quite often puzzled as to just how to understand what they meant. The language here becomes fraught with ambiguity and conceptual obscurity. But it is not my purpose here either to lay out all the possible things they might have meant, or to decide just which of the many understandings of their words we ought to attribute to them. Again, I doubt if there is any definitive answer to those questions. We shall be exploring a variety of possible meanings in detail in Chapters 5, 6, and 7.

Throughout this section it is important to keep in mind that what was primarily at stake here was the attempt to show that the apparent contradiction of the kinetic theory with underlying micro-mechanics could be avoided. That is not the same thing at all as showing that the theory is correct, nor of explaining why it is correct. We will see here, however, how some of the fundamental problems of rationalizing belief in the theory and of offering an account as to why it is correct received their early formulations.

Maxwell's probabilism. In a train of thought beginning around 1867, Maxwell contemplated the degree to which the irreversibility expressed by the Second Law is inviolable. From the new kinetic point of view, the flow of heat from hot to cold is only the mixing of molecules faster on the average with those slower on the average. Consider a Demon capable of seeing molecules individually approaching a hole in a partition and capable of opening and closing the hole with a door, his choice depending on the velocity of the approaching molecule. Such an imagined creature could sort the molecules into fast on the right and slow on the left, thereby sorting a gas originally at a common temperature on both sides into a compartment of hot gas and a compartment of cold gas. And doing this would not require overt mechanical work, or at least the amount of this demanded by the usual Second Law considerations. From this and related arguments, Maxwell concludes that the Second Law has "only a statistical certainty."

Whether a Maxwell Demon could really exist, even in principle, became in later years a subject of much discussion. L. Brillouin and L. Szilard offered arguments designed to show that the Demon would generate more entropy in identifying the correct particles to pass through and the correct particles to block than would be reduced by the sorting process, thereby saving the Second Law from the Demon's subversion. Later, arguments were offered to show that Demon-like constructions could avoid that kind of entropic increase as the result of the Demon's process of knowledge accrual.

More recently, another attack had been launched on the very possibility of an "in principle" Maxwell Demon. In these works it is argued that after each molecule has been sorted, the Demon must reset itself. The idea is that the Demon, in order to carry out its sorting act, must first register in a memory the fact that it is one sort of particle or the other with which it is dealing. After dealing with this particle, the Demon must "erase" its memory in order to have a blank memory space available to record the status of the next particle encountered. R. Landauer and others have argued that this "erasure" process is one in which entropy is generated by the Demon and fed into its environment. It is this entropy generation, they argue, that more than compensates for the entropy reduction accomplished by the single act of sorting.

In his later work, Maxwell frequently claims that the irreversibility captured by the Second Law is only "statistically true" or "true on the average." At the same time he usually seems to speak as though the notions of randomness and irregularity he invokes to explain this are only due to limitations on our knowledge of the exact trajectories of the "in principle" perfectly deterministic, molecular motions. Later popular writings, however, do speak, if vaguely, in terms of some kind of underlying "objective" indeterminism.

Boltzmann's probabilism. Stimulated originally by his discussions with Loschmidt, Boltzmann began a process of rethinking of his and Maxwell's results on the nature of equilibrium and of his views on the nature of the process that drives systems to the equilibrium state. Various probabilistic and statistical notions were introduced without it being always completely clear what these notions meant. Ultimately, a radically new and curious picture of the irreversible approach of systems (and of "the world") toward equilibrium emerged in Boltzmann's writings.

One paper of 1877 replied specifically to Loschmidt's version of the Reversibility Objection. How can the *H*-Theorem be understood in light of the clear truth of the time reversibility of the underlying micro-mechanics?

First, Boltzmann admits, it must be clear that the evolution of a system from a given micro-state will depend upon the specific micro-state that serves to fix the initial conditions that must be introduced into the equations of dynamical evolution to determine the evolution of the system. Must we then, in order to derive the kinetic equation underlying the Second Law of Thermodynamics, posit the existence of specific, special initial conditions for all gases? Boltzmann argues that we can avoid this by taking the statistical viewpoint. It is certainly true that every individual micro-state has the same probability. But there are vastly more micro-states corresponding to the macroscopic conditions of the system being in (or very near) equilibrium than there are numbers of micro-states corresponding to non-equilibrium conditions of the system. If we choose initial conditions at random, then, given a specified time interval, there will be many more of the randomly chosen initial states that lead to a uniform, equilibrium, micro-state at the later time than there will be initial states that lead to a non-equilibrium state at the later time. It is worth noting that arguments in a similar vein had already appeared in a paper of Thomson's published in 1874.

In his 1877 paper, Boltzmann remarks that "one could even calculate, from the relative numbers of different state distributions, their probabilities, which might lead to an interesting method for the calculation of thermal equilibrium." He develops this idea in another paper also published in 1877.

Here the method familiar to generations of students of elementary kinetic theory is introduced. One divides the available energy up into small finite intervals. One imagines the molecules distributed with so-and-so many molecules in each energy range. A weighing factor is introduced that converts the problem, instead, into imagining the *momentum* divided up into the equal small ranges. One then considers all of the ways in which molecules can be placed in the momentum boxes, always keeping the number of molecules and the total energy constant. Now consider a state defined by a distribution, a specification of the number of molecules in each momentum box. For a large number of particles and boxes, one such distribution will be obtained by a vastly larger number of assignments of molecules to boxes than will any other such distribution. Call the probability of a distribution the number of ways it can be obtained by assignments of molecules to boxes. Then one distribution is the overwhelmingly most probable distribution. Let the number of boxes go to infinity and the size of the boxes go to zero and one discovers that the energy distribution among the molecules corresponding to this overwhelmingly probable distribution is the familiar Maxwell–Boltzmann equilibrium distribution. (See Figure 2-3.)

It is clear that Boltzmann's method for calculating the equilibrium distribution here is something of a return to Maxwell's first method and away from the approach that takes equilibrium to be specified as the unique stationary solution of the kinetic equation. As such it shares "precariousness" with Maxwell's original argument. But more has been learned by this time. It is clear to Boltzmann, for example, that one must put the molecules into equal momentum boxes, and not energy boxes as one might expect, in order to calculate the probability of a state. His awareness of this stems from considerations of collisions and dynamics that tell us that it is only the former method that will lead to stationary distributions and not the latter. And, as we shall see in the next section, Boltzmann is also aware of other considerations that associate probability with dynamics in a non-arbitrary way, considerations that only become fully developed shortly after the second 1877 paper appeared.

Combining the definition of H introduced in the paper on the kinetic equation, the calculated monotonic decrease of H implied by that equation, the role of entropy, S, in thermodynamics (suggesting that S in some sense is to be associated with $-H$), and the new notion of probability of a state, W, Boltzmann writes for the first time the equation that subsequently became familiar as $S = -K \log W$. The entropy of a macrostate is determined simply by the number of ways in which the macrostate can be obtained by arrangements of the constituent molecules of the system. As it stands, much needs to be done, however, to make this "definition" of entropy fully coherent.

One problem – by no means the only one – with this new way of viewing things is the use of "probability." Botzmann is not oblivious to

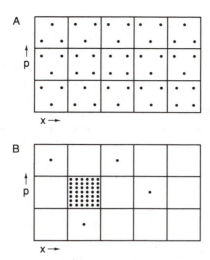

Figure 2-3. Boltzmann entropy. Box A (and Box B) represent all possible values of position x and momentum p a molecule of gas might have. This "molecular phase space" is divided up into small sub-boxes. In the theory, these sub-boxes are of small size relative to the size of the entire molecular phase space but large enough so that many molecules will generally be in each box for any reasonable micro-state of the gas. An arrangement of molecules into boxes like that of Fig. A can be obtained by many permutations of distinct molecules into boxes. But an arrangement like that of Fig. B can be obtained only by a much smaller number of permutations. So the Boltzmann entropy for arrangement A is much higher than that for arrangement B. The equilibrium momentum distribution (the Maxwell–Boltzmann distribution) corresponds to the arrangement that is obtainable by the maximum number of permutations, subject to the constraints that total number of molecules and total energy of molecules remain constant.

the ambiguities latent in using that term. As early as 1881 he distinguished between "probability" as he means it, taking that to be "the time during which the system possesses this condition on the average," and as he takes Maxwell to mean it as "the ratio of the number of [innumerable similarly constituted systems] which are in that condition to the total number of systems." This is an issue to which we will return again and again.

Over the years, Boltzmann's account of irreversibility continues to evolve, partly inspired by the need to respond to critical discussion of his and Maxwell's ideas, especially in England, and partly due to his own ruminations on the subject. To what extent can one stand by the new, statistical reading of the H-Theorem, now taken to be read as describing the "overwhelmingly probable" course of evolution of a system from "less probable" states to "more probable" states? Once again, we are using states here not to mean micro-states, all of which are taken as equally probable, but states defined by numbers of particles in a given momentum range.

Most disturbing to this view is a problem posed by E. Culverwell in 1890. Boltzmann's new statistical interpretation of the H-Theorem seems to tell us that we ought to consider transitions from micro-states corresponding to a non-equilibrium macro-condition to micro-states corresponding to a condition closer to equilibrium as more "probable" than transitions of the reverse kind. But if, as Boltzmann would have us believe, all micro-states have equal probability, this seems impossible. For given any pair of micro-states, S_1, S_2, such that S_1 evolves to S_2 after a certain time interval, there will be a pair S_2', S_1' – the states obtained by reversing the directions of motion in the respective original micro-states while keeping speeds and positions constant – such that S_2' is closer to equilibrium than S_1', and yet S_2' evolves to S_1' over the same time interval. So "anti-kinetic" transitions should be as probable as "kinetic" transitions.

Some of the discussants went on to examine how irreversibility was introduced into the kinetic equation in the first place. Others suggested that the true account of irreversibility would require some kind of "friction" in the system, either in form of energy of motion of the molecules interchanging with the aether or with the external environment.

In a letter of 1895, Boltzmann gave his view of the matter. This required, once again, a reinterpretation of the meaning of his kinetic equation and of its H-Theorem, and the introduction of several new hypotheses as well. These latter hypotheses, as the reader will discern, are of an unexpected kind, and, perhaps, unique in their nature in the history of science.

In this new picture, Boltzmann gives up the idea that the value of H will *ever* monotonically decrease from an initial micro-state. But it is still true that a system, in an initial improbable micro-state, will probably be found at any later time in a micro-state closer to equilibrium. As Culverwell himself points out, commenting on Boltzmann's letter, the trick is to realize that if we examine the system over a vast amount of time, the system will nearly always be in a close-to-equilibrium state. Although it is true that there will be as many excursions from a close-to-equilibrium micro-state to a state further from equilibrium as there will be of the reverse kind of transition, it is still true that *given* a micro-state far from equilibrium as the starting point, at a later time we ought to expect the system to be in a micro-state closer to equilibrium. (See Figure 2-4.)

But the theory of evolution under collision is now *time-symmetric*! For it will also be true that given a micro-state far from equilibrium at one time, we ought, on probabilistic grounds, to expect it to have been closer to equilibrium at any given past time. The theory is also paradoxical in

Figure 2-4. Time-symmetric Boltzmann picture. In this picture of the world, it is proposed that an isolated system "over infinite time" spend nearly all the time in states whose entropy S_{system} is close to the maximum value S_{max} – that is, in the equilibrium state. There are random fluctuations of the system away from equilibrium. The greater the fluctuation of a system from equilibrium, the less frequently it occurs. The picture is symmetrical in time. If we find a system far from equilibrium, we ought to expect that in the future it will be closer to equilibrium. But we ought also to infer that in the past it was also closer to equilibrium.

that it tells us that equilibrium is overwhelmingly probable. Isn't this a curious conclusion to come to in a world that we find to be grossly distant from equilibrium?

Boltzmann takes up the latter paradox in his 1895 letter. He attributes to his assistant, Dr. Schuetz, the idea that the universe is an eternal system that is, overall, in equilibrium. "Small" regions of it, for "short" intervals of time will, improbably, be found in a state far from equilibrium. Perhaps the region of the cosmos observationally available to us is just such a rare fluctuation from the overwhelmingly probable equilibrium state that pervades the universe as a whole.

It is in 1896 that Zermelo's application of the Poincaré Recurrence Theorem is now invoked to cast doubt on the kinetic explanation of irreversibility and Boltzmann responds to two short papers of Zermelo's with two short pieces of his own. Boltzmann's 1896 paper points out that the picture adopted in the 1895 letter of a system "almost always" near equilibrium but fluctuating arbitrarily far from it, each kind of fluctuation being the rarer the further it takes the system from equilibrium, is perfectly consistent with the Poincaré Recurrence Theorem.

The 1897 paper repeats the picture of the 1896 paper, but adds the cosmological hypothesis of Dr. Schuetz to it. In this paper, Boltzmann makes two other important suggestions. If the universe is mostly in equilibrium, why do we find ourselves in a rare far-from-equilibrium portion? The suggestion made is the first appearance of a now familiar "transcendental" argument: Non-equilibrium is essential for the existence of a sentient creature. Therefore a sentient creature could not find itself existing in an equilibrium region, probable as this may be, for in such regions no sentience can exist.

Even more important is Boltzmann's answer to the following obvious question: If the picture presented is now time-symmetrical, with every piece of the path that represents the history of the system and that slopes toward equilibrium matched by a piece sloping away from equilibrium to non-equilibrium, then why do we find ourselves in a portion of the universe in which systems approach equilibrium from past to future? Wouldn't a world in which systems move from equilibrium to non-equilibrium support sentience equally well? Here the response is one already suggested by a phenomenological opponent of kinetic theory, Mach, in 1889. What we *mean* by the future direction of time is the direction of time in which our local region of the world is headed toward equilibrium. There could very well be regions of the universe in which entropic increase was counter-directed, so that one region had its entropy increase in the direction of time in which the other region was moving away from equilibrium. The inhabitants of those two regions would each call the direction of time in which the entropy of their local region was increasing the "future" direction of time! The combination of cosmological speculation, transcendental deduction, and definitional dissolution in these short remarks has been credited by many as one of the most ingenious proposals in the history of science, and disparaged by others as the last, patently desperate, ad hoc attempt to save an obviously failed theory. We shall explore the issue in detail in Chapters 8 and 9, even if we will not settle on which conclusion is correct.

6. The origins of the ensemble approach and of ergodic theory

There is another thread that runs through the work of Maxwell and Boltzmann that we ought to follow up. As early as 1871, Boltzmann describes a mechanical system, a particle driven by a potential of the form $\frac{1}{2}(ax^2 + by^2)$, where a/b is irrational, where the path of the point in phase space that represents the motion of the particle will "go through the entire surface" of the phase space to which the particle is confined by its constant total energy. Here, by phase space we mean that abstract multiple dimensional space each point of which specifies the total position-momentum state of the system at any time. Boltzmann suggests that the motion of the point representing a system of interacting molecules in a gas, especially if the gas is acted upon by forces from the outside, will display this same behavior:

The great irregularity of thermal motion, and the multiplicity of forces that act on the body from the outside, make it probable that the atoms themselves, by virtue of the motion that we call heat, pass through all possible positions and velocities consistent with the equation of kinetic energy. (See Figure 2-5.)

Given the truth of this claim, one can then derive such equilibrium features as the equipartition of energy over all available degrees of freedom in a simple way. Identify the equilibrium value of a quantity with

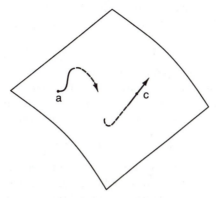

Figure 2-5. The Ergodic Hypothesis. Let a system be started in any micro-state, represented by point a in the phase space. Let b represent any other micro-state possible for the system. The Ergodic Hypothesis posits that at some future time or other, the system started in state a will eventually pass through state b as well. But the posit is in fact provable false.

its average over an infinite period of time, in accordance with Boltzmann's general approach of thinking of the "probability" of the system being in a given phase as the proportion of time the system spends in that phase over vast periods of time. If the system does indeed pass through *every* phase in its evolution, then it is easy to calculate such infinite time aver-ages by simply averaging over the value of the quantity in question for each phase point, weighting regions of phase points according to a measure that can easily be derived. (We will see the details Chapter 5.) Here we see Boltzmann's attempt to eliminate the seeming arbitrariness of the probabilistic hypotheses used earlier to derive equilibrium features.

Maxwell, in a very important paper of 1879, introduces a new method of calculating equilibrium properties that, he argues, will give a more general derivation of an important result earlier obtained by Boltzmann for special cases. The methods of Boltzmann's earlier paper allowed one to show, in the case of molecules that interact only upon collision, that in equilibrium the equipartition property holds. This means that the total kinetic energy available to the molecules of the gas will be distributed in equal amounts among the "degrees of freedom" of motion available to the molecules. Thus, in the case of simple point molecules all of whose energy of motion is translational, the x, y, and z components of velocity of all the molecules will represent, at equilibrium, the same proportion of the total kinetic energy available to the molecules. Maxwell proposes a method of calculating features of equilibrium from which the equiparti-tion result can be obtained and that is independent of any assumption about the details of interaction of the molecules. It will apply even if they exert force effects upon one another at long range due to potential interaction.

Suppose we imagine an infinite number of systems, each compatible with the macroscopic constraints imposed on a given system, but having every possible micro-state compatible with those macroscopic constraints. We can characterize a collection of possible micro-states at a time by a region in the phase-space of points, each point corresponding to a given possible micro-state. If we place a probability distribution over those points at a time, we can then speak of "the probability that a member of the collection of systems has its micro-state in a specified region at time t." With such a probability distribution, we can calculate the average value, over the collection, of any quantity that is a function of the phase value (such as the kinetic energy for a specified degree of freedom). The dynamic equations will result in each member of this collection or ensemble having its micro-state evolve, corresponding to a path among the points in phase-space. In general, this dynamic evolution will result in the probability that is assigned to a region of phase points changing with time, as systems have their phases move into and out of that region.

There is one distribution for a given time, however, that is such that the probability assigned to a region of phase points at a time will not vary as time goes on, because the initial probability assigned at the initial time to any collection of points that eventually evolves into the given collection will be the same as the probability assigned the given collection at the initial time. So an average value of a phase-function computed with this probability distribution will remain constant in time. If we identify equilibrium values with average values over the phase points, then for this special probability assignment, the averages, hence the attributed equilibrium values, will remain constant. This is as we would wish, because equilibrium quantities are constant in time. This special probability assignment is such that the average value of the energy per degree of freedom is the same for each degree of freedom, so that our identification results in derivation of the equipartition theorem for equilibrium that is dependent only upon the fundamental dynamical laws, our choice of probability distribution, and our identification of equilibrium values with averages over the ensemble. But the result is independent of any particular force law for the interaction of the molecules. (See Figure 2-6.)

Maxwell points out very clearly that although he can show that the distribution he describes is one that will lead to constant probabilities being assigned to a specified region of the phase points, he cannot show, from dynamics alone, that it is the *only* such stationary distribution. One additional assumption easily allows that to be shown. The assumption is that "a system if left to itself . . . will, sooner or later, pass

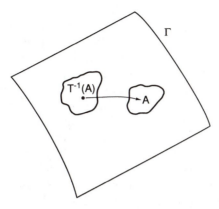

Figure 2-6. Invariant probability distribution. The space Γ represents all possible total micro-states of the system, each represented by a point of the space. A probability distribution is assigned over the space. In a time Δt, the systems whose points were originally in T^{-1} (A) have evolved in such a way that their phase points have moved to region A. Only if, for each "measurable" A, the total probability assigned to T^{-1} (A) is equal to that of A will a specified region of the phase space have a constant probability assigned to it as the systems evolve.

through every phase consistent with the energy." (How all this works we will see in detail in Chapter 5.)

Furthermore, Maxwell asserts, the encounters of the system of articles with the walls of the box will "introduce a disturbance into the motion of the system, so that it will pass from one undisturbed path to another." He continues:

It is difficult in a case of such extreme complexity to arrive at a thoroughly satisfactory conclusion, but we may with considerable confidence assert that except for particular forms of the surface of the fixed obstacle, the system will sooner or later, after a sufficient number of encounters pass through every phase consistent with the equation of energy.

Here we have introduced the "ensemble" approach to statistical mechanics, considering infinite collections of systems all compatible with the macroscopic constraints but having their micro-states take on every possible value. And we have the identification of equilibrium quantities with averages over this ensemble relative to a specified probability assigned to any collection of micro-states at a given time. We also have another one of the beginnings of the ergodic theory, that attempt to rationalize, on the basis of the constitution of the system and its dynamical laws, the legitimacy of the choice of one particular probability distribution over the phases as the right one to use in calculating average values.

In the 1884 paper, Boltzmann takes up the problem of the calculation of equilibrium values (a paper in which the term "ergodic" appears for the first time). Here he studies the differences between systems that will stay confined to closed orbits in the available region of phase space, and those that, like the hypothesized behavior of the swarm of molecules in a gas, will be such that they will wander all over the energetically available phase space, "passing through all phase points compatible with a given energy." In 1887 he utilizes the Maxwellian concept of an ensemble of macroscopically similar systems whose micro-states at a time take on every realizable possibility and the Maxwellian notion of a "stationary" probability distribution over such a micro-state.

The justifiable roles of (1) collections of macroscopically similar systems whose micro-states take on every realizable value, of (2) probability distributions over such collections, of stationary such probability distributions, of (3) the identification of equilibrium values with averages of quantities that are functions of the micro-state according to such probability measures, and of (4) the postulates that rationalize such a model by means of some hypothesis about the "wandering of a system through the whole of phase space allowed by energy" become, as we shall see, a set of issues that continue to plague the foundations of the theory.

III. Gibbs' statistical mechanics

1. Gibbs' ensemble approach

J. Gibbs, in 1901, presented, in a single book of extraordinary compactness and elegance, an approach to the problems we have been discussing that although taking off from the ensemble ideas of Boltzmann and Maxwell, presents them in a rather different way and generalizes them in an ingenious fashion.

Gibbs emphasizes the value of the methods of calculation of equilibrium quantities from stationary probability distributions reviewed in the last section. He looks favorably on the ability of this approach to derive the thermodynamic relations from the fundamental dynamical laws without making dubious assumptions about the details of the inter-molecular forces. He is skeptical that at the time, enough is known about the detailed constitution of gases out of molecules to rely on hypotheses about this constitution, and his skepticism is increased by results, well known at the time, that seem in fact to refute either the kinetic theory or the standard molecular models. In particular, the equipartition theorem for energy gives the wrong results even for the simple case of diatomic molecules. Degrees of freedom that ought to have their fair share of the energy at equilibrium seem to be totally ignored in the sharing out of energy that actually goes on. This is a problem not resolved until the underlying classical dynamics of the molecules is replaced by quantum mechanics.

In Gibbs' method we consider an infinite number of systems all having their micro-states described by a set of generalized positions and momenta. As an example, a monatomic gas with n molecules can be described by $6n$ such coordinates, 3 position and 3 momentum coordinates each for each point molecule. In a space whose dimensionality is the number of these positions and momenta taken together, a point represents a single possible total micro-state of one of the systems. Given such a micro-state at one time, the future evolution of the system having that micro-state is given by a path from this representative point, and the future evolution of the ensemble of systems can be viewed as a flow of points from those representing the systems at one time to those dynamically determined to represent the systems at a later time.

Suppose we assign a fraction of all the systems at a given time to a particular collection of phase points at that time. Or, assign to each region of phase points the probability that a system has its phase in that region at the specified time. In general, the probability assigned to a region will change as time goes on, as systems have their phase points enter and leave the region in different numbers. But some assignments of probability will leave the probability assigned a region of phases constant in time under the dynamic evolution. What are these assignments?

Suppose we measure the size of a region of phase points in the most natural way possible, as the "product" of its position and momentum sizes, technically the integral over the region of the product of the differentials. Consider a region at time t_0 of a certain size, measured in this way. Let the time go to t_1 and see where the flow takes each system whose phase point at t_0 was a boundary point of this region. Consider the new region at t_1 bounded by the points that represent the new phases of the "boundary systems." It is easy to prove that this new region is equal in "size" to the old.

Suppose we calculated the probability that a system is in a region at a time by using a function of the phases, $P(q,p)$, a probability density, such that the probability the system is in region A at time t is

$$\int \ldots \int_A P(q,p)\ dq, \ldots dqndp, \ldots dpn$$

What must $P(q,p)$ be like that the probability assigned to region A is invariant under dynamic evolution? We consider first the case where the systems have any possible value of their internal energy. The requirement on $P(q,p)$ that satisfies the demand for "statistical equilibrium" – that is, for unchanging probability values to be attributed to regions of the phase space – is simply that P should be a function of the q's and p's that stays constant along the path that evolves from any given phase point. If we deal with systems of constant energy, then any function P that is constant for any set of systems of the same energy will do the trick.

Gibbs suggests one particular such P function as being particularly noteworthy:

$$P = \exp(\psi - \varepsilon/\theta)$$

Where θ and ψ are constants and ε is the energy of the system. When the ensemble of systems is so distributed in probability, Gibbs calls it canonically distributed. When first presented, this distribution is contrasted with others that are formally unsatisfactory (the "sum" of the probabilities diverges), but is otherwise presented as a posit. The justification for this special choice of form comes later.

Now consider, instead of a collection of systems each of which may possess any specific energy, a collection all of whose members share a common energy. The phase point for each of these systems at any time is confined to a sub-space of the original phase space of one dimension less than that of the full phase space. Call this sub-space, by analogy with surfaces in three-dimensional space, the energy surface. The evolution of each system in this new ensemble is now represented by a path from a point on this surface that is confined to the surface.

Given a distribution of such points on the energy surface, how can they be distributed in such a way that the probability assigned to any region on the surface at a time is always equal to the probability assigned to that region at any other time? Once again the answer is to assign probability in such a way that any region ever transformed by dynamical evolution into another region is assigned the same probability as that latter region at the initial time. Such a probability assignment had already been noted by Boltzmann and Maxwell. It amounts to assigning equal probabilities to the regions of equal volume between the energy surface and a nearby energy surface, and then assigning probabilities to areas on the surface in the ratios of the probabilities of the "pill box" regions between nearby surfaces they bound. Gibbs calls such a probability distribution on an energy surface the micro-canonical ensemble.

There is a third Gibbsian ensemble – the grand canonical ensemble – appropriate to the treatment of systems whose numbers of molecules of a given kind are not constant, but we shall confine our attention to ensembles of the first two kinds.

2. The thermodynamic analogies

From the features of the canonical and micro-canonical ensembles we will derive various equilibrium relations. This will require some association of quantities calculable from the ensemble with the familiar thermodynamic quantities such as volume (or other extensive magnitudes), pressure (or other conjugate "forces"), temperature, and entropy. Gibbs is quite cautious in offering any kind of physical rationale for the associations he makes. He talks of "thermodynamic analogies" throughout his work, maintaining only that there is a clear formal analogy between the quantities he is able to derive from ensembles and the thermodynamic variables. He avoids, as much as he can, both the Maxwell–Boltzmann attempt to connect the macro-features with specific constitution of the system out of its micro-parts, and avoids as well their attempt to somehow rationalize or explain why the ensemble quantities serve to calculate the thermodynamic quantities as well as they do.

He begins with the canonical ensemble. Let two ensembles, each canonically distributed, be compared to an ensemble that represents a system generated by the small energetic interaction of the two systems represented by the original ensembles. The resulting distribution will be stationary only if the θ's of the two original ensembles are equal, giving an analogy of θ to temperature, because systems when connected energetically stay in their initial equilibrium only if their temperatures are equal.

The next analogy is derived by imagining the energy of the system, which also functions in the specification of the canonical distribution, to be determined by an adjustable external parameter. If we imagine every system in the ensemble to have the same value of such an energy fixing parameter, and ask how the canonical distribution changes for a small change in the value of that parameter, always assuming the distribution to remain canonical, we get a relation that looks like this:

$$d\varepsilon = -\theta d\overline{\eta} - \overline{A}_1 da_1 \ldots - \overline{A}_n da_n$$

where $\eta = \psi - \varepsilon/\theta$, the a_i's are the adjustable parameters, and the A_1's are given by $A = -d\varepsilon/da_i$. A bar over a quantity indicates that we are taking its average value over the ensemble. If we compare this with the thermodynamic,

$$d\varepsilon = TdS - A_1 da_1 - \ldots - A_n da_n$$

we get the analogy of θ as analogous to temperature, and $-\overline{\eta}$ as analogous to entropy (where the $A_i da_i$ terms are such terms as PdV, and so on).

Next, Gibbs takes up the problem of how systems will be distributed in their energies if they are represented by a canonical ensemble. The main conclusion is that if we are dealing with systems with a very large number of degrees of freedom, such as a system of gas molecules, then we will find the greatest probability of a system having a given energy centered closely around a mean energetic value.

This leads to the comparison of the canonical with the micro-canonical ensemble, and eventually to a picture of the physical situations that the two ensembles can plausibly be taken to represent. The conclusion Gibbs comes to is that a canonical ensemble, with its constant, θ, analogous to temperature, best represents a collection of identically structured systems all in perfect energetic contact with an infinite heat bath at a constant temperature. Here one would expect to find the systems closely clustered (if they have many degrees of freedom) about a central value of their energy, but with systems existing at all possible energies distributed around this central value in the manner described by the canonical distribution.

The micro-canonical ensemble, on the other hand, seems to be the appropriate representative of a collection of systems each perfectly isolated energetically from the external world. In the former case, the thermodynamic result of each system at a constant energy when in contact with the heat bath is replaced by the idea of the systems having varying energies closely centered around their thermodynamically predicted value. In the micro-canonical case the thermodynamic idea of an energetically isolated system as in a perfect, unchanging equilibrium state is replaced by that of systems whose components fluctuate from their equilibrium values, but that are, with overwhelming probability, near equilibrium and at equilibrium "on the average" over time. An examination of the fluctuation among components in an ensemble whose members are micro-canonically distributed shows that these fluctuations will be describable by a canonical distribution governed by the equilibrium temperature of the isolated system.

One can find thermodynamic analogies for the micro-canonical ensemble as one can for the canonical ensemble. The quantity "analogous" to entropy turns out to be log V, where V is the size of the phase space region to which the points representing possible micro-states of the system are confined by the macro specification of the system. Analogous to temperature is $d\varepsilon/d\log V$, although, as Gibbs is careful to point out, there are difficulties in this analogy when one comes to treating the joining together of systems initially isolated from one another.

Gibbs is also careful to point out that there are frequently a number of distinct quantities that converge to the same value when we let the number of the degrees of freedom of the system "go to infinity." When we are dealing with systems with a finite number of degrees of freedom, one of these quantities may be most "analogous" to a thermodynamic quantity in one context, whereas another one of them might, in its functional behavior, be more analogous to that thermodynamic feature in some other context.

Gibbs also points out that the values calculated for a quantity using the canonical ensemble and its thermodynamic analogy, and those calculated from the micro-canonical ensemble with its appropriate analogy, will also coincide in the limit of vast numbers of degrees of freedom. Because it is usually easier to analytically evaluate quantities in the canonical ensemble framework, this suggests this framework as the appropriate one in which to actually carry out calculations, even though Gibbs himself declares that the canonical ensemble "may appear a less natural and simple conception than that which we have called a micro-canonical ensemble of systems."

In his discussion of the thermodynamic analogies for the microcanonical ensemble, Gibbs also points out clearly the distinction between average values calculated by means of an ensemble and most probable values calculated by means of the same ensemble. Once again for finite systems these may well differ from one another, even if the usual convergence of values in the limit of systems of vast numbers of degrees of freedom seems to obliterate the distinction when it comes to calculating the values of thermodynamic quantities by means of the analogies.

3. The theory of non-equilibrium ensembles

The Gibbsian program supplies us with a method for calculating equilibrium values of any thermodynamic quantity. Given the constraints imposed on a system, set up the appropriate canonical or micro-canonical ensemble of systems subject to those constraints. Calculate from this ensemble the appropriate analogue to the thermodynamic quantity and identify the value obtained as the thermodynamic value for the system in equilibrium.

But how are we, within this ensemble view, to understand the inevitable approach to equilibrium of systems initially not in equilibrium? In Chapter XIII of his book, Gibbs shows, for a number of standard cases such as that of two systems originally isolated and at equilibrium that are then allowed to energetically interact, that if we assign to the systems the appropriate ensembles, the features of equilibrium so described will match the thermodynamic prediction. For example, in the case cited, after the interaction the component of the new combined system that was originally at the higher temperature will have transferred heat to the component of the new combined system that was originally at the lower temperature.

But what right do we have to assume that the modified system will be appropriately described by a canonical or micro-canonical ensemble? The problem here is this: We have already described by means of an ensemble the initial state of the system when it begins its transition to a new equilibrium by, say, having the constraints to which it originally was subject changed, making its old equilibrium condition no longer an equilibrium condition. But each system of the original ensemble is determined in its dynamical evolution by the underlying dynamical laws during the period after the constraints have been changed. So it is not fair to simply choose a new ensemble description of the system at the later time. Instead, the appropriate ensemble description that evolves by dynamics from the ensemble description of the system at the earlier time must be the ensemble description of the system at the later time. Can we in any way justify describing the system at the later time by means of the appropriate canonical or micro-canonical ensemble, rationalizing this by an argument showing that such a description is the appropriate dynamically evolved ensemble from the original ensemble?

Gibbs takes up this problem in two chapters of the book. Chapter XI is devoted to showing that the canonical ensemble is, in a certain sense, the ensemble that "maximizes entropy." Chapter XII takes up the dynamical evolution of an ensemble, as the member systems of the ensemble follow the evolution dictated to them by dynamical laws.

We have already seen how, for the canonical ensemble, Gibbs offers, by means of the thermodynamic analogies derived from considerations of the variation of the ensemble with varying external constraints, $\bar{\eta}$ or $-\overline{\log P}$, which is $-\int P \log P \, dp \, dq$ as the ensemble analogue to the thermodynamic entropy. In Chapter XI, Gibbs proves a number of "extremal" theorems about this ensemble analogue to entropy. The major conclusions are that (1) if an ensemble of systems is canonically distributed in phase, then the average index of probability, $\bar{\eta}$, is less than in any other distribution of the ensemble having the same average energy; and (2) a uniform distribution of a given number of systems of fixed energy within given limits of phase gives a smaller average index of probability of phase than any other distribution. Together, these two results give us, in a certain sense, the result that the canonical and micro-canonical distributions are, for their respective cases of fixed average energy or fixed energy, the ensembles that maximize ensemble entropy.

Next, Gibbs takes up the problem of the dynamic evolution of ensembles. Suppose, for example, we have a micro-canonical ensemble determined by some fixed values of constraints. We change the constraints. Can we expect the micro-canonical ensemble to evolve, in a manner determined by the dynamic evolution of each system in the ensemble in accordance with the dynamical laws, into the micro-canonical ensemble relative to the newly changed constraints?

Now the ensembles we want to be the endpoint of evolution are the unique ensembles that are such that two constraints are met: (1) the number of systems between any values of functions of the phases that are constants of motion are constant; and (2) the value of $\bar{\eta}$, the mean of the index of probability, is minimized. Can we show, then, that the ensemble whose constraints have been changed will have its $\bar{\eta}$ value evolve with time toward the minimum value? That would show that the ensemble is evolving to the unique, stationary ensemble consistent with the constants of motion that are the standard equilibrium ensemble. That is, can we use the "excess" of the $\bar{\eta}$ value of the ensemble at a time over the $\bar{\eta}$ value of the equilibrium ensemble as an indicator of how far the ensemble is from equilibrium, and then prove the monotonic decrease of this value for the evolving ensemble?

Alas, a simple application of the same results that were required to show the canonical and microcanonical ensembles stationary in the first place shows that any ensemble, no matter how it evolves, will always have a constant value of $\bar{\eta}$.

But, says Gibbs, one ensemble may in some sense approach another even if some function of the first does not approach that function of the other. He offers an explanation by analogy. Imagine two mutually insoluble fluids, A and B, in initially different but connected regions of a container. Stir the fluids. Because one never dissolves in the other, the volume occupied by fluid A is always constant and equal to its original volume. And the density of A at any point is always 1 or 0, just as it was when the mixing began. But take any small, fixed, spatial region of the container and look at the portion of that region occupied by the A fluid. Initially this will, except for a small number of regions on the boundary of the fluids, be the fraction 1 or 0. But as the mixing continues, we will expect, after a sufficiently long time, that the portion of any region that is occupied by the A fluid will be the same fraction of that region as the portion of original fluid that was A. (See Figure 2-7.)

We can expect an analogous result with ensembles. We divide the phase space into small but non-zero regions. When the constraints are changed, the ensemble initially fills a box entirely or the box is entirely unoccupied. As dynamic evolution goes on, the original compact ensemble "fibrillates" into a complex "strung out" ensemble ranging over the entire region of now, newly accessible phase space. Eventually we expect the portion of each box occupied by the ensemble to be the same, and equal to the original portion of the expanded phase space occupied by the ensemble when it started. In this sense the ensemble has become "uniform" over the expanded region of phase space, and, is therefore "like the appropriate equilibrium ensemble, which is genuinely, uniformly so spread, and not just spread uniformly in this new coarse-grained" sense. If we let time go to infinity and then let the boxes go to zero in

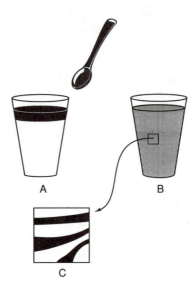

Figure 2-7. Mixing of insoluble fluids. A glass is filled 90% with clear water and 10% with a black ink as in A, the water and ink being insoluble in one another. The glass is stirred. The apparent result is uniform light gray inky water, as in B. But a magnified look at a small region of the fluid in B, as shown in C, reveals that each point is either totally in clear water or totally in black ink. The portion of any small volume filled by black ink is, however, 10%, if the ink is throroughly mixed into the water.

the limit, we will have a mathematical representation of this "spreading to uniformity" of the initial ensemble. We would miss this spreading if we first let the boxes go to zero size, and then let time go to infinity, because that would give us, once again, the result of constancy of the measure of "spread-outness" of the initial ensemble. (See Figure 2-8.)

With this new coarse-grained notion of spread-outness of an ensemble, we can introduce a new measure of the deviation of an ensemble from the equilibrium ensemble. Take the density of the ensemble in box i at a time as P_i, and define the "coarse-grained entropy" as $-\Sigma_i P_i \log P_i$. Whereas $\int P \log P d\Gamma$ is provably invariant, we can expect $\Sigma_i P_i \log P_i$ to decrease to its minimal value as time goes on. Gibb does point out that there will be "exceptional" ensembles that will not approach uniformity in even this coarse-grained sense. He doesn't offer any proof, however, that such "exceptional" ensembles will really be rare "in the world." Nor is there any proof that the approach to equilibrium, in this new sense, will be monotonic, nor that the time scale will be the one we want to replicate for the ensembles in the actual approach to equilibrium in time of non-equilibrium systems in the world. (See Figure 2-9.)

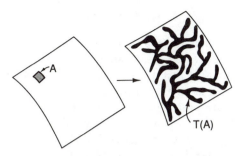

Figure 2-8. Coarse-grained spreading of an initial ensemble. Region A represents the collection of points in phase space corresponding to a collection of systems. Each system has been prepared in a non-equilibrium condition that is macroscopically the same. But the preparation allows a variety of microscopic initial states. As the systems evolve following the dynamics governing the change of microscopic state, A develops into T(A). The size of T(A) must be equal to that of A by a law of dynamics, but whereas A is a simple region confined to a small portion of the available phase space, T(A) is a complex fibrillated region that is spread all over the available phase space "in a coarse-grained sense." A uniform distribution over the available phase space is what corresponds to equilibrium in the theory. T(A) is not really uniformly spread over the phase-space, but may nevertheless be considered to represent a spreading of the initial ensemble that represents approach to equilibrium.

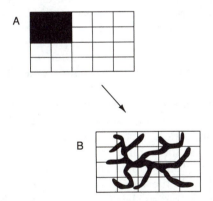

Figure 2-9. Coarse-grained entropy increase. Suppose an ensemble starts with the situation as noted in A – that is, each coarse-graining box is either totally full of ensemble points that are spread uniformly throughout the occupied boxes, or else totally empty of points. In B, the initial ensemble has become fibrillated and coarsely spread out over the available phase space. One can certainly show that the Gibbs coarse-grained entropy in B must be higher than it is in A, for by the definition of coarse-grained entropy it is minimal for the situation in which boxes are totally full or totally empty. But that, as the Ehrenfests noted, is a long way from showing that coarse-grained entropy will always increase monotonically, that it will approach the appropriate entropy for the new equilibrium condition after the constraints have been changed, or that it will approach this value in the appropriate time scale to represent observed approach to the new equilibrium.

Gibb does point out the provable fact that the initial coarse-grained entropy can, at least, not be greater than the entropy of future states. But it is also the case that the fact that one can expect in the future ensembles of greater and greater coarse-grained entropy is matched by similar statements about the past. The "increase of coarse-grained entropy toward the future," if we have a right to expect it, is matched by a similar expectation of increase of coarse-grained entropy in the past of a system. That is, if we trace the dynamic evolution of the systems that constitute a non-equilibrium ensemble at a given time back into the past, we will generate ensembles that are more uniform in the coarse-grained sense than the ensemble at the specified time.

It is worth quoting Gibbs' response to this time-symmetry paradox in full, as it represents a theme we will see recur as the paradox of time-symmetry is faced by later investigators:

But while the distinction of prior and subsequent events may be immaterial with respect to mathematical fictions, it is quite otherwise with respect to events in the real world. It should not be forgotten, when our ensembles are chosen to illustrate the probabilities of events in the real world, that while the probabilities of subsequent events may be often determined from probabilities of prior events, it is rarely the case that probabilities of prior events can be determined from those of subsequent events, for we are rarely justified in excluding from consideration the antecedent probability of the prior events.

We shall try to fathom what this means later on.

In Chapter XIII, Gibbs applies his results to the discussion of the appropriate ensemble description of various non-equilibrium processes. Here, Gibbs focuses not on the evolution of an ensemble representing an isolated system, but on the course of change to be expected as external constraints are varied. First, he considers constraints being abruptly varied. Here, changing the parameter does not effect the systems directly but does change the distribution from an equilibrium one to one that is now non-equilibrium in the newly accessible phase-space. In this case, one expects the ensemble to change in such a way that it finally, in the "coarse-grained" sense, approximates the equilibrium ensemble for the newly specified parameters. And one expects, therefore, the coarse-grained entropy to increase. In the case of slow enough change of parameter, one expects the ensemble to vary in such a way that even the coarse-grained entropy is constant.

Gibbs then takes up the problem of the bringing into energetic contact of systems originally energetically isolated, and argues, once again, that the resulting change of the ensemble representing the composite system can be expected to evolve, in a coarse-grained sense, in such a way that the resulting ensemble eventually comes to approximate the equilibrium ensemble appropriate to the new joint system subject to its overall external constraints. For example, heat having flowed from the originally higher temperature body to the one originally at lower temperature, the two components will be at equal temperatures as represented by the appropriate "modulus" constant of the ensemble.

IV. The critical exposition of the theory of P. and T. Ehrenfest

In 1912, P. And T. Ehrenfest published a critical review of the then current state of kinetic theory and statistical mechanics entitled "The Conceptual Foundation of Statistical Mechanics." Published in the *Encyclopedia of Mathematical Sciences*, the article provided a brilliantly concise and illuminating, if sometimes controversial, overview of the status of the theory at that time. This piece can be considered the culmination of the early, innovative days of kinetic theory and statistical mechanics, and the beginning of the critical discussion of the foundations of the theory that accompanied and still accompanies its scientific expansion and development. The piece is directed first to an exposition of the original kinetic theory and its changes in response to the early criticism, along with a discussion of remaining unsolved problems in the "statistical-kinetic" approach. It then moves to a critical exposition and critique of Gibbs' statistical mechanics.

The general argument of the piece is this: The development of kinetic theory from Krönig and Clausius to Maxwell and Boltzmann culminated in Boltzmann's kinetic equation and his proof of the *H*-Theorem. The *H*-Theorem led to the criticisms of the theory summed up in the Reversibility and Recurrence Objections of Loschmidt and Zermelo. These led Boltzmann to a reinterpretation of the description of system evolution given by his *H*-Theorem and kinetic equation that is perfectly consistent with the underlying dynamical theory. Despite this there are those who remained unconvinced that Boltzmann had avoided the fundamental problems of consistency posed by the Reversibility and Recurrence Objections. In order to show that he had, one must resolve many ambiguities in Boltzmann's expression of his views. This elimination of obscurities proceeds primarily by "transforming the terminology of probability theory step by step into hypothetical statements about relative frequencies in clearly defined statistical ensembles." On the other hand, the revised Boltzmannian theory of non-equilibrium does require for its justification postulates whose plausibility and even consistency remain in doubt.

Furthermore, "a close look at this process of development shows that the systematic treatment, which W. Gibbs attempts to give ... covers only a small fraction of the ideas (and not even the most important ones) which have come into being through this process."

1. The Ehrenfests on the Boltzmannian theory

The older formulation of the theory. The Ehrenfests first offer a brief survey of the evolution of assumptions about "equal frequencies of occurrence" in the early development of kinetic theory. Krönig asserted that the motions of the molecules being very irregular, one could, "in accordance with the results of probability theory" replace the actual motion by a regular motion – for example, assuming the molecules all have the same speed. Clausius made a number of assumptions about equal frequencies of molecular conditions, some explicit and some only tacit. He assumed, for example, that the spatial density of the molecules was uniform, that the frequencies of molecular speeds did not vary from place to place, and that all directions of molecular motion occured with equal frequency. He also, tacitly, assumed the *Stosszahlansatz*.

These are all postulates about "equal frequency" of conditions. What about frequency postulates where the frequencies are not all equal – the non-*Gleichberechtigt* cases? Clausius had already assumed that at equilibrium there was a definite distribution of molecular velocities, even if he didn't know what it was. Maxwell posited his famous distribution law, guided, it seems, by the Gaussian theory of errors. Boltzmann succeeded in generalizing this distribution law to the cases of gases in external potentials and to the cases of polyatomic gases from Maxwell's law, which holds for monatomic gases.

But how can one derive the equilibrium velocity distribution law? Maxwell and Boltzmann, treating dynamic evolution of the system, showed the equilibrium distribution to be stationary. Boltzmann, by means of the H-Theorem, showed it to be the uniquely stationary distribution. The H-Theorem requires a mechanical assumption in its proof, but it also requires the kinetic equation, which assumes the truth of the *Stosszahlansatz*. So we can reduce the case of non-equal frequency assumptions to an equal-frequency assumption, the *Stosszahlansatz*. But clearly, then, the kinetic equation and the H-Theorem are the central components of the theory that we must critically explore.

Now Boltzmann had defined the quantity H and shown that if the evolution of the distribution function obeyed his kinetic equation, H must monotonically decrease. But Loschmidt's Reversibility Objection and Zermelo's Recurrence Objection show that this alleged monotonic decrease in H from any non-equilibrium distribution cannot be valid. This led to the new "Statistical Mechanical" or "Kineto-Statistical" approach to the theory.

The "modern" formulation of the theory of equilibrium. The revised approach to the theory of equilibrium, according to the Ehrenfest account, is one that takes as its basic arena the phase space of points, each of which represents a possible micro-state of the system compatible with its macroscopic constraints. The evolution of a system under the dynamical laws is represented by a path from a point in the phase space. If we measure volume in phase space in the standard way noted in Section III,1, a theorem of dynamics, Liouville's Theorem, will show us that the volume assigned to a set of points will be the invariant volume of the regions that set evolves into under dynamic evolution. If we restrict our attention to points all of which have a common energy – that is, to an energy hypersurface in the phase space – and to flows confined to that surface, a new measure of "area" on the surface – the micro-canonical measure discussed earlier – will play the role of invariant measure under dynamical flow. If the probability distribution over all phase space is an arbitrary function of the constants of motion of the systems, it will be constant along any path. And it will be stationary in time, always assigning the same probability to any region of the phase space as the systems evolve. In the energy-surface constrained case, a similar probability function, normalized by the surface measure, will similarly be a stationary probability distribution.

At this point, the Ehrenfests introduce their notion of an ergodic system as one whose path in phase space "covers every part of a multi-dimensional region densely." They attribute to Maxell and Boltzmann the definition of an ergodic system as one whose phase point "traverses every phase point which is compatible with its given total energy." (Although, as historians point out, it isn't at all clear that either Maxwell or Boltzmann ever really defined the relevant systems in exactly this way.) If the systems obey this latter condition, then every system in the ensemble traverses the same path if we follow it over an unlimited time. Therefore, every system will have the same time average – defined by $\lim_{t\to\infty}\frac{1}{t}\int_0^t f(p,q,t)dt$ – of any quantity that is a function of the phase value of the system,

But, the Ehrenfests maintain,

The existence of ergodic systems, *i.e.* the consistency of their definition, is doubtful. So far, not even one example is known of a mechanical system for which the single G-path, that is the path in the phase space under dynamic evolution, approaches arbitrarily closely each point of the corresponding energy surface. Moreover, no example is known where the single G-path actually traverses all points of corresponding energy surface.

But it is just these assumptions that are necessary for Maxwell and Boltzmann's claim that in the gas model, all motions with the same total energy will have the same time average of functions of the phase.

Next, special distribution functions are introduced: those over phase space that are a function of energy alone, of which the canonical distribution is a special case; and the standard micro-canonical distribution over the energy surface. It is these distributions that are constantly used to derive results in gas theory, despite the fact that they were originally introduced as justified by the assumption of ergodicity, even though ergodicity is not usually even mentioned by those who have taken up the use of these distributions in the foundations of their work.

At this point, the Ehrenfests offer a plausible reconstruction of what "may have been" the starting point of Boltzmann's investigation – that is, of his suggested plan to use the micro-canonical ensemble average as the appropriate representative of equilibrium values. Begin with the empirical observation that a gas goes to equilibrium and stays there. So the average behavior of a gas over an infinite time ought to be identical to its behavior at equilibrium. Can we calculate "time averages of functions of the phase over infinite time," in the limiting sense of course? In particular, can we show that the infinite time average of a gas corresponds to the Maxwell–Boltzmann distribution? Take any phase quantity f – that is, $f(p,q)$. Then consider the following equalities:

(1) Ensemble average of f = time average of the ensemble average of f
(2)   = ensemble average of the time average of f
(3)   = time average of f

Equality (1) follows from the choice of a stationary probability distribution over the ensemble. Equality (2) follows from the legitimacy of interchanging averaging processes. Equality (3), however, is crucial. To derive it, we must invoke the postulate of ergodicity, for it is that postulate that tells us that the time average of f is the same for each system in the ensemble. This is so because the dynamic path from each point is always the same path if it is true that each system eventually traverses each dynamical point. The Ehrenfests point out that the derivation, as they have reconstructed it, looks rather more like the work of Maxwell, Lord Rayleigh, and J. Jeans than it does like that of Boltzmann, because Boltzmann chooses to derive from ergodicity the fact that the time spent by the system in a region of phase space goes – given ergodicity and in the limit as t goes to infinity – to the size of the region. But, they say, and as we shall see in detail in Chapter 5, that conclusion is equivalent to the earlier one that proves the identity of ensemble phase averages with infinite time averages given the posit of ergodicity.

The Ehrenfests conclude with the observation that given the postulate of ergodicity, the facts about infinite time averages and about relative duration of a system in states corresponding to a specified region of phase space in the infinite time limit follow from dynamical considerations alone. No "probabilistic hypotheses" are called for. But if, on the other hand, we dispense with the posit of ergodicity, it is unclear what rationales the probabilistic hypotheses of the standard distribution of probability over the ensemble have in the first place.

The statistical approach to non-equilibrium and the H-Theorem.
At this point, the discussion moves on to an attempt to make fully unambiguous the statistical reading of the kinetic equation and the *H*-Theorem. In the process the Ehrenfests offer their demonstration of the consistency of these important non-equilibrium results with the underlying dynamics.

First, apparatus is set up that will allow the Ehrenfests to clear up a number of misunderstandings resulting from Boltzmann's loose language. Consider the phase space appropriate for a single molecule – that is, the space such that points in it represent the position and momentum, in all degrees of freedom, of one of the molecules of the system. Coarse-grain this phase space into small boxes, each of equal extent in position and momentum. Now consider the position and momentum of each molecule of the gas at a given time. This results in a certain number of molecules having their representative points in this small phase space, called μ-space, in each i-th box. Call the number in box i, a_i, and a given specification of the a_i's a state-distribution, Z. To each Z corresponds a region of points in the original "big" phase space, now called Γ-space – that is, the region of all those phase points representing the total molecular state of the gas with a given corresponding state-distribution, Z.

First, note that the region of Γ-space corresponding to the Z that is the Maxwell–Boltzmann distribution is overwhelmingly large and dominates the allowed region of Γ-space open to the gas. If we assume ergodicity, this will justify Boltzmann's claim that for the overwhelming amount of time (in the infinite time limit) spent by the gas, it is in the Maxwell-Boltzmann distribution.

Now, consider any function F of Z. Considered as a function of p,q it will be discontinuous, changing its value when p,q change in such a way that an a_i changes its value. One such function is $H(Z) = \Sigma_i a_i \log a_i$, and it is this H that will play a crucial role in our discussion of the *H*-Theorem. Next, let us discretize time, looking at the gas only at a number of discrete time points separated by an interval Δt from each other.

With this apparatus we can now present a tidied-up version of Boltzmann's claims about the behavior of H over time. We can say that $H(Z)$ will remain for overwhelmingly long time intervals near its minimum value, H_0. If H_1 is a value of H much above the minimum, then the sum of the time intervals in which H has a value at H_1 or above decreases rapidly with the increasing value of H_1. If we now move to discrete times, and look at the value of $H(Z)$ at each time, we can argue that whenever Ha, Hb, and Hc follow one another in time and all are well above H_0, when we look at neighboring time moments we will usually find situations in time like this:

$$Hb$$
$$Ha \qquad Hc$$

Much less often, but with equal frequency, we will find:

$$Ha \qquad\qquad\qquad\qquad Hc$$
$$Hb \qquad\qquad\qquad Hb$$
$$Hc \quad \text{or} \quad Ha$$

Only very rarely will we observe the following pattern to be the case:

$$Ha \qquad Hc$$
$$Hb$$

So, from a state with an H much above its minimum value we will almost always find that the immediately succeeding state is one with a lower value of H. (And almost always the immediately preceding value of H will be lower as well!)

In what sense, then, could the H-Theorem be true? Pick a Z_a, at t_a such that $H(Z_a)$ is well above H_0. Corresponding to this Z_a is a region of points in Γ-space. From each point, a dynamic path goes off, leading to states that at later times, $t_a + n\Delta t$ have a value H_n. The claim is that there will be values of $H, H_1, \ldots H_n$ such that nearly all the systems will have at time t_i, H values at or near H_i. The set of values $H_1, \ldots, H_n$ we call the concentration curve of the bundle of H-curves. We then assert that this curve monotonically decreases from its initial high $H_a(Z_a)$ value, converges to the minimum value H_0, and never departs from it. At any time t_n, the overwhelming majority of curves of H will be at a value near H_n, but, as the Recurrence Theorem shows us, only a set of measure zero of curves of individual systems will themselves behave in a manner similar to the concentration curve. Next, we claim that the curve of the H-Theorem, derived from the kinetic equation that presupposes the *Stosszahlansatz*, is identical to this concentration curve. Note that neither of the claims made (that the concentration curve will show this monotonic decreasing behavior and that it will, in fact, be replicated by the curve of the H-Theorem) has been proven. The claim here is only that

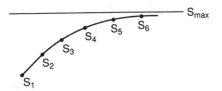

Figure 2-10. The concentration curve of a collection of systems. A collection of systems is considered, each member of which at time 1 has entropy S_1. The systems evolve according to their particular micro-state starting at the initial time. At later times, 2, 3, 4, 5, 6, . . ., the collection is reexamined. At each time, the overwhelming majority of systems have entropies at or near values S_2, S_3, S_4, S_5, S_6, . . ., which are plotted on the "concentration curve." This curve can monotonically approach the equilibrium value S_{max}, even if almost all the systems, individually, approach and recede from the equilibrium condition in the manner depicted in Figure 2-4.

a consistent reading of Boltzmann's claims has now been produced. An additional supplement, suggested by the Ehrenfests, is that a given observationally determined state of a gas must be posited to have one particular Z state overwhelmingly dominate the class of Z states compatible with that observational constraint. This is needed to connect the observational thermodynamic initial state with the posited definite initial Z_a in the earlier treatment of the H-Theorem. (See Figure 2-10.)

The picture presented here is time-symmetrical and perfectly compatible with the Objection from Reversibility. It is also compatible with the Objection from Recurrence. A further variant of the Reversibility Objection – that to each state there is the reverse state, and that if the former leads to an increase of H in a time interval, the latter must lead to a decrease of H in the same time interval of equal size – is noted to be fallacious, because there is no general argument to the effect that in a given time interval a state and its time reverse will lead to equal and opposite changes in the value of H for their respective systems.

Finally, the Ehrenfests take up the question of how to formulate properly the *Stosszahlansatz*, essential to the derivability of the kinetic equation and hence of the H-Theorem, in light of the new, statistical, understanding of the non-equilibrium situation. Boltzmann, they say, could be understood as making claims in his revised, statistical, view of irreversibility: (1) The *Stosszahlansatz* gives, for each time interval, only the most probable number of collisions, and the H-Theorem only the most probable value in the change in H; (2) the actual number of collisions (and the actual change in H) fluctuates around this most probable value. Here, once again, we must resolve obscurities and ambiguities by replacing "probability" language with a careful formulation in terms of relative frequencies in well-defined reference classes. While, the Ehrenfests say, statement (2) remained (at that time) to be clarified, J. Jeans, responding to criticisms of the *Stosszahlansatz* by S. Burbury, had made the meaning of statement (1) precise.

Consider a gas in state Z_a at a given time. How many collisions of a given kind will occur in time Δt? This depends on the specific phase of the gas at the time, and is under-determined by the Z-state. Look at the points in Γ-space corresponding to the Z-state. We can formulate the statistical *Stosszahlansatz* as the claim that the sub-region of the region of points in space corresponding to the Z-state in which the number of collisions is given by the *Stosszahlansatz* is the overwhelmingly largest part of the given region of Γ-space.

But to get the kinetic equation we need more. For we apply the *Stosszahlansatz* at every moment of time in deriving that equation. This leads to the Hypothesis of Molecular Chaos as a posit. Take the subset of the original region in Γ-space for which the *Stosszahlansatz* gave the right number of collisions in the time interval. Look at the systems those systems have evolved into at the end of that time interval, characterized by a new region of points in Γ-space. In order to derive the kinetic equation, we must assume that the overwhelming fraction of those points also represent systems whose collisions in the next interval of time will also be in accord with the *Stosszahlansatz*. And this hypothesis about the dominance of the *Stosszahlansatz* obeying region must be continuously repeated for the set into which that subset evolves, and so on.

We can get the result we want by considering the points reached by trajectories started in our initial region. These will correspond to different distributions of state, $Z_{B'}$, $Z_{B''}$, and so on. But each such $Z_{B'}$ region of points started in our initial region will correspond to only part of the Γ-space region corresponding to a given $Z_{B'}$ distribution, that part occupied by points on trajectories that started in our initial Z region. What we must assume is that in each such partial region corresponding to a Z-state, the largest region is that for which the *Stosszahlansatz* will hold, just as we assumed it for the whole region of Γ-space corresponding to a given initial Z-state.

Although this new Hypothesis of Molecular Chaos may need subtle modification (because of molecular correlation induced by finite size of molecules, and so on), it shows, according to the Ehrenfests, that a statistical formulation of the *Stosszahlansatz*, and therefore a statistical derivation of the kinetic equation and *H*-Theorem, exists. And the derivation is immune to the Objections of Recurrence and Reversibility. Note, once again, that no claim as to the provability or derivability of the needed posit from underlying dynamics is being made here.

2. The Ehrenfests on Gibbs' statistical mechanics

Whereas the Ehrenfests view Boltzmann, aided by his predecessors and Maxwell, as the seminal thinker in the field, the researcher whose work defines the field, even if it is sometimes ambiguously stated and needful of clarification and supplementation, the Ehrenfests' critical view of Gibbs is rather unfavorable. In successive sections they offer an exposition and critique of Gibbs' approach: (1) to the theory of equilibrium; (2) to the theory of the irreversible approach to equilibrium; and (3) to the association of Gibbs' statistical mechanical quantities with thermodynamic quantities by means of the "thermodynamic analogies."

Critique of Gibbs' equilibrium theory. Gibbs, according to the Ehrenfests, first recapitulates Boltzmann's investigations into the most general stationary ensembles, and then fixes his attention on two "very special" cases – the micro-canonical distribution, which is equivalent to Boltzmann's ergodic surface distribution, and the canonical distribution.

Gibbs is able to compute many average values of functions of the phase quantities when the systems are canonically distributed in phase. For a given modulus (θ) of a canonical distribution, the overwhelmingly large number of systems have the same total energy, E_0, given the vast number of degrees of freedom of the system. For this reason, "it is plausible that in general the average over the canonical distribution will be very nearly identical with the average taken over the micro-canonical or even ergodic ensemble with $E = E_0$." On the other hand, that distribution will be almost identical to the one obtained by taking all those phase points in Γ-space that correspond to a number a_i of molecules in box i in coarse-grained Γ-space such that $\Sigma_i a_i \varepsilon_i = E_o$, where ε_i is the energy of a molecule "centered" in the i-th box, and then distributing probability uniformly over those phase points in Γ-space. But average values calculated by this last probability function will be values of phase quantities when the molecules are in the Maxwell–Boltzmann distribution. So, essentially, average values calculated using the canonical ensemble can be expected to be the same as values calculated by assuming that the gas is in one of its "overwhelmingly most probable" Maxwell–Boltzmann distributed states. "Thus, from the point of view of Boltzmann's presentation, the introduction of the canonical distribution seems to be an analytical trick" – that is, merely a device to aid in calculation.

And, the point is, Gibbs does not even mention the subtle problems of the posits of ergodicity needed to justify the claims that these methods are the appropriate methods for calculating equilibrium.

Critique of Gibbs on non-equilibrium and the approach to equilibrium. First, say the Ehrenfests, Gibbs defines a function $\sigma = \int \rho \log \rho \, dq dp$ and shows that this function takes on its minimum value subject to appropriate constraints (definite average energy for canonical case, definite energy for micro-canonical case) in the canonical and micro-canonical distributions. But this quantity is probably invariant under dynamical evolution, the proof being a simple corollary of Liouville's Theorem. Then Gibbs suggests dividing the Γ-space into small boxes, taking the average ensemble density in each box, P_i, and defining $\Sigma = \Sigma_i P_i \log P_i$ as a new measure of departure of an ensemble from its equilibrium distribution.

Gibbs concludes that every non-equilibrium ensemble will stream in such a way that the P_i's will become stationary in the infinite time limit and such that in the limit of infinite time, the value of $\Sigma(t)$ will assuredly be less than or equal to its value at the initial moment. His argument for this consists in the analogy with the mixing of insoluble liquids. Further, his arguments can be taken to show that in the infinite time limit, the P_i's will all be the same on a given energy surface.

Now one certainly can demonstrate the inequality noted here, for if we start with an ensemble all of whose coarse-grained boxes are either totally filled or totally empty, then $\Sigma(t)$ cannot increase in time, and may very well decrease.

But, say the Ehrenfests, Gibbs has not shown, by any means, all that needs to be shown. First, there is the question of the time the ensemble will take to spread out in such a way that $\Sigma(t)$ is at or near its minimum value. The Ehrenfests suggest that this can be expected to be many cycles of the time needed for a system that leaves a coarse-grained box to return to that box – that is, many of the " 'enormously large' Poincaré–Zermelo cycles." Nor does Gibbs' argument show in any way that the decrease of $\Sigma(t)$ with time will be monotonic. It could very well sometimes decrease and sometimes increase, even if the limiting inequality is true. Finally, Gibbs fails to show that the limiting value $\Sigma(t)$ will in fact be that value corresponding to the appropriate equilibrium ensemble for the now modified constraints. As far as Gibbs' proofs go, $\lim_{t \to \infty} \Sigma(t)$ could be arbitrarily larger than the fine-grained entropy of the appropriate canonical or micro-canonical ensemble. In his treatment of the thermodynamic analogies, Gibbs simply assumes, without noticing he is making the assumption, that one can identify the ultimate ensemble as that which approximates the equilibrium ensemble although, as just noted, he really has not demonstrated this, for he has only shown, at best, "a certain change in the direction of the canonical distribution."

What, finally, is the response of the Ehrenfests to Gibbs' observation that his "increase of coarse-grained entropy" result holds both forward and backward in time from an ensemble stipulation at a given time? Gibbs observed that we often infer future probabilities from present probabilities, but that such inference to the past was illegitimate. In their footnote 190, the Ehrenfests respond: "The penultimate paragraph of Chapter XII [the paragraph in which Gibbs makes the relevant observations quoted above], so important for the theory of irreversibility, is incomprehensible to us."

Critique of Gibbs on the thermodynamic analogies. As the final component of their critique of Gibbs' theory, the Ehrenfests offer a systematic comparison of how the components of the statistical-kinetic theory are to be associated with those of thermodynamics in the alternative approaches of Boltzmann and Gibbs.

First, they note that the Maxwell–Boltzmann distribution of molecules as they are distributed in μ-space is formally comparable to the canonical distribution function description of the distribution of systems in Γ-space in the canonical ensemble. If we think of slowly modifying the constraints on a system, we get, in the Boltzmann picture, a description of how the parameters determining the Maxwell–Boltzmann distribution vary with this parameter variation. This parallels Gibbs' description of how the parameters governing the canonical distribution will have their variation connected with variation of constraints in his approach. And both resulting equations will *formally* resemble the familiar equation governing the relation of thermodynamic quantities in the phenomenological theory.

For a system in equilibrium, Boltzmann thinks of the system as having a representative point in Γ-space, which characterizes its micro-state, as overwhelmingly likely to be in that dominating region of points that correspond to the Maxwell–Boltzmann distribution. From this distribution, one can then calculate the appropriate surrogates to the thermodynamic quantities. Gibbs, on the other hand, identifies averages over all phases in the appropriate canonical ensemble with the thermodynamic quantities.

How do Boltzmann and Gibbs treat the problem of two systems, initially each in equilibrium but energetically isolated, and then brought into a condition where they can exchange energy with one another? Boltzmann describes each system, prior to interaction, as having its appropriate Maxwell–Boltzmann distribution over its molecules. Once in interaction, they are in a non-equilibrium initial state. The kinetic equation and *H*-Theorem show that on the average (and with overwhelming frequency) the combined system will evolve to the new equilibrium Maxwell–Boltzmann state appropriate to the combined system. Gibbs will describe each component system initially by a canonical ensemble.

To represent the systems in thermal contact we pick a new ensemble each member of which is a combination of one system from each of the earlier ensembles. Unless the initial such ensemble is canonical, it will "smear out," in the coarse-grained sense, until the new ensemble "approximates" the appropriate canonical ensemble for the combined system.

How do they treat reversible processes? For Boltzmann, one can, given the slow variation in the parameters, treat the system as always in a Maxwell–Boltzmann distribution. Using the thermodynamic analogy relation obtained earlier, one can derive that entropy is (tentatively) generalizable to $-H = -\Sigma_i a_i \log a_i$ where a_i is the occupation number of the i-th box in coarse-grained μ-space. For Gibbs, slow change of parameters means that we can view the system as represented by a canonical ensemble at every moment. The analogue to entropy is given by $-\int \rho \log \rho \, dpdq$, where ρ is the density function of the canonical ensemble in Γ-space.

How do they treat the increase of entropy in an irreversible process? For Boltzmann, if the system does not have a Maxwell–Boltzmann distribution at a time, then by the H-Theorem or the kinetic equation, it will, in almost all motions, assume smaller values of H at later times. For Gibbs, it is the quantity $\Sigma_i P_i \log P_i$, defined by coarse-graining Γ-space and taking the P_i's as average ensemble density in the i-th box, whose decrease with time represents the changing ensemble and which, in the infinite time limit, must assume a smaller value than its initial value.

Concluding remarks. Summing up their critiques both of Boltzmann and Gibbs, the Ehrenfests remark that their conceptual investigation into the foundations of kinetic-statistical mechanical theory required that they "emphasize that in these investigations a large number of loosely formulated and perhaps even inconsistent statements occupy a central position. In fact, we encounter here an incompleteness which from the logical point of view is serious and which appears in other branches of mechanics to a much smaller extent."

But these foundational and conceptual difficulties have not prevented physicists from applying the basic modes of thought over wider and wider ranges of phenomena, generalizing from the theory of gases to radiation theory and other areas. How optimistic can one be that they will prove warranted in their optimistic use of the theory? Here the Ehrenfests are guarded in their prognosis, not least because of the notorious difficulties with the fundamental consequence of equilibrium theory – the equi-partition theorem – giving apparently wrong results in the case of the theory of poly-atomic molecules and absurd results (divergence of the distribution function with increasing frequency) in the case of radiation. These difficulties were not resolved until the classical dynamical underpinning of the theory was replaced with a quantum micro-dynamics.

V. Subsequent developments

In the three-quarters of a century since the Ehrenfest review article appeared, there has been an intensive and massive development of kinetic theory and statistical mechanics. To hope to even survey the multiplicity of ramifications of the theory would be impossible here. It will be helpful for our further exploration of foundational problems, however, to offer an outline of some of the high points in the development of the theory subsequent to its innovative first period described earlier.

Two of the following sections – 1 and 3 – will simply note some of the gains that have been made in extending the theory of equilibrium and of non-equilibrium to allow them to cover cases outside the range of the early theory. Whereas the early theory is primarily devoted to understanding dilute gases, the theory can be extended widely both to cover more difficult cases within the realm of kinetic theory proper (such as gases in which more than binary collisions are relevant, and so on) and to realms of phenomena such as magnetization of matter and the distribution of energy in radiation that although treatable in general statistical mechanics, go beyond the case of many-molecule gas systems in various ways.

The other two following sections – 2 and 4 – treat material more directly relevant to our project. Here I will offer the barest outline of the direction in which attempts have been made to continue the program of rationalizing or justifying the methods used in statistical mechanics, in particular in rationalizing and justifying the choice of probabilistic postulates such as the traditional ensemble probability distributions in equilibrium theory and the Hypothesis of Molecular Chaos in non-equilibrium theory. Such programs of rationalization and justification are also intimately connected with the problem of explaining *why* the posited probability assertions are true, if indeed they are. Or why, if they are not, the theory that posits them works as well as it does. Because the topics of these two sections will occupy our attention for three full chapters in Chapters 5, 6, and 7, all that is offered here is an outline to be fleshed out in greater detail.

1. The theory of equilibrium

From classical to quantum micro-theory. We have already seen how a major failure in the predictive reliability of equilibrium statistical

mechanics cast doubt upon the validity of the entire theory. One fundamental result of the theory is that at equilibrium the thermal energy of a
substance will be equally distributed over all of the degrees of freedom
available to its components. But even in the case of diatomic molecules,
the energy fails to be so distributed when the vibrational and rotational
degrees of freedom of the molecule are added to its translational degrees
of freedom. In the case of radiation, the equi-partition theorem leads to
totally incoherent results, because the possibility of standing waves in a
cavity having unlimitedly short wavelength leads to a distribution function that diverges with high frequency.

One of the origins of quantum mechanics is in Planck's study of the
thermodynamics and statistical mechanics of radiation. He was able to
get the empirically observed distribution curve by assuming that energy
is transferred from radiation to matter, and vice versa, only in discrete
packets, each having its energy proportional to the frequency of the
relevant radiation. This was generalized by Einstein to the view that
energy exists in such "quanta" even in the radiation field. Combined by
Bohr with the existence of stationary states of electrons in atoms and the
quantization of emission and absorption of radiant energy by them, we
get the older quantum theory. Soon, difficulties of applying that theory
to more general cases than the hydrogen atom, along with the desire
for a general quantum kinematics on the part of Heisenberg and an
exploration of "wave-particles" duality by de Broglie and Schrödinger,
give rise to the full quantum theory.

From the point of view of this theory, the underlying dynamics of the
components of a system (molecules of the gas, frequency components of
the radiation field, and so on) is not governed by classical mechanics at
all but by the quantum mechanical laws. This requires a total reformulation of kinetic theory and statistical mechanics. Where we previously
dealt with ensembles of micro-states, taking micro-states to be specifica-
tions of the positions and momenta of each molecule, now we must deal
with ensembles of quantum micro-states, where these are represented by
rays in a Hilbert space or, more generally, by density matrices.

We shall, of course, not divert ourselves into the mysteries encumbent
in a study of the meaning of the foundational concepts of quantum
mechanics. Nor shall we even devote a great deal of attention to the
ways in which a statistical mechanics founded upon an underlying
quantum mechanical micro-theory differs in its structure and its predictions from a statistical mechanics founded upon an underlying classical
mechanical micro-theory. But we will note here one curious apparent
consequence of the theory and a few ways in which the change in the
micro-theory does impinge on the study of the fundamental statistical
assumptions of statistical mechanics.

The curious consequence of the theory arises out of its statistical mechanics of systems made up of a multiplicity of similarly constituted particles. For reasons that are, to a degree, made available by some results of quantum field theory, it turns out that in considering possible states of systems made of particles with half-integral spin, we must construct our ensembles only over state-functions that are anti-symmetric under permutation of the particles. In constructing ensembles for particles whose spin is integral, we must restrict ourselves to symmetric wave functions. In both cases, the simple way of counting micro-states in the classical theory is abandoned. This is the method that takes micro-states obtained from one another by a mere permutation of particles from one phase to another as distinct micro-states. Statistically, the result is the creation of two new distribution functions: the Fermi–Dirac distribution for half-integral spin particles, and the Bose–Einstein distribution for particles with integral spin. The Maxwell–Boltzmann distribution remains only as an approximating distribution for special cases.

Philosophically, the new way of counting "possible micro-states compatible with the macroscopic constraints" leads us, if we insist upon thinking of the particles as really being particles – which, if quantum mechanics is correct is a position hard to maintain – to say very curious things both about the "identity through time" of a particle, about the alleged Principle of Identity of Indiscernibles, and about the possibility of apparently non-causal correlations among the behaviors of components of a compound system. Because I believe these are puzzles that can only be coherently discussed within the context of a general study of the foundations of quantum mechanics, it will be best to leave them to the side.

More relevant to our purposes are some special features of ensembles in the quantum mechanical statistical theory that do interfere with the smooth treatment of the problem of the nature and status of the fundamental statistical posits of statistical mechanics. In classical statistical mechanics, we can have an ensemble that has no "recurrence" over time in its distribution, even if the individual systems in the ensemble are all governed by the Poincaré Recurrence Theorem. In quantum statistical mechanics, this is not true. If we want a non-recurrent ensemble, we can only obtain it in some special way – for example, by going to the idealization of a system with an infinite number of constituents. Further, we can show that what we will call ergodicity and mixing results of various kinds obtainable in classical statistical mechanics cannot hold in the theory with a quantum micro-foundation.

Finally, as we shall see, a number of investigators have tried to found an explanatory theory of the nature of the fundamental statistical posits of the theory by arguing that these rest upon constraints in our ability to so prepare systems as to construct ensembles that fail to behave in the appropriate kinetic way. Their arguments frequently draw analogies between limitations on our preparation of systems, which result from purely classical considerations, with the alleged limitations on measurement and preparation familiar from the Uncertainty Relations in quantum mechanics. When we explore this approach to the rationalizaion of non-equilibrium ensemble theory, we will have to deal with these alleged similarities to the quantum mechanical situation.

Extending the equilibrium theory to new domains. In standard statistical mechanics, the methodology for deriving the thermodynamic features of equilibrium is routine, if sometimes extraordinarily difficult to carry out in practice. One assumes that equilibrium properties can be calculated by establishing the appropriate canonical ensemble for the system in question. The form of this ensemble is determined by the external constraints placed upon the system, and by the way the energy of the system is determined, given those constraints, by the generalized position and momentum coordinates of its component parts. Then one uses the now familiar argument that in the limit of a very large number of components, one can use the micro-canonical ensemble or the canonical ensemble to represent the problem with the assurance that the thermodynamic results obtained will be the same. Actually, the true rationale for this requires subtle argumentation, some of which we will note later.

Formally, what one calculates is the function Z, the *Zustandsumme* or partition function, where Z is defined by

$$Z = \int e^{-E(p,q)/T} dp_1 \ldots dp_n dq_1 \ldots dq_n$$

in the classical case, and by the analogous sum

$$Z = \Sigma_i e^{-Ei/T}$$

in the quantum mechanical case. E is the energy and T the temperature of the system.

By a Gibbsian thermodynamic analogy, one can identify a thermodynamic quantity, the Helmholtz free energy, F, as $F = -T \log Z$. Then, from the transformation laws of the thermodynamic quantities, one can derive the other thermodynamic functions. The end result of the process is a characterization of the equilibrium thermodynamic properties of the system in terms of the parameters constraining the system and the features of its internal constitution, such as the number of constituent components in the system. Here, the standard Gibbsian identification of ensemble averages with equilibrium thermodynamic features is presupposed.

Because the energy function can be a complex function of the positions and momenta of the components, depending in crucial ways, for example, on spatial separations of particles acting on one another by potential-derived forces and, in the case of electromagnetic forces, by forces that may depend on relative velocity as well, the actual calculation of the partition function as an analytically expressible function of the constraints and structural features may be impossible. This leads to a wealth of approximative techniques, series expansions, and so on. Cases such as L. Onsager's exact calculation of the thermodynamic functions for the Ising model of a two-dimensional lattice of spinning particles interacting by their magnetic moments are rare triumphs for the theory.

But if one is satisfied with approximative solutions, the wealth of systems whose equilibrium properties can be calculated by this orthodox method is great: dilute and dense gases of molecules; one-, two-, or three-dimensional arrays of spinning particles interacting by their magnetic moments; radiation confined to a cavity (which can be viewed as being "composed" of its monochromatic component frequencies); interacting systems of matter and radiation; plasmas of charged particles, and so on. All fall prey to the diligent application of sophisticated analytical and approximative techniques.

Phases and phase changes. One of the most characteristic macroscopic features of matter is the ability of macroscopically radically different forms of matter to coexist at equilibrium. Typical examples are the gas-liquid, gas-solid, and liquid-solid phases of matter at boiling, sublimation, and freezing points. But examples abound from the macroscopic magnetic states of matter and other thermodynamically describable situations as well.

We shall see in Section 2 how the study of the so-called "thermodynamic limit" throws light upon the existence of phases and the nature of the change from one phase to another. Here I wish only to note the existence of a theory that supplements the general principles of statistical mechanics and has proven enormously successful in shedding light upon the nature of the change from one phase to another, at least upon the nature of the transition to the solid phase. The special feature of solidity is long-range order. Whereas in a gas one expects at most short-range deviation from pure randomness in the state of the molecules the deviation to be expressible in a correlation of the state of nearby molecules because of their dynamic interaction, in a crystal one finds an astonishing regularity that holds to macroscopic distances.

Reflecting on the fact that small arrays and large arrays show the same kind of correlational order in the solid phase, a theory – renormalization theory – is invented that provides great insight into the nature of the solid phase and the transition to it from, say, the random gas phase. In particular, this theory is able to explain why it is that quite dissimilar systems (atoms interacting by van der Waals forces and spinning electrons interacting by means of their magnetic moments, for example) can show phase transitions to the solid state that are quite similar in their macroscopic characteristics. Renormalization theory is able to explain why it is that general features of the system such as its dimensionality and the number of degrees of freedom that are available to its constituents are what are important in the characterization of the nature of the phase transition. The specifics of the intercomponent forces are much less important.

2. Rationalizing the equilibrium theory

Ergodicity. We have seen how, from the very beginning of the ensemble approach to the theory, there has been a felt need to *justify* the choice of the standard probability distribution over the systems represented (in Gibbs) by the canonical and micro-canonical ensembles. In Chapter 5, we shall spend a great deal of time exploring exactly what it would mean to justify or rationalize such a probability distribution, as well as ask in what sense such a rationalization would count as *explanation* of why the probability distribution holds in the world. It will put matters into perspective, though, if I here give the barest outline of the progress of so-called ergodic theory subsequent to the Ehrenfest review. The details of this history and an explanation of exactly what was shown at each stage and how it was shown will be provided in Chapter 5.

Maxwell, Boltzmann, and Ehrenfest all propose one version or another of an Ergodic Hypothesis. In some versions, one speaks of the dynamical path followed from the micro-state at a time of a perfectly isolated system. In other versions, one speaks of the path followed by the representative point of a system subjected to repeated small interference of a random sort from the outside. In some versions of the hypothesis, one speaks of the point as traversing a path that eventually goes through each point in the accessible region of phase space. In other cases, the path is alleged to be dense in the accessible region, or to come arbitrarily close to each point in the region over an unlimited time.

What is supposed to be provided by the hypothesis that is otherwise not demonstrable? Although the micro-canonical ensemble is provably a stationary probability distribution, we do not, without an Ergodic Hypothesis, know that it is the unique such stationary probability distribution. Ergodicity is supposed to guarantee this. Ergodicity is supposed to guarantee that the infinite time limit of a function of the phase is equal to the average of that phase function over all accessible phase points,

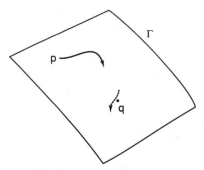

Figure 2-11. The Quasi-Ergodic Hypothesis. As a suggestion to replace the Ergodic Hypothesis, which, in its strong versions, is provably false, the Quasi-Ergodic Hypothesis is proposed. Let a system be started in a micro-state represented by some point p in the phase space Γ. Let g be any other representing point in the phase space. Then if Quasi-Ergodicity holds, the trajectory started from p will eventually come arbitrarily close to g. That is, given any initial micro-state of a system and any other micro-state allowed by constraints, the system started in the first micro-state will eventually have a micro-state as close as one likes to the given, second, micro-state.

where the average is generated using the micro-canonical probability distribution. Finally, it is supposed to follow from ergodicity that the limit, as time goes to infinity, of the fraction of time spent by the representative point of a system in a given region of phase space is proportional to the size of that region of phase space, when the size is measured using the unique invariant measure.

First, A. Rosenthal and M. Plancherel show that the strongest version of Ergodic Hypothesis – that the path of an isolated system will eventually actually go through each point in the accessible region of phase space – is provably false. This leads attention to be directed to quasi-ergodicity – that is, to the claim that the path will be, instead, dense in the phase-space region. It seems to be as difficult to prove any realistic model of a system quasi-ergodic as it was to prove it ergodic in the stricter sense. In any case, as we shall see, quasi-ergodicity, even if it holds, proves to be insufficient to prove such things as equality of time and space averages. (See Figure 2-11.)

Eventually, J. von Neumann and G. Birkhoff proved theorems that give necessary and sufficient conditions on the dynamical structure of an isolated system for the alleged *consequences* of ergodicity to hold. That is, the study of the path is dropped in favor of trying to prove facts about the system sufficient to show directly the uniqueness of the stationary distribution, the equality of time and phase average, and the proportionality of time in a region to size of region. But their necessary and sufficient condition – metric indecomposability – is itself extremely difficult to establish for realistic models.

For some years there have been attempts at performing "end runs" around ergodicity, offering weaker rationales for the probability distribution, rationales that are, at least, dependent only upon conditions that can be shown to follow from the dynamics of the system. Others drop the rationalization program entirely, proposing instead that the fundamental probability hypothesis be taken as an ineliminable basic posit of the theory.

Finally, as the culminating result of a long-term mathematical research program initiated by Birkhoff and von Neumann and carried on by E. Hopf, A. Kolmogorov, N. Krylov, V. Arnold, and others, Ya. Sinai is able to prove metric indecomposability for certain models that bear an interesting relation to realistic models of molecules in a box. But, at the same time, other important mathematical results at the hands of Kolmogorov, Arnold, and J. Moser provide rigorous limitations on the range of systems that could possibly satisfy the Birkhoff–von Neumann ergodic condition.

As noted, we shall follow out this history in detail in Chapter 5. More importantly, we shall there explore in detail just what one has a right to expect from ergodicity results, when they are available, in the manner of justifications of the probabilistic posits, rationalizations of them, or explanations of why they are so successful in predicting thermodynamic results.

The thermodynamic limit. We have seen continual reference in our historical survey to the important fact that the systems under consideration are constituted out of a vast number of micro-constituents. This undeniable fact has been invoked to make such claims as the assertion that fluctuations from equilibrium will be extremely rare over time and that the canonical and micro-canonical ensemble calculations will give the same values for thermodynamic quantities. Can this use of the large numbers of degrees of freedom of the system be rigorously justified? The attempt to put such claims on a firm foundation is known as the study of the thermodynamic limit.

Here we deal, if we are working with the typical case of a gas of molecules confined to a box, with systems whose volume is taken as very large, whose number of molecules is taken as enormous, but whose density is that of the system under consideration. This suggests the obvious idealization of dealing with systems in the limit as the volume and number of particles become infinite, with the density being held constant in the limiting process. The theory of the thermodynamic limit has been highly fruitful. Results may be rigorously obtained in interesting idealized cases. Furthermore, there is in this theory a nice sense of "control" over the limiting process in that one can not only prove things about the limit but get estimates on deviation from the limit in finite cases. This is something usually impossible to obtain in the situation of ergodic theory where infinite time limits can be obtained, but where, often, no hold on how much deviation from the limit one can expect in finite time intervals is available.

What are some of the results one desires from the study of the thermodynamic limit?

(1) We presuppose in statistical mechanics that the effect on the behavior of the system of the interaction of the molecules with the box at the edges of the gas is ignorable in calculating, for example, total energies due to intermolecular interaction. But what genuine right do we have to believe that this is so? Can we show that in the thermodynamic limit, the contribution of the molecule-box interaction to the potential energy stored in the gas goes to zero?

(2) If we demand that entropy be a function that is maximal for the standard equilibrium distribution when that distribution is compared with all distributions consistent with the specified constraints, and if we demand that it have the appropriate additivity properties for independent components of a system, then the unique statistical mechanical definition of entropy can be derived. Now in thermodynamics we take entropy to be an extensive quantity. A portion of gas in an equilibrium state that is like a portion in a similar state but twice the size is taken to have twice the entropy. Can we show in the thermodynamic limit that this will be so? That is, assuming the usual equilibrium ensemble distribution and the usual definition for entropy in terms of the distribution, can we show that in the limit of the vast system, the entropy will have the appropriate extensive nature?

(3) We generally assume that in the thermodynamic limit, an ensemble will describe a system whose fluctuations away from the dominating equilibrium state will become negligible. Whereas each ensemble provides a probability distribution over micro-states that is such that "most probable" distributions in Γ-space overwhelmingly dominate in probability those distributions in Γ-space that are far from equilibrium, what assurance do we have that the probabilities of the fluctuational Γ-space distributions will go to zero in the thermodynamic limit?

This problem is intimately related to the problem of the equivalence of ensembles. It is just this assumption of vanishing fluctuations that allows one to conclude that calculations done canonically and microcanonically will give the same thermodynamic results. Thermodynamic quantities are derivable from one another by appropriate transformations in the phenomenological thermodynamical theory. But different ensembles may be appropriate for deriving the different thermodynamical quantities in statistical mechanics. Only the proof of vanishing fluctuations – fluctuations whose characteristic nature varies from ensemble to ensemble – will allow us to legitimately select ensembles indiscriminately at our need or convenience in calculating the values of thermodynamic quantities while being assured that the resulting quantities will bear the appropriate thermodynamic transformational relations to one another.

(4) The existence of multiple phases in equilibrium with one another can plausibly be claimed to be modeled in statistical mechanics by a partition function that is non-analytic – that is, that fails at certain points to be a smooth function of its variable. But a simple proof is available to the effect that in a finite system the partition function is provably analytic everywhere. The thermodynamic limit approach to this problem is designed to show that such smooth behavior of the partition function can indeed break down in the thermodynamic limit, leading in that idealized case to a non-everywhere analytic partition function representing multiple phases. An alternative but similar approach to this problem recognizes in the path to the thermodynamic limit an analogy to the ergodic path to the infinite time limit from orthodox ergodic theory. This second approach attempts to correlate phase-change with a failure of this kind of "system size ergodicity."

But how can the results about the thermodynamic limit be obtained? Let us first note that the task here is not to justify the equilibrium ensemble distribution. This is presupposed. What additional presuppositions need be made in order to derive the desired results? In the case of a gas of interacting molecules, it is the nature of the interaction that is crucial. Only for certain energetic interactions of the molecules will the thermodynamic limit results be obtainable.

First, it is required that a stability condition be met. This demands that there be a lower bound to the potential energy produced by the interaction potential attracting one molecule to another. Next, there is a requirement on the interaction that the positive part of the potential converges sufficiently quickly to zero with increasing molecular separation. For some interactions, proving these results is not difficult. For other idealized interactions (such as the Coulomb interaction), various additional assumptions and clever analytical tricks are necessary. For gravitational interaction, as we shall see, the stability condition is not met, but for most realistic cases, one simply neglects the relatively insignificant gravitational attraction of molecules for one another.

Given these presuppositions, there are basically two ways to proceed. The first looks at the formulas connecting thermodynamic features to ensemble-calculated features for finite systems, and studies how those functions behave in the thermodynamic limit. Here, one is often able to prove the four results discussed as the aim of this theory.

There is an alternative to the approach that defines the thermodynamic functions for finite systems and then investigates how these functions behave in the thermodynamic limit. In this other program, one attempts to characterize an infinite system from the very beginning, seeking the appropriate representation to describe an idealized system of infinite extent and with an infinite number of components, but whose density is a specified finite value. When it comes to defining thermodynamic properties for such a system, one looks for the appropriate ensemble definition for quantities such as entropy density, rather than total entropy, which will be divergent.

One problem in characterizing such systems and their statistical mechanics is finding the appropriate restriction on initial micro-states to guarantee that wildly divergent behavior doesn't ensue. For such infinite systems, the possibility of solutions to dynamic evolution failing to exist after a finite time arises – say, by an infinite amount of energy being transferred to a single component in that finite time. Such possibilities do not arise in the finite-system case.

Once such constraints are discovered, the next task is to say what an equilibrium state is for such an idealized infinite system. One approach uses ergodicity as the defining criterion for a state being equilibrium. Although such equilibrium states for typical systems have been found, there remain difficulties in being sure all such states have been described. It is a curiosity of this approach that systems that when finite are never ergodic, such as the ideal gas, are ergodic in the infinite-system limit.

Other ingenious characterizations of equilibrium have also been proposed. A suggestion of Dobrushin, Lanford, and Ruell (the DLR condition) is that we think of equilibrium for a classical infinite system as being that state characterized by the fact that for any finite region, if we keep the particles outside of the region fixed, the particles inside the region will be describable by a Gibbs grand canonical ensemble. For quantum systems, an ingenious proposal from Kubo, Martin, and Schwinger (the KMS condition) provides a clear characterization of equilibrium state that can be extended to classical systems as well. Another way to characterize equilibrium in this approach to the thermodynamic limit is to think of equilibrium states as states that are stable under all local perturbations.

3. The theory of non-equilibrium

Solving the Boltzmann equation. The Maxwell transfer equations and the Boltzmann kinetic equation provide alternative, equivalent, means for describing the approach to equilibrium of a non-equilibrium dilute gas. But whereas the H-Theorem – at least if the presuppositions of the posit of Molecular Chaos can be established – gives us reason to expect, in the probabilistic sense, an approach to equilibrium from a non-equilibrium initial state, detailed knowledge of the nature of that approach requires a solution to the equation. Although Maxwell and Boltzmann were able to obtain limited results for the (unrealistic) inverse fifth-power forces – the so-called Maxwell potential – the solutions of the kinetic equation for realistic potentials were a long time in coming.

In 1916–17, S. Chapman, working from Maxwell's transfer equations, and D. Enskog, working from the Boltzmann equation, were able to determine a class of special solutions to the equations. They did not look for a general solution to the equation from an arbitrary initial condition, an analytically hopeless task to solve. Instead, relying upon the phenomenological equations of hydrodynamics, they simply assumed that linear transfer coefficients existed, and looked only for solutions that presupposed such linear coefficients. The implicit assumption here is that even from the most wildly non-equilibrium initial condition, the gas will very quickly reach a state that although still non-equilibrium, is one whose future evolution obeys the familiar phenomenological regularities of hydrodynamics. The special class of solutions Chapman and Enskog found are usually characterized as normal solutions to the equations. Although the discovery of such solutions can hardly be taken to give a full account of why the phenomenological regularities hold, given that their form is presupposed, the theory has positive and novel results that outrun the resources of the phenomenological theory. One can, for example, compute numerical values for the linear transport coefficients, values that are simply inserted into the phenomenological theory. One can also derive such empirically confirmed results as the dependence of the coefficients on various thermodynamic parameters, such as the variation in viscosity of a dilute gas with its temperature.

The problem of non-equilibrium ensembles. In Chapter 6, we shall explore in great detail the structure of various attempts to generalize non-equilibrium statistical mechanics beyond its origins in Maxwell's transfer equations and in Boltzmann's kinetic equation. As we shall see, there is no single, coherent, fully systematic theory that all will agree to that constitutes the correct core of non-equilibrium theory. There is, instead, a collection of approaches to the fundamental problems. Although various approaches are plainly related to one another, exact or rigorous demonstrations of equivalence of approach are rare. Even the question of the limits of the statistical mechanical approach is not a settled issue.

It is simply not known in any definitive way what general classes of non-equilibrium phenomena will be susceptible to the apparatus of ensembles and their evolution, or of kinetic equations derived from this kind of ensemble dynamics.

Whereas equilibrium theory is founded upon a choice of the equilibrium ensemble, non-equilibrium theory presents a somewhat different problem. Plainly, our approach to the foundations of non-equilibrium will require an ensemble way of dealing with systems. If a kinetic equation or some other dynamical description of approach to equilibrium is to be possible at all, it can only be in a probabilistic sense – whatever that turns out to mean. But what should a theory of non-equilibrium ensembles look like?

The first thing to notice here is the joint role played by the choice of initial ensemble and by the dynamical laws of ensemble evolution that are entailed by the micro-dynamics of the individual systems in the ensemble. If an initial ensemble is picked at one time, the evolutionary course of that ensemble is not open to our choice, for it is fixed by the underlying dynamics. This is true despite the fact that, as we shall see in Chapter 6, the method for deriving an appropriate description of dynamical evolution often involves the positing of some "perpetual randomness" assumption that in one way or another, generalizes the Posit of Molecular Chaos. How the ensemble evolves will depend upon its initial structure. We will need to justify or rationalize any such assumptions about its structure in order to complete our descriptive theory. But if we derive our description of the evolution of this initial ensemble by superimposing on the micro-dynamical laws some posit of a statistical nature, we shall have to show that the posit is actually *consistent* with the constraints on evolution imposed by the micro-dynamical laws.

How should an initial ensemble be chosen? This immediately leads to a fundamental preliminary problem: What are to be the macroscopic-phenomenological quantities by which the system and its evolution are to be described? In the case of equilibrium theory, the answer to this question has always been clear. One chooses the energy (or its mean if one was doing canonical ensembles) or, in special cases, the energy plus the small number of additional knowable constants of motion. Then one generates the appropriate ensemble. In some cases of non-equilibrium statistical mechanics, where we are dealing with systems near equilibrium or started from an equilibrium that is then destroyed by a parameter being changed, the equilibrium methods guide us to our choice of non-equilibrium initial ensemble as well. But in the more general case we really have no systematic guide to what macroscopic parameters can be fixed by the experimenter, and hence no systematic guide to what kinds of initial ensembles we must consider. This issue will be crucial when the alleged reduction of thermodynamics to statistical mechanics is discussed in Chapter 9.

Once we have decided what the macroscopic constraints are to be, the further problem of choosing a probability distribution for the systems over the allowable phase space of systems arises. Here, the method of equilibrium theory that picks out the standard invariant ensemble is not available to us, although, once again, it provides direction in limited kinds of cases. We shall explore some general proposals for systematically picking the probability distribution given the macroscopic constraints.

What methods, given an initial ensemble, or even leaving the choice of this ensemble indeterminate, can we find for deriving the appropriate dynamical evolution equations that will lead to our macroscopic non-equilibrium behavior?

Ideally, one would like there to be methods that, given a sufficient characterization of the initial ensemble, could, using the micro-dynamics alone, allow one to derive a description of ensemble evolution from which generalized kinetic equations could be derived. Such a derivation does exist in a very limited realm. Even there it serves only to rationalize or justify a kinetic equation (in this case, the Boltzmann equation) derived by other means. The actual route to kinetic equations in the general case comes about through the positing of some probabilistic principle that, superimposed on the dynamical laws, permits the derivation of a kinetic equation in a manner similar to the way in which the Posit of Molecular Chaos permits the derivation of the Boltzmann equation.

One group of such methods appears in the form of a direct generalization of the Posit of Molecular Chaos. Here, the functions that describe the correlations of the motions of molecules of various kinds (positions, momenta, and so on) are posited to be simple functionals of the functions that express lower-order correlations. The result is a hierarchy of equations for the correlation functions that can be transformed, by a posit or *Ansatz*, into closed equations for the lower-order correlations. The method is a clear generalization of the basic posit of a "perpetual *Stosszahlansatz*" of the kind noted by the Ehrenfests.

By use of this method, one can seek for kinetic equations for dense gases, plasmas, and so on. It is worth noting here that even for the simple cases that go beyond the Boltzmann equation for a moderately dense gas, it has proven to be extraordinarily difficult to construct an appropriate kinetic equation. And, in fact, it is usually not only impossible to *solve* the kinetic equation derived, in the sense that Chapman and Enskog solved the Boltzmann equation, but even to prove an appropriate *H*-Theorem for the more general cases.

A very important class of approaches that has proven to be of value in deriving kinetic equations is applicable to cases where the system can be viewed as a large number of energetically, almost independent, sub-systems that interact only to an energetic degree much smaller than their intrinsic energy. Many systems are of this nature – an almost ideal gas with small intermolecular potential energy, radiation coupled by a small "dust mote" that absorbs and reradiates, almost harmonic oscillators coupled by a small mutual inharmonic force, and so on.

Here, one can deal with states that are the invariant states of the uncoupled system, and one can assume that the introduction of the small coupling serves to give rise to transitions of the system from one distri-bution over the invariant states of the uncoupled system to another. The technique is subtle, for often one must deal not with values of functions that are determined by the invariant uncoupled states, but with averages of such functions over those states. But once again, an assumption that transitions from one such state to another are determined by fixed con-stant probabilities over time becomes another way of representing the system as a Markov process, and leads, by means of what is called a Master Equation, once more to a kinetic equation describing a monotonic trend toward equilibrium.

Other approaches to the problem of generalizing from the Boltzmann equation follow a plan derived in conception from Gibbs' treatment of the non-equilibrium ensemble. Gibbs suggested that we could deal with non-equilibrium, at least in part, by "coarse graining" the Γ-phase space and defining the new coarse-grained entropy. The approach to equilib-rium would be represented by an increase in coarse-grained entropy, corresponding to the initial ensemble "spreading out" in such a way that although its volume remained constant, the proportion of each coarse-grained box occupied by it became more and more uniform with time, heading toward an ensemble in which each coarse-grained box was equally full and equally empty of systems.

But Gibbs did not offer a postulate sufficient to show that this monotonic increase in coarse-grained entropy would in fact occur, a deficiency clearly pointed out by the Ehrenfests. One can supply such a postulate, how-ever. As we shall see in Chapter 6,III,3, the assumption that the system evolves as a kind of Markov process does the trick. The assumption, in essence, is that the fraction of systems in a box at one time that will be in a given box at the next specified later time is constant. Again, this assumption, like the generalized Posit of Molecular Chaos, amounts to a continuous rerandomization assumption about the dynamic evolution of the ensemble, an assumption whose consistency with the underlying dynamical laws remains in question.

In Chapter 6, we shall follow in some detail a number of approaches toward finding some general principles for construing non-equilibrium ensembles and for deriving, by means of various posits, kinetic equations of various sorts from them. This is an area of physics lacking in any single, unified, fully coherent approach. There is, rather, a wide variety of approaches. That they are related to one another is clear, but there is in general no systematic way of rigorously or definitively elucidating their relationships to one another.

4. Rationalizing the non-equilibrium theory

It is one thing to be able to describe initial ensembles in a systematic way and to be able to posit appropriate randomization hypotheses so as to be able to derive a kinetic description of non-equilibrium systems. It is quite another thing to justify the correctness of one's description. And it is another thing again to be able to explain why the description is success-ful, if it indeed is. The whole of Chapter 7 is devoted to just these issues but, once again, it will be useful to give here the briefest possible outline of what some approaches to the justification and explanation problem might be.

First, there are a number of approaches to this problem that lie outside the "mainstream" or "orthodox" approaches. One of these non-standard approaches argues for the reasonableness of the principles that give rise to kinetic equations and irreversibility through an interpretation of the probabilities cited in statistical mechanics as subjective probabilities. The rules for determining probabilities are taken as originating in general principles of inductive inference under uncertainty.

A second non-orthodox approach, whose origins can be traced back quite far in the history of statistical mechanics, seeks the rationale in the non-isolability of systems from the external world. Here it is claimed that at least part of the explanation of the success of the statistical mechanical methods arises from the fact that systems idealized as energetically isolated from the rest of the world are actually not so isolated. It is, in this view, the perpetual interaction of the system with its environment, small as that may be, that accounts for the success of the statistical method and the randomization posits. A third non-orthodox approach seeks the resolu-tion of the difficulties in the positing of time-asymmetric fundamental dynamical laws.

The more common, orthodox, directions in which justification and explanation are sought look for their resources to identifiable physical features of the world arising out of the structure of the system in question and the structure of the micro-dynamical laws in order to ground the derivation of the kinetic description.

In one unusual case, the kinetic equation can be derived from a posited structure of the initial ensemble by itself using solely the micro-laws of dynamical evolution. In other cases, it is derived by means that rely both on the structure of the initial ensemble and on posited randomizing features of dynamic evolution. In both cases one would like to find some grounds for believing that all initial ensembles will have the requisite feature. Here, what one seeks are physical reasons for a limitation on the kinds of actual ensembles of systems that we can prepare. The resources to which one turns include the structure of the individual systems – the underlying micro-dynamical laws – plus additional constraints upon the ability of an experimenter, restricted in some sense to the macroscopic manipulation of the system, to constrain a system's initial conditions. Whether such resources will, by themselves, serve to rationalize the restriction on initial ensembles or whether, instead, some additional fundamental probabilistic postulates resting on different grounds will be needed as well is, as we shall see in Chapter 7, a matter of important controversy and is fundamental to the study of the foundations of the theory.

When one's derivation of a kinetic equation rests either in whole or in part upon randomization posits imposed on dynamic evolution, a rather different task of justification and explanation needs to be faced. Because the dynamic evolution of an ensemble is fixed by the underlying micro-dynamics – at least in the orthodox approaches in which isolation from the outside is retained for the system and in which one accepts the standard micro-dynamics as a given and not as some approximation to a more stochastic or probabilistic underlying theory – one must reconcile the posited randomization imposed on the evolution with the evolution as constrained by the laws of the micro-dynamical theory.

Here, one might try to rely solely upon some demonstration that the evolution described by the micro-dynamics is adequately representable by the probabilistic description, the justifiability of the alternative representation resting upon the structure of the laws of dynamical evolution alone. Or one might offer a justificatory or explanatory argument that utilizes both the structure of the micro-dynamical laws and some posited structure feature of the initial ensemble. We shall see both approaches used in our detailed reconstruction of these arguments in Chapter 7.

It is in this part of the program of rationalizing the non-equilibrium theory that the use is made of a series of important advances in our understanding of the micro-dynamical laws and the structure of the possible solutions. In particular, various demonstrations of the radical instability of solutions under minute variations of initial conditions for the equations, and of the connection of such demonstrable instabilities with various "mixing" features of the ensembles whose evolution is governed by these equations, play a central role in the part of foundational studies. Here, powerful generalizations of the ergodicity results that we shall study in Chapter 5 come into play. We shall explore these more general results – which play in the foundational study non-equilibrium a role both like in some respects and unlike in other respects the role played by ergodicity results in the equilibrium theory – in detail in Chapter 7.

VI. Further readings

For a history of thermodynamics, see Cardwell (1971).

Two clear books on thermodynamics emphasizing fundamentals and concepts are Buchdahl (1966) and Pippard (1961). Truesdell (1977) is historically and conceptually oriented.

On the approach to thermodynamics that directly embeds probability into the theory, see Tisza (1966).

For the theory of non-equilibrium systems close to equilibrium, two good sources are Kubo, Toda, and Hashitsuma (1978) and de Groot and Mazur (1984). Introductions to the concepts used to deal with non-equilibrium thermodynamics far from equilibrium can be found in Prigogine (1980) and (1984). See also, Truesdell (1984).

Brush (1983) is a brief and clear history of kinetic theory. Brush (1976) is a compendium of essays covering many areas of the history of kinetic theory and statistical mechanics. Brush (1965) contains many of the seminal works in the history of kinetic theory and statistical mechanics selected from the original sources and translated into English where necessary. Maxwell's Demon is exhaustively treated in Leff and Rex (1990).

An excellent introduction to statistical mechanics emphasizing foundational issues is Tolman (1938). Two later works covering both older and newer topics and emphasizing fundamental issues are Toda, Kubo, and Saito (1978) and Balescu (1975). Munster (1969) is a treatise covering both foundational issues and applications.

Gibbs (1960) is essential reading.

Ehrenfest and Ehrenfest (1959) stands as a masterpiece of critical analysis of the theory up to 1910. For quantum statistical mechanics, Toda, Kubo, and Saita (1979) and Tolman (1938) are excellent.

An excellent survey of the important developments in statistical mechanics in the period prior to its publication is O. Penrose (1979), a retrospective piece that can direct the reader to much of the original literature.

Toda, Kubo, and Saito (1979) and Balescu (1975) contain very accessible treatments of the phase-transition problem and descriptions of the exactly solvable models.

The issue of the thermodynamic limit is taken up with mathematical rigor in Ruelle (1969) and in Petrina, Gerasimenko, and Malyshev (1989). See also, O. Penrose (1979), Section 2.

For further readings on the foundational problems of the rationalization of the probabilistic posits in equilibrium and non-equilibrium theory, see the "Further readings" at the ends of Chapters 5, 6, and 7 of this book.

3

Probability

As is already clear from the preceding historical sketch of the development of foundational problems in statistical mechanics, the concept of probability is invoked repeatedly in the important discussions of foundational issues. It is used informally in the dialectic designed to reconcile the time-asymmetry of statistical mechanics with the time-reversibility of the underlying dynamics, although, as we have seen, its introduction cannot by itself resolve that dilemma. Again, informally, it is used to account for the existence of equilibrium as the macro-state corresponding to the "overwhelmingly most probable" micro-states, and to account for the approach to equilibrium as the evolution of micro-states from the less to the more probable. More formally, the attempts at finding an acceptable derivation of Boltzmann-like kinetic equations all rest ultimately on attempts to derive, in some sense, a dynamical evolution of a "probability distribution" over the micro-states compatible with the initial macro-constraints on the system. The picturesque notion of the ensemble, invoked in the later work of Maxwell and Boltzmann and made the core of the Gibbs presentation of statistical mechanics, really amounts to the positing of a probability distribution over the micro-states of a system compatible with its macro-constitution, and a study of the changes of such a distribution over time as determined by the underlying dynamics.

But what is probability? The theory of probability consists of a formal mathematical discipline of astonishing simplicity at its foundation and of astonishing power in what can be derived in it. Along with the formal theory, there is a wide and deep literature devoted to trying to explicate just what elements are described within the formal theory. As we shall see, the so-called problem of the interpretation of probability – the problem of understanding in a general way what probabilities consist of – is one that remains replete with controversy. What seems initially to be a simple matter, requiring only conceptual clarification in matters of detail, is actually a puzzling and difficult area of philosophy.

Although it will be the aim of this chapter to survey a number of the most crucial assumptions and puzzles in several modes of the interpretation of probability, it will not be my purpose here to explore this to the depth it deserves as a problem in its own right. Such an effort would take us too far afield from our main goal – the pursuit of the foundational problems in statistical mechanics. But a survey of these general approaches to interpretation will provide an essential background for studying the more specialized problems in the application of probability to physics that are encountered in statistical mechanics. We shall also see how some of these general interpretive issues may very well depend for their resolution on some of the particular results obtained in the study of statistical mechanics, because, as we shall see in Chapter 7,IV,1 and 2, it is sometimes unclear where the dividing line between purely conceptual issues and issues about the nature of the world as revealed in our best physics lies.

This chapter will first outline some essential formal aspects of probability theory. Next, it will survey some of the philosophical work on the interpretation of probability. The final section will deal with the special problems encountered when probability is introduced as a fundamental element in statistical mechanics.

I. Formal aspects of probability

1. The basic postulates

One begins with a collection, E, called a set of elementary events. Next, there must be a collection of subsets of E, called events, that is closed under set union, intersection, and difference, and that contains E. In general, this collection, F, will not contain every subset of E. Probability is then introduced as a function from the members of F into the real numbers, P. P assigns the value 1 to E – that is, $P(E) = 1$. Most importantly, P obeys an additive property: If A and B have no members in common, then P assigns to their union the sum of the values assigned to A and B – that is,

$$A \cap B = \phi, \text{ then } P(A \cup B) = P(A) + P(B)$$

When the set E is infinitely large, F can contain an infinite number of subsets of E. In this case, the additivity postulate here is usually extended to allow for denumerably infinite unions of subsets of F. If a collection of A_i's consists of sets, all of which have no members in common whenever $i \neq j$, then it is posited that $P(\cup_i A_i) = \Sigma_i P(A_i)$. This is called countable additivity or σ-additivity. This postulate is sometimes questioned in subjective probability theory, although denying it leads to peculiar behavior for probabilities even when they are interpreted subjectively. But in all of the standard applications of probability theory to statistical mechanics, it is assumed to hold.

Next, one needs the notion of conditional probability, introduced in this approach by definition. If $P(A) \neq 0$, then the probability of B given that A, the conditional probability of B on A, written $P(B/A)$, is just $P(B \cap A)/P(A)$. This can be understood intuitively as the relative probability of B to the condition of A being assumed to be the case. Two events are probabilistically independent if $P(B/A) = P(B)$ and $P(A/B) = P(A)$. This can easily be generalized to hold for collections of events where each event is independent of all the others in the collection, rather than independence being merely between pairs of events.

A sequence of "trials" that constitute an "experiment" is an important notion. Here, one thinks of a sequence in which some elementary event is the outcome of each running of the trial. If the probability of an event on a given trial is independent of the outcomes of all other trials, the sequence is said to be a Bernoulli sequence. If the probability of an event in a trial differs, perhaps, from its probability conditional upon the outcome of the immediately preceding trial, but if this probability $P(A_n/A_{n-1})$ is then the same as the probability of A_n conditioned on all other additional outcomes, the sequence is called a Markov sequence. In a Bernoulli sequence, no knowledge of other outcomes leads to a change of one's probability attributed to the outcome of a given trial. In a Markov sequence, knowing the outcome of what happened just before the specified trial may lead one to modify one's probability for a given trial, but knowledge of earlier past history is irrelevant.

Finally, we need the notions of random variables and distribution. A random variable is merely a function that assigns a value to each of the elementary events in E in such a way that the set of all elementary events that have a value less than a specified amount is in F, and hence is assigned a probability. One can define the distribution function for the random variable as the function whose value at a is just the probability that the random variable has a value less than a. If this function is differentiable, its derivative, intuitively, gives the rate at which the probability that the random variable will have a value less than a is increasing at a. From the distribution function for a random variable, one can define its expectation. Intuitively, this is just the "sum of the product of the value of a random variable times the probability of that random variable." Naturally, for continuously distributed random variables, this requires formalization in terms of that generalization of the sum that is the integral. Crudely, the expectation of a random variable is its mean value.

The basic postulates of probability are of extraordinary simplicity. Naturally, much refinement is needed to make things precise in the general cases of infinite collections of events and so on. Although the postulational basis is simple, it is extraordinarily rich in consequences. We will need to note only a few of the most important of these consequences in the next section.

2. Some consequences of the basic postulates and definitions

A group of important consequences of the basic postulates of probability theory deal with sequences of trials that are independent – Bernoulli sequences. Suppose such a sequence of trials is performed, with a fixed probability for the outcomes of each trial and with the independence condition holding, so that for any collection of trials the probability of a joint outcome is merely the product of the probabilities of the outcomes of each trial, a condition easily seen to be equivalent to the earlier definition of independence.

One can take the outcomes of the first n trials, characterized by a random variable defined over the elementary event outcomes, add them up, and divide by n, getting a "sample mean" of the outcomes in a particular series of trials. The laws of large numbers relate these sample means in the limit as the number of trials "goes to infinity" to the expected value of the random variable as calculated by using the probability distribution for this random variable as fixed by the probabilities of the various outcomes on any one trial. This probability distribution, by the definition of the sequence as a Bernoulli sequence, is the same for each trial.

The Weak Law of Large Numbers tells us that

$$\lim_{n\to\infty} P(|y_n - \mu| > \varepsilon) = 0$$

where y_n is the sample mean of the random variable in question and μ is the expected value of it on a single trial. What this says is this: Suppose an infinite number of Bernoulli sequences of trials are run, and the sample mean calculated for each sequence at each n. For any deviation of this mean from the expected value, the proportion of sequences such that the sample mean at n, a specific trial number, will differ by more than ε from the expected mean can be made arbitrarily small by picking an n far enough out in the sequence.

This result is compatible, however, with the following: An infinite number of such infinite number of trials are run. In each, the deviation of the sample mean from the expected value becomes, as the Weak Law demands, sparser and sparser as the number of trials increases. But in each run, or in a large proportion of them, the sample mean never gets within ε of the expected value, *and stays there forever*. That this is not to be expected is given us by the Strong Law of Large Numbers. This says $P(\lim y_n = \mu) = 1$. What this comes down to is that when we look at this infinite collection of repeated trials, as the number of trials goes "to infinity" in each sequence, we can take it that the probability of a sequence getting a sample mean that settles down to within ε of the expected value forever has probability one, and the set of those that never do this has probability zero.

One can go much further. The Central Limit Theorem, for example, tells us that there is a limiting particular distribution function to which the distribution of such sample means in this infinite collection of infinite trials will converge.

These facts about large numbers of trials (or more correctly about the behavior of sample means of sequences of trials in the limit) will play a role in some of our attempts to outline just what probability is. For the moment the reader should only note how the notion of probability plays a crucial role in the statement of these results. The "convergence of the sample mean to the expected value" noted in all these results is some form of convergence "in probability." For future reference, it is important to note also how the crucial premise that allows these results to be derived is that the sequence is one of independent trials. In Chapter 5, we shall see in discussing the role of the Ergodic Theorem in attempts to provide a rationale for equilibrium statistical mechanics, that it is possible to provide a close analogue of these "large number" results even in some important cases where, intuitively, independence of trials is not the case, at least in the physical sense of trials whose outcomes are causally independent of one another. Indeed, we shall see that something like a law of large numbers can be derived in some cases even where the outcome of each successive trial is completely causally determined by the outcome of the initiai trial.

One very simple consequence of the postulates and of the definition of conditional probability is Bayes' Theorem. Suppose $A_1 \ldots A_n$ constitute events in F so that each elementary event in E is in one and only one A_i. Then, for any event B in F, the following holds:

$$P(A_k/B) = \frac{P(B/A_k)P(A_k)}{\Sigma_i P(B/A_i)P(A_i)}$$

The theorem relates the probability of one of the A_i's conditional upon B to a quantity calculable from the unconditional probabilities of the A_i's and the probability of B conditional upon the specific A_i in question. We shall note in Section II,5 the central role this theorem plays in one view about the way probability assessments are to vary dynamically over time.

3. Some formal aspects of probability in statistical mechanics

The role of probability in statistical mechanics is one fraught with puzzles, many of which will be components of the central issues to be discussed in this book. Here, I want only to note a few basic aspects of the formalism by which probability is usually introduced into the theories we will survey.

The class of elementary events relevant to statistical mechanics is phase-space, the class of points representing each possible micro-state of the dynamical system in question. For the classical statistical mechanics with which we will be dealing, each point represents the full specification of generalized position and momentum coordinates for each dynamical degree of freedom of the system. The relevant phase-space for quantum statistical mechanics is of vital importance for statistical mechanics in practice, but not something we will have to deal with very often. The class of events, F, is just a class of sets of micro-dynamical states. Random variables will then be functions that assign numbers to micro-states in such a way that when the probability function over the relevant class of sets of micro-states, F, is defined, well-defined probability distributions for the random variables in question also exist.

To define one's probabilities, start with a *measure*, a generalization of the ordinary notion of volume, on the phase-space in question. The choice of this measure ultimately constitutes the determination of the relevant values a probability is to take. Justifying the choice of this measure and explaining why it works often constitute central parts of the foundational question. The standard measures are all derived from the measure on phase-space as a whole that works by taking a volume in phase-space to have its size be essentially the product of all its extensions in coordinate and momentum values. Usually one will be dealing with a lower-dimensional sub-space of the phase-space, a sub-space determined by the restriction of possible micro-states of the system to those compatible with some class of macro-quantities. The appropriate measure on this sub-space will not in general be the simple volume measure of the phase-space restricted to this lower dimension, but an appropriate modification of it. In the case of equilibrium statistical mechanics, this appropriate measure is a measure of the size of the regions in the sub-space provably invariant over time under the action of the dynamic evolution of the systems represented by trajectories from their initial representing phase-point. In non-equilibrium cases, as we shall see in Chapter 7, such measures also play a role, but one whose rationalization is rather less well understood.

The measure of regions then allows one to assign probability to appropriate regions. Here, it is important to note crucial facts about regions "of zero size," regions that are assigned zero probability. It is easy to show from the basic postulates that the empty set in F – the set containing no elementary events – must get probability zero. But having probability zero is not a property of this set alone. In general, many non-empty sets in F will have probability zero. In the ordinary measure for regions of the real line, for example, all sets of rational numbers only get measure zero. In a many-dimensional space with its ordinary volume measure, all subspaces of lower dimension will receive measure zero. In statistical mechanics, the role of sets of micro-states of probability zero is an important one, and we shall have a good deal to say about what can, and what cannot, be inferred from some class of micro-states being of probability zero.

Formally, the choice of a probability assignment to sets of micro-states that is such that probability zero is assigned to any set with zero size in the measure – an assignment of probability that is, in mathematicians terms, "absolutely continuous" with respect to the standard measure – makes the formal method for assigning probabilities simple. When the probability assignment is absolutely continuous with respect to the measure, we can derive all probabilities assigned to sets by positing a non-negative, measurable, function, f, over the micro-state points of the relevant subspace of phase-space. The probability assigned to a region A in the phase-space is then obtained by "adding up" the f values for all the points in A, or, more rigorously, by integrating the function with respect to the measure over the region A. That is, a single probability density function can be defined that "spreads the probability" over the appropriate region of phase-space so that the total probability assigned to a region is obtained by a measure of the amount of the total probability over the whole phase-space that is spread in the region in question. Most commonly in statistical mechanics it is "uniform" spreading that is posited, the probability being assigned to a region of micro-states just being proportional to the size of that region in the chosen measure. Of course, the grounds for choosing such a measure in the first place, and the explanation of why such a choice works as well as it does (when it does work), remain deep conceptual and physical issues.

II. Interpretations of probability

A single formal theory can be applied in many different domains. The same differential equation, for example, can apply to electromagnetic fields, sound, heat, or even to phenomena in the biological or social world. The postulates and basic definitions of formal probability theory are clear. But what is the domain that we are describing when we take the formal apparatus of set of elementary events, set of subsets of it called events, and additive function on these called the probability function to be characterizing what we take intuitively to be probability? As we shall see in the brief survey next, this has become one of those issues that seems at first easily resolvable, but that becomes slippery and elusive when a complete and clear explication is demanded.

1. Frequency, proportion, and the "long run"

Consider a finite collection of individuals with a number of properties such that each individual has one and only one member of the set of properties. The relative frequency with which a property is instanced in the population is plainly a quantity that obeys the postulates of probability theory. So, are probabilities in the world nothing but relative frequencies in ordinary finite populations?

One objection to this is its limitation of probability values to rational numbers only, whereas in many of our probabilistic models of phenomena in the world, we would want to allow for probability to have any value in real numbers from zero to one. But surely we can generalize beyond actual relative frequencies to actual proportions. Imagine a particle moving in a confined region that we think of as partitioned into non-overlapping sub-regions. Surely we understand the notion of the proportion of some finite time interval the particle spends in any one of the partitioning sub-regions. These proportions can, in general, have non-rational values. And it is transparent that the proportions will obey the laws of formal probability theory. The particle will spend all of its time somewhere in the region $P(E) = 1$, and for exclusive regions the proportion of time the particle spends in the joint region of A and B will just be the sum of the proportion of time spent in A and that spent in B.

But there is a standard objection to any proposal to take such clearly understood notions as finite relative frequency or ordinary proportion as being identifiable with *probability*. Our intuition tells us that although frequencies and proportions should in some sense cluster around probabilities, we cannot, in general, expect probabilities and proportions to be equal. The probability of heads on a fair coin toss is one-half, but only a small proportion of coin tossings have heads come up *exactly* half the time – and it will never be so if the number of tossings is odd. A simple identification of probabilities with frequencies or proportions, it is often argued, is too naive, too direct, and too simple-minded a theory to capture the subtle relation that probabilities hold to actual frequencies and proportions.

One familiar attempt to resolve this problem is to move to idealized "infinitely large" reference classes. Here, the attempt is made to use the close association of probability with observed frequency mediated by the laws of large numbers in probability theory to find something in the world more strictly identifiable with probability than the frequencies or proportions in finite reference classes. For example, it is sometimes argued that in the case of something like coin tossings, where only one of a finite number of outcomes is possible, the probability of an outcome ought to be identified not with the frequency of that outcome in some limited set of trials, but with the "limit of relative frequency in the long run" – that is, as the number of tosses increases without limit or "to infinity."

But such long-run relative-frequency accounts of probability are problematic in several ways, at least if they are intended as providing some definition of probability in terms that advert to actual features of the world and that themselves do not invoke probabilistic notions. One problem is that the limit of a series – the natural way to define probability in this approach being to use the usual mathematical definition of a quantity as a limit – can vary depending upon the order of the terms in the series. Although the finite relative-frequency approach to probability requires no implicit order among the events whose probability is to be determined, the limit of relative frequency in the long-run approach does demand such an order of events. But for events that are orderable in time or in some other natural way, this is not an insuperable conceptual problem.

More disturbing is the degree of unrealistic idealization that has been introduced in this "more sophisticated" approach to a definition of probability. Although the finite classes and sub-classes of the finite relative frequency view are taken as existing in the world, in what sense do the infinite sequences of events needed to define probability in the long-run approach really exist? Will there ever really be an infinite number of tossings of fair coins in the world? If not, then have we abandoned the ideal of finding something "real" and "in the world" to be the element described by the formal probability theory? If we take such infinite sequences not as real elements of the world, but as some "ideal" element, then exactly how is this ideal related to what is real (presumably actual relative frequencies in actual finite reference classes), and, given this idealization, what are we to take its probabilities to be?

Most important of all is the realization that going to long-run limits of relative frequencies doesn't really solve the problem it was introduced to handle – the possibility of deviation between relative frequency and what is taken, intuitively, to be probability. First, note that the association of probability with relative frequency guaranteed by the laws of large numbers holds only when the sequence is a Bernoulli sequence, a sequence of probabilistically independent trials. But this notion is one that requires an introduction of a probabilistic characterization of the physical situation in its definition, leading us to be quite skeptical that there is some way in which, by using infinite sequences of trials, we could "define" the notion of probability in terms that themselves do not presuppose the understanding of probabilistic notions.

A further consideration in this vein is even more important. Even given that the sequence has the proper probabilistic independence between trials, even the Strong Law of Large Numbers fails to identify the limit of relative frequency with the probability in an individual trial. We are told only that the probability that they will be equal is one, or that the probability of their being unequal goes to zero. But, as we have noted, probability zero is not to be identified with impossibility. Even if an infinite sequence existed, and even if we could be assured that it was a Bernoulli sequence, the laws of probability still leave open the possibility that the limit of relative frequency, even if it existed, could differ from the probability of the relevant outcome in an individual trial. It would seem, then, that we can associate limits of relative frequency with probabilities only relative to a modality that is itself probabilistic in nature. For this reason as well as for the others noted, many have become skeptical that the invocation of long runs, construed either realistically or in terms of some idealization, will do the job of mitigating the fundamental objection to the finite relative frequency (or finite proportion) account of probability – that is, the possibility of a divergence between the appropriate proportion, finite or long-run, and the probability we are seeking to identify with some feature of the world.

2. Probability as a disposition

One very important group of attempts to solve some of the problems encountered in the relative frequency (or proportion) and long-run relative frequency approaches is the one, beginning with seminal work of K. Popper, that offers an account of probability as a disposition or propensity. Although the frequentist takes probability as something "in the world" but attributable, at least primarily, to collections of trials or experiments, the dispositionalist thinks of a probability attribution as being fundamentally an attribution of a property to a single trial or experiment. For the frequentist, probability adheres to the individual trial only insofar as that trial is subsumable into some general kind or class, and frequentists generally take the correct probability attribution to an individual trial as something that is only relative to thinking of that trial as in some specific class or other. But such problems concerning the uniqueness or non-uniqueness of the correct probability attribution to an individual trial (or, on its relativization to thinking of that trial as a member of some general kind) are sometimes puzzles for the dispositionalist as well.

The property – or rather magnitude, for probability is taken to be a "quantity" that inheres in the individual event to some "degree" or other ranging from zero to one – that probability is taken to be is a dispositional property. Here, probability is assimilated to such properties as the solubility or the fragility of objects. Intuitively, a distinction is made between categorical properties and dispositional properties. The former are, in some sense, the "occurrent" properties of the object, the latter present in the form only of "conditional" properties, attributions that have their presence because of what the object *would* do were certain conditions met. Thus, a dry piece of salt is categorically cubical, but soluble only in the sense that if it *were* put in a solvent it *would* dissolve. Needless to say, that distinction between the categorical and dispositional properties is a hard one to pin down. Many, in fact, would doubt that any such distinction that holds in a context-independent way can be made, arguing that what is for the purposes of one discussion categorical can be for other purposes viewed as dispositional.

Dispositional theories of probability vary widely in what the disposition is attributed to and in their detailed analysis of how the attribution of probability as a disposition is to be precisely construed. We will have to survey the most general features of these accounts only briefly.

As we have seen, the stipulation or definition of a dispositional property is usually by means of some counter-factual conditional: how the object would behave were certain test conditions satisfied. What would such an account of probability as a disposition look like? Usually this is thought of in terms of the propensity of the trial to reveal a certain relative frequency (or proportion) of outcomes on repetitions of it. Crudely, what it means to attribute a probability of one-half for heads on a coin toss is that were one to toss the coin (an infinite number of times?), a relative frequency of heads of one-half would be the result. Here, at least one problem of the ordinary relative frequency view is addressed. Even if there were no repeated trials – indeed, even if the coin were never tossed at all – it could still have a definite probability for an outcome upon tossing, for the relevant tossings are posited only in the mode of possibility.

There are many alleged problems with the dispositional account and, once again, we will only be able to note a few of them in outline. Intuitively, many think there are two distinct situations where probabilities are attributed to a kind of trial: the cases where the outcome on a single trial is determined by some underlying "hidden" parameters whose values remain unavailable to us, and the "tychistic" case where, as is frequently alleged of trials in quantum mechanics, there are no such underlying hidden parameters whose values actually determine the outcome of the trial in question. How does the dispositionalist address these two cases?

In the deterministic case, some dispositionalists would deny that any non-trivial probabilities are involved, the probability of the outcome being one or zero depending on whether it is determined that it occur or that it not occur. For these dispositionalists, the appearance of probability in deterministic cases is just that, an illusion. For them, where genuine indeterministic chance is not exemplified in the world, it is wrong to think of the outcomes as having non-trivial probabilities.

Other dispositionalists would be loath to drop the idea that even in such cases there would still be real, non-trivial probabilities. How would these be defined? By the relative frequency of the outcome that would result from repeated trials of the *kind* of experiment in question. But what would this frequency be? It would, given the determinism, be fixed by the *actual* distribution over the initial conditions that would have held. This leads to two problems. First, it seems to some that the dispositionalist account here may be parasitical on an actual relative frequency (or proportion) account, making it no real change over the latter. More importantly, wouldn't such an account be subject to the same fundamental objection as that made against the actual relative frequency account – that it would fail to do justice to our intuition that frequency and probability can diverge from one another, even in the "long run"?

What about the other case, where the outcome of a given trial is truly undetermined by any hidden parameter values? This is the case that many dispositionalists find most congenial, indeed some of them asserting that only in such cases does the trial have real non-trivial probabilities for its outcomes.

Here we have no actual distribution of underlying hidden parameters that fully determine the outcomes of the trials to rely on, nor do we have the worry that in any individual trial the "real" probability must be zero or one because the outcome, whatever it is, is fully determined by the values of the hidden parameters even if they are unknown to us. We do have, of course, the actual relative frequencies (or their limits or proportions) of outcomes in collections of trials that are (now, in all respects) like the trial in question.

Are these frequencies or proportions to be taken to be the probabilities? No. For the dispositionalist, the real probability – the dispositional feature of the individual tychistic situation – is connected to the manifested frequency, but not in a definitional manner. The real propensities are supposed to be, in some sense, causally responsible for the frequencies. The probabilities as dispositions in the individual case "generate" the relative frequencies in some sense of that term. And the manifested frequencies or proportions may be taken as evidence of just what the real dispositional probability values are. Manifested frequencies clustering about some value are an indication to us that the real probability has its value somewhere in the vicinity of the revealed frequency. But the manifested frequencies or proportions are not taken to be "constitutive" or definitive of what the probabilities are.

For many dispositionalists, the real probabilities are to be, once again, defined by the use of counter-factual locutions. The probability has a certain value if in a long run of repeated trials of the experiment in question the proportion or frequency of the value specified by the probability would now be obtained. The type of trial is now fixed as a kind by all of its features, there being no "hidden" features relative to which a further subdivision of kinds can be made into which the trial could be placed.

Yet that can't be quite right either. For, as we have seen, even the strongest laws of large numbers don't identify long-run proportion with individual case probability. Wouldn't this "slack" – a slack characterizable only in probabilistic terms – still hold even if it were possible but we were talking about non-actual runs of trials? One can imagine how to try and improve the situation from a dispositionalist stance. Think of a counter-factual being true if in some possible world like ours, but differing only in the satisfaction of the "iffy" part of the conditional, the "then" part holds. Now imagine the repeated trials being carried out in a vast number of such other possible worlds. In each such world, a certain proportion of outcomes would result. Take the disposition of probability as having a certain value if the distribution over all these possible worlds of the proportions of outcomes is as would be described by the laws of large numbers for the appropriate probability. So, in vast numbers of such worlds the proportion would cluster strongly around the proportion identical to the probability, but we could still allow for possible worlds where the frequency diverged from the probability, assuming such worlds were themselves infrequent enough.

But few are willing to have quite such a "realistic" attitude toward these other possible worlds. At this point, one begins to wonder if this "explication" of probability really is an explanation of what "in the world" probability is. Has it become, rather, a repetition of the facts about probability as described by the formal theory cast into a guise that seems to be providing an interpretation but that is now diverging from the original intent of finding some feature of the world, the actual world, that can be taken as described by the formal terms of probability theory?

3. "Probability" as a theoretical term

A group of important attempts to understand probability, still in the objectivist vein that we have been exploring in the last two sections, make a deliberate effort to avoid some of the problems encountered by frequentist and dispositionalist, problems they run into at least in part because of their desire to define probability in non-probabilistic terms. This alternative approach is in part inspired by views on the nature of theoretical terms in science. Early empiricist and operationalist views on the meaning of terms referring to non-directly observable entities and properties in science, wanting to legitimize the usage of such terms as they appeared in physical theories but wary of the introduction into discourse of terms whose meaningfulness was dubious because of their lack of associability with items of experience, frequently proposed that any term in science not itself clearly denoting an item of "direct experience" be strictly definable in terms of such items from the "observational vocabulary." But later empiricistically minded philosophers became dubious that our typical terms of physical theory could be so explicitly defined. Couldn't such terms be legitimized, though, so long as they played an essential role in a theoretical structure of sentences that was tied down to direct experience at some level, if only through a network of logical implications? A term, then, could be legitimate, and its meaning could be made clear, if it functioned in the overall network of scientific assertion, so long as that network as a whole led to testable empirical consequences.

The application of this general approach to probability would be to argue that attributions of probability to the world were also to be understood as functioning in a complex network of assertions, some of the components of which, including the attributions of probabilities to outcomes of experimental setups, were tied to the immediate data of observation only in an indirect way.

One component of the overall structure that fixes the meaning of probability attributions would be the rules of inference that take us upward from assertions about observed frequencies and proportions to assertions of probabilities over kinds in the world, and downward from such assertions about probabilities to expectations about frequencies and proportions in observed samples. These rules of "inverse" and "direct" inference are the fundamental components of theories of statistical inference.

Needless to say, there is no reasonable plausibility to the claim that there is a body of such rules that is universally accepted. The appropriate structure for a theory of inference to statistical or probabilistic generalizations from observed frequencies and proportions is one fraught with controversy at very fundamental levels. Even the apparently simpler problem of inferring from an accepted probabilistic assertion about a population to a legitimate conclusion to be drawn about a sample of one or more individual cases is wrapped in controversy. But the claim we are looking at here is only this: Once one has adopted some principles for making such inferences – from sample to probabilistic assertion and from probabilistic assertion to expectations for samples from the population in question – the very adoption of those rules of upward and downward inference is constitutive, at least in part, of what the probabilistic assertion means for you. The principles of "warranted assertion" that state what one will take to be legitimized inferences to and from probability assertions to assertions about frequencies and proportions in finite sample classes do at least part of the job of determining what we mean by the probabilistic assertions themselves. And fitting the probabilistic assertions into a network of inference in this way, we can "fill in" their meaning without requiring us to give an explicit definition of "probability" in terms of proportion or its limit, or in terms of these even in other possible worlds (as the dispositionalist suggests).

One puzzle with this approach is that it seems, so far, to leave out some crucial elements. Suppose two statistical schools rely on differing upward and downward rules of inference, but whose rules, when combined, lead to the same inferences from proportions in samples to proportions anticipated in other samples. Do they merely disagree on the meaning of the "intervening" statistical assertions, because they accept different such assertions on the same sample evidence, or do they have a disagreement about which upward rule of inference is correct? The latter seems to require some notion of another constraint on the truth of the statistical assertions, something that goes beyond its being the assertion warranted on the upward rule chosen. And doesn't it seem plausible to claim that the truth of the statistical assertion has at least something to do with the actual frequency or proportion of the feature in question in the total population, even if, as we have seen, a naive identification of the probability with that proportion can be challenged?

One direction in which to seek additional elements in the world pinning down the meaning we give to objectivist probability assertions is to look toward the origin of the actual frequencies and proportions in the total population in the general features of the world described by our background scientific theories. The idea here is that although some frequencies and proportions have a merely accidental or contingent aspect, others can, in one way or another, be shown to be "generated" out of the fundamental nature of the world as described by our most general and foundational lawlike descriptions of the world.

The ideas here are connected with those that arise in the philosophy of science when one seeks for a notion of a law of nature as opposed to a mere true generalization. Whereas "$F = ma$" is supposed to have a lawlike status, "all the coins in my pocket are copper" does not. The former generalization is inductively inferable from a sample of the cases it covers, the latter knowable only by an exhaustive inspection of each coin in the pocket. The former grounds counter-factual inferences about possible but non-actual situations, the latter hardly allows one to infer "if something else were a coin in my pocket it would be copper." What constitutes the lawlike force of some generalizations? Reliance on such notions as laws being true not only in the actual world but in all "physically possible worlds" just seems to beg the question.

One general approach focuses on the place the generalization holds in our hierarchical system of general beliefs about the world. Some beliefs of a generalized nature play a fundamental role in that they are the simple axiomatic beliefs from which other beliefs of great generality and explanatory importance can be derived. The idea is that lawlikeness is not a feature of a generalization having some semantically differentiable content when compared with ordinary generalizations, but rather that the lawlike generalizations are those fundamental axiomatic generalizations that ground our overall explanatory structure of the world, or those more restrictive generalizations whose lawlikeness accrues from their being derivable from the more fundamental laws.

The connection of these ideas with objective probability goes through a claim to the effect that a quantity that appears in such a fundamental lawlike generalization, and that has the formal requisites to obey the laws of probability theory, is what counts as an "objective probability in the world." The idea is that we connect up this quantity, obeying the formal postulates of probability theory and placed in the fundamental laws of nature to the world of experience, by taking its specification of probabilities to be our guide to proportions in samples of the population according to the usual probabilistic inferences. So although such a theoretical probability is a guide to frequencies in the world, and may be, at least in part, inferred to have the value it does by our observation of frequencies in the world, there is no strict identification of probability with frequency or proportion, even in the whole population in question or in that population in some dominant class of possible worlds either.

The wave-function of quantum mechanics, or rather that function multiplied by its complex conjugate, provides, in specific physical situations, a formally probabilistic distribution over the possible outcomes of an experiment. From this we can formulate our expectations not only of what proportion to find in some run of the experiment, but even of the way in which proportions themselves will be distributed over repeated runs of repeated experiments. Naturally we expect to find closer concentration of proportions and frequencies around theoretically posited probabilities as the number of trials in an experimental run increases. But we don't claim that the probability is the actual frequency even in the totality of trials, or even in that overwhelmingly large number of totalities of trials we could have run.

Something like this idea of probability will indeed fit well with much of the way in which probability functions in statistical mechanics. Needless to say, this "probability as a theoretical quantity" view is not without its puzzles as well. Some skepticism that we have genuinely fixed a unique meaning for "probability" still affects those who find general fault with the holistic idea that meanings can really be fixed by putting a term into a theoretical web that connects somewhere or other with experience, faults engendered by problems of referential indeterminism that come along with such holistic accounts of meaning. Again, the connection between theoretical assertions about probability and assertions about relative frequencies in finite samples is not completely clear here. Have probabilistic notions been presupposed when we say that we have interpreted the probabilistic assertion at the theoretical level by drawing inferences – inferences of what kind? – about proportions in finite populations from it?

One puzzle to consider is that the virtue of this approach – its ability to distinguish probability from proportion even in the total population – may be in some sense a vice as well. Could it be the case that probability and actual proportion in the entire population radically differed from one another? Of course we might doubt, in the case of wide variance of probability and frequency, that what we took to be the probabilities in the world really had the value we supposed. But from the point of view of the theoretical notion of probability we have been looking at, could it not be the case that that probability really did have the original value even though the proportion in the population as a whole radically differed from this value? Such an outcome would itself be "improbable" in that view, but not in any way impossible. But do we want to let the notion of "what probability is in the world" become *that* detached from the frequencies or proportions that actually manifest themselves? Perhaps the idea could be filled out that probability is that proportion of the total population posited as an "ideal" by the simple but very general postulates that we take to be the fundamental laws of nature.

As we shall see in Section III of this chapter, and as we shall again see a number of times, especially in Chapter 8, where the details of probability in statistical mechanics are discussed, this problem of radical divergence of probability from proportion or frequency is not merely one that occurs in an "in principle" discussion of the meaning of probability in general. Some vexing issues within statistical mechanics proper, and in its interpretation, hinge on the question of how we ought to respond to the puzzle of accepting a theory that posits a probability for an outcome while seeming to simultaneously maintain that the actual proportionate outcome of the feature in question in the world we inhabit radically diverges from the probability. The germ of this issue has already been noted in our discussion in Chapter 2 of Boltzmann's response to the problem of explaining the apparent non-equilibrium nature of the world as a whole, in the face of his contention that equilibrium is the over-whelmingly most probable state of a system.

Additional questions abound concerning the relation of probability construed as a theoretical property and the other features of the world described by our theories. Suppose we watch a piece of salt dissolve in water. Asked why the salt dissolved we could say, "because it is soluble," but we believe a much more interesting answer is available in terms of the composition of the salt out of ions, the nature of water on the molecular scale, and so on. Similarly, we might "explain" an observed proportion in nature by reference to a probability that leads to some degree of expectation for that proportion to occur in such a sample. But, once again, we feel that some other, deeper, answer in terms of the laws of nature governing the phenomenon in question, the distribution of initial conditions of hidden parameters in nature, and so on is available. What is the relation of "probability" to these more familiar features of the world as described by our scientific theories?

Some have suggested that we think of probability as a "temporary place-holder" in our theories, to be eliminated by future, deeper, theories that will dispense with it in terms of the underlying physics of the situation. So, it is suggested, "soluble" holds a place for descriptions of atomic constitution of solute and solvent and their mutual interaction, and "probability" is, similarly, a temporary and dispensable place-holder in our theory.

Much of this book will be concerned with the issues surrounding the question of just what underlying facts of nature ground the attributions of probability encountered in statistical mechanics. Whether one con-cludes that ultimately "probability" ought to be somehow defined in terms of these deeper physical elements of nature (as some have sug-gested), or, instead, that it be treated as a term "holding a place for them in theory" to be replaced by them as our deeper understanding progresses (as others have suggested), or even that it is, in its own terms, an ineliminable component of our ultimate theory (which also suggests it-self as the right expectation to others), one still must get clear just what the nature of the world is, on the microscopic scale and as described by our physics, that plays the role relative to probability for thermal pro-cesses analogous to the role played by ionic constitution in grounding notions such as solubility for solids. Getting an agreed answer to this question at present will be too much to expect. The issues surrounding the importance of fundamental dynamical laws, of distributions of initial conditions, of the interaction of a system with its environment, of the nature of the means by which systems are constructed and prepared, and of the basic cosmological facts about the universe considered as a global whole in "grounding" the fundamental probability attributions needed to get the standard conclusions out of the theory remain replete with controversy.

4. Objective randomness

At the time attempts were being made to define objective probability in terms of limits of frequencies in long-run sequences of events, it was noticed that the order of outcomes in the sequence could be important for reasons that went beyond the dependence of the value of the limit on the order. A sequence of two outcomes coded by zeros and ones, for example, might have the limiting frequency of zeros being one-half. But suppose the zeros and ones alternated in regular order. Would we want to say that the probability of a zero outcome in a trial was one-half? If we knew the immediately previous outcome, wouldn't we be sure that the zero would or would not occur? Clearly, at least for the application of our knowledge of frequencies and their limits, some assurance that the outcomes occurred "randomly" in the sequence was essential. For some, randomness became a requisite of there being a genuine probability in the sequence. For others, it was thought to be merely a condition of its applicability. But both camps required some understanding of what randomness *is*.

Whereas some argued that questions of randomness and order ought best be considered a matter of the "subjective" knowledge of the observer, others looked for objectivistic characterizations of a sequence being random in the world. R. von Mises suggested that randomness consisted in the same limit of relative frequency holding in the original sequence and in any sub-sequence derived from it by a characterization that did not refer to the actual outcomes of the trials. For this rule to select at least some sequences as random, some restriction of the rule for selecting out sub-sequences needs to be imposed. A. Church made the important suggestion that a sequence is random if the same limiting relative frequency occurs in any sub-sequence selected by an effectively computable function generating the indexes of the selected trials. This ingenious characterization proved a little too weak, because it included as random sequences in which, for example, there were more zeros than ones in every finite sequence initiating an infinite sequence, even though the probability of zeros in the sequence as a whole was one-half. Strengthened versions of it do a better job, however.

Other ingenious characterizations of objective randomness have been developed. Some rely on the intuition that "almost all" sequences ought to be random, and that the orderly sequences ought to be sparse in the sequences as a whole. A. Wald and P. Martin-Löf have suggested that a property of randomness be defined as any property that "almost all" (sets of measure one) sequences be characterizable in a specific way (i.e. in some limited language), and that the random sequences be those with all the random properties. On this definition, the random sequences are, indeed, of measure one in the set of all sequences. But the definition moves rather far from our intuitive grasp of randomness.

Martin-Löf has developed another approach. Here, one looks at the way statisticians reject hypotheses. A hypothesis is rejected at a certain "significance level" if the outcome observed is sufficiently improbable given that the hypothesis is true. Effectively characterizable tests for randomness are described, and it is then shown that they can all be combined into a universal effectively characterizable test for randomness. The random sequences are those not rejected by this universal test at any significance level.

Another ingenious notion of randomness utilizes the notion of how long a computer program it would take to program an effective computer to generate the sequence in question. Although the length of a computer program will depend upon the programming language chosen, A. Kolmogorov has shown that the problem can be discussed in terms of a universal programming language. Intuitively, for finite sequences, the randomness of the sequence is measured by the relative length of the shortest program that can generate it, random sequences being those that, in essence, can be generated only by simply stipulating the outcomes in their order in the sequence. To a degree, the results can be extended to infinite sequences, although the natural way of so extending it fails, and several, inequivalent, alternative ways can be formulated that extend the notion in different ways. Furthermore, getting agreement between our intuition as to what is to count as random and what gets defined as random by the method of computational complexity requires our thinking of the sequence as itself being generated by some mechanism characterized in a probabilistic way. Once again this makes it implausible that we can define probability in non-probabilistic terms using the notion of objectively random sequences.

The multiplicity of definitions of randomness do not all coincide, but it is not a great surprise to discover that the initial vague intuitive notion had a number of distinct (although related) formal explications. We shall not be concerned much with these notions of randomness here. In Chapter 7,II,3, however, we will discuss in some detail other notions of randomness that have their origin in the way in which a collection of systems, each of which has a strictly deterministic evolution, can be characterized as displaying randomizing behavior when the systems are given descriptions only at the macroscopic level. That is, when we coarse-grain the phase space in the manner suggested by Gibbs, it turns out that for systems suitably constructed, the evolution of a system from coarse-grained box to coarse-grained box may generate a sequence of numbers, characterizing the box it is in at a given time, that has features closely related to the features we would expect of sequences generated in purely stochastic ways. One of these features will be the fact that "almost all" such sequences of occupation numbers for the members of the ensemble started in different initial conditions will be of the "random" sort. We shall also briefly mention later work that has shown that systems can be "chaotic" in their behavior as described by the macroscopic equations of evolution as well. That is, that the equations of hydrodynamics, for example, can have solutions that can be characterized as chaotic in nature. Work has also been done relating chaos in this sense (which comes down to ineliminable large scale deviation of future behavior on infinitesimal variation of initial state) to randomness of sequences of the kind noted in this section.

5. Subjectivist accounts of probability

Whereas objectivists look for some feature of the world that can be identified as the probability formally captured by the familiar axioms of probability theory, subjectivists focus on how our acceptance of a given probability assertion will govern our beliefs and our behavior. What is it to hold that a certain outcome in a kind of trial has a specified probability? Isn't it to hold a "degree of partial belief" that when the trial is undertaken that outcome will result? And what is it to have such a "degree of partial belief"? Can't we understand that in terms of the familiar phenomenon of betting behavior? In the face of reward and punishment for having our expectations come out confirmed or disconfirmed by experience, we will act as though some outcome will occur depending both on the gains and losses we would suffer were the outcome to actually occur or not, and on our degree of certainty that the outcome will in fact turn out to be the case. The higher the probability we hold the outcome to have, the lower the odds we will demand from the bookie (human or nature) before we will bet on that outcome being the case. Turning this around, can't we simply understand the notion of probability as being a measure of our confidence in an outcome as defined by the minimum odds at which we would bet (act as if) that outcome were really going to be the case?

Much rich and profound work has been done, from several perspectives, in trying to understand probability from this subjectivist point of view. Probability is here viewed as a measure of degree of belief fixed in its functional role in our network of psychological states by its place in an account of our action in the face of risk. One problem any such subjectivist account faces is to explain to us why, given probability as such "partial belief," probability ought to conform to the usual formal axioms of probability theory. For the objectivist, this conformity follows either directly or indirectly from the facts about proportions. But why should our partial beliefs obey the axioms, in particular the addition postulate?

One ingenious and philosophically underexplored group of arguments begins with the work of R. Cox. In these approaches, probability is a number assigned to propositions, perhaps conditional one on the other, so that we are looking for $P(i/h)$, the "probability of the inference i on the hypothesis h." The usual Boolean algebra of propositional logic on the propositions is assumed. Next, axioms are posited such as: (1) the probability of i on h determines the probability of (not i) on h; (2) the probability of i and j on h is determined by the probability of i on j and h and the probability of j on h; (3) the probability of $i \cup j$ on h (when $i \cap j = \phi$) is a continuous and monotonic function of the probabilities of i on h and j on h. These axioms of functional dependence, plus some assumptions about the smoothness (differentiability) of the functional dependence, lead to a generalization of the usual probability axioms. It is then argued that the standard form of the axioms can be obtained by making some conventional stipulations. Here, then, the formal aspects of probability are taken to arise out of our intuitions that some partial beliefs (probabilities) ought to depend in a functional way only on a limited class of others.

More familiar and more widely explored alternative justifications of the usual axioms focus directly on our betting behavior. The so-called "Dutch Book" arguments have us reflect on making bets against a bookie on the outcome of some trial. We pick our degrees of partial belief in the outcomes, and the bookie offers us the minimum odds we would accept appropriate to those degrees of partial belief. An ingenious but simple argument can show us that unless our subjective probabilities conform to the usual postulates of probability theory, the bookie can offer us an array of bets that we will accept, but that guarantee that when all stakes are collected he wins and we lose, no matter what the outcome of the trial. In order for us to avoid having such a "Dutch Book" made against us, our probabilities must be coherent – that is, accord with the usual postulates. Even if our probabilities are coherent, we might still be put into a position where we cannot win, no matter what the outcome, but we might lose or break even. If we wish to avoid that we must make our probabilities "strictly coherent" – that is, coherent and with a zero probability credited only to outcomes that are impossible. Normally in the physical situations we won't want strict coherence because classes of events that are non-empty but of probability zero are a familiar, if puzzling, part of our usual physical idealizations of phenomena.

An alternative and more general approach considers an agent offered choices of lottery tickets in which gains and losses are to occur to the agent conditional on an outcome of a trial occurring or not occurring. If the agent's preference ordering among the lottery tickets is transitive, and if some other conditions of the choices ordered being rich enough are met, then one can show that the agent acts "as if" he had a subjective probability over the trial outcomes that obeyed the standard axioms of probability and that was uniquely determinable by his preferences, and as if he had a valuation or utility or desirability function over the rewards and losses unique up to a linear transformation. It goes something like this: Suppose the agent always prefers x to z whenever x is preferred to y and y to z. Suppose also that if the agent prefers x to y, he will also prefer a given probabilistic mixture of outcomes involving x to the identical mixture with y substituted for x. And suppose there is a sufficiently rich collection of preferences. Then we can assign a probability to each outcome, p, and a utility to the gains or losses incumbent upon the outcome occurring or not occurring, u_1 and u_2, such that if each lottery ticket is assigned an "expected value," $pu_1 + (1 - p)u_2$, then the preference ordering among lottery tickets instanced in the agent can be reproduced by ordering the lottery tickets in terms of their expected value. It will be "as if" the agent assigned a unique probability to the trial outcomes consistent with the laws of probability theory, and a utility to the gains and losses, unique up to a linear transformation (which amounts to picking one gain or loss as "neutral" and fixing a scale unit for units of desirability), and then ordered his preference for the lottery tickets by calculating their expected values.

This approach to subjective probability is important as a component of various "functionalist" accounts as to what partial beliefs and desirabilities are, as well as a crucial component of the theory of rational decision making. Naturally the full theory is a very complicated business, involving many difficulties in associating the "psychological" states to behavior, and in formulating a theory of normative rational action immune to intuitive paradox. But the overall strategy for showing why probabilities, if they are to be taken as measures of partial belief, ought to obey the usual probability axioms (as a consequence of "rationality of choice as evinced by transitiveness") is clear.

An alternative derivation of subjective probabilities starts with the notion of comparative believability – that is, with the idea of a relation among propositions that, intuitively, can be thought of as one proposition being "as worthy of belief" as another. Intuitively plausible axioms are imposed on this notion of comparative believability. For example, we may demand transitivity, so that if A is as believable as B and B as believable as C, it follows that A is as believable as C. Additional constraints can be placed on the notion of comparative believability that are sufficient to guarantee that one's believability structure can be represented by a probability assignment to the propositions. Such a probability representation assigns real numbers between zero and one (inclusive) to the propositions. And the assignment is such that A will be as believable as B just in case the probability number assigned to A is at least as great as that assigned to B.

The subjective theory of probability is concerned not only with our holding partial beliefs, but with our changing them in the face of experience. How should we modify our distribution of partial beliefs in the face of new evidence? The usual rule suggested is conditionalization. We have, at a time, not only probabilities for propositions, but conditional probabilities as well, the probability of h given e, whenever the probability of e is non-zero. Suppose we then observe e to be the case. Conditionalization suggests that we take as the new probability of h the old conditional probability it had relative to e. This rule has been nicely generalized by R. Jeffrey and others to handle cases where we don't go to e as a certainty, but instead take the evidence as merely modifying the older probability of e. Conditionalization is a conservative strategy. It makes the minimal changes in our subjective probability attributions consistent with what we have learned through the evidence, and it generates a new probability distribution as coherent as the one we started with.

Much effort has gone into giving a rationalization for changing probabilities by conditionalization as persuasive as the standard rationales for having subjective probabilities at a time obey the usual formal axioms. An argument reminiscent of the Dutch Book arguments and due to D. Lewis has the agent confronting a bookie and making bets on the outcome of one trial, and then additional bets on other outcomes of other trials should the first trial result in one specific outcome or the other. Only if the second set of bets is made on odds consistent with conditionalization on the outcome of the first trial can a compound Dutch Book situation be avoided by the agent. P. Teller has shown that conditionalization follows from demanding that one's new probability distribution has $P(h) = P(k)$ after the evidence e has come in if $P(h) = P(k)$ before the observation and if h and k both implied the evidence assertion, e. And B. van Fraassen has shown that conditionalization is the only method of changing one's probability distribution that will make the probability dynamics invariant under reclassifications of the elementary events into event classes. A variety of other rationalizations for conditionalizing can also be given. On the other hand, because conditionalization is a conservative procedure, intuitively changing one's subjective probabilities only to the extent that the change is directly forced by the change in the probability attributed to the evidence induced by the result of the trial, there could be circumstances where the subjectivist would doubt its reasonable applicability, say if the new evidence indicated the agent's earlier irrational state of mind when the initial probability distribution was chosen.

In Chapter 7,III,3, we shall see that some of those who advocate what they call a subjectivist approach to probability in statistical mechanics seem to evade the strictures on conditionalizing with respect to all known evidence in some of their uses of probability attributions.

Suppose one accepts the legitimacy of probability as interpreted as a measure of partial belief, in the manner we have outlined. What, from this subjectivistic perspective, is the place of the version of probability as "a feature of the objective world" in one of the guises previously outlined? Here, defenders of subjective probability take a number of different positions. Some would allow for their being two kinds of probability (often labeled by subscripts as probability$_1$ and probability$_2$), objective and subjective, just as most defenders of objective probability are perfectly happy to countenance a legitimate subjectivist interpretation of the formalism as well.

The advocate of objective probability as actual short or long-run frequency or proportion will usually think of subjective probabilities as estimates on our part of the real proportions of the world. Naturally he will seek principles of rational inference to and from proportions in samples to those in populations, and hence to and from proportions in samples to subjective probabilities. The believer in objective probability as a propensity of an individual trial situation, it has been suggested by D. Lewis, will hold to a "Principal Principle," that if the chance of an outcome on the trial is taken to have a certain value, the subjective probability of that outcome must have the same measure. At least this will be so if the propensity chance is one that supposes the absence of underlying hidden variables that would, if known, change the propensities assigned. Subjective probabilities are, for some of the probability-as-a-theoretical-feature proponents, the basic interpretive device by which the theory structure fixing objective probabilities gets its connection to the world. It has been suggested, for example, by S. Leeds, that we can give an interpretation to the state-function of quantum mechanics by simply taking the rule that the values thought of as probabilities upon measurement computed from it are to be understood as subjective probabilities. From this point of view, subjective probabilities generated by a theory so interpreted partially fix any meanings for objective probabilistic expressions appearing in the theoretical network, leaving it open, of course, that these objective probabilities may be further pinned down by their association with non-probabilistically characterized features of the world, or even eliminated in terms of them. (The "fairness" of the coin that leads to subjective probabilities of one-half for heads and tails upon tossing, for example, would be replaced by its structural symmetry, and perhaps the distribution in the world of initial conditions over tossings, which underlie its being a fair coin.)

Other subjective probability theorists find no room whatever for the notion of objective probability as we have been construing it. For them, probability *is* subjective probability. Not that there aren't, of course, frequencies in the world, maybe even long-run limits of them. And, of course, there are also all those other features of the world, such as the balance of the coin, the distribution of initial conditions, the quantum state of the electron, that are causally connected to the frequencies we observe. But, from this point of view, there is no need to think of probability itself as anything over and above degree of partial belief. Our partial beliefs may very well be determined for us by various beliefs we have about frequencies and structures in the world, but there is no need to think of probability as being anything itself "in the world" generated by or identifiable with these familiar objective features of things.

A very interesting proposal in this vein, initiated by B. de Finetti, tries, from a subjectivist perspective, to explain to us why we might be tempted to think of "objective probability in the world" when no such concept can be made coherent (as the usual objectivist intends it, at least). Imagine a sequence of tossings of a coin the objectivist thinks of as biased – that is, of having some definite propensity to come up heads that might not have the value one-half. The objectivist thinks of us as learning from experience. We observe a long run of trials, and on the basis of the observed proportion of heads in the trials come to an estimate of the real propensity of the coin to come up heads, the objective probability of heads. How can a subjectivist understand this learning from experience without positing objective probabilities?

De Finetti asks us to consider someone with a subjective probability distribution over finite sequences of outcomes of trials of the coin tossing. Suppose this objective probability distribution is exchangeable – that is, suppose the subjective probability given to a sequence of heads and tails is a function only of the proportion of heads in the sequence and is independent of the order of heads and tails. Then, de Finetti shows, this agent's probability distribution over the sequences can be represented "as if" the agent took the sequences to be generated by independent trials of tossing a coin that is biased to some particular propensity for heads, but with an unknown bias. It will be "as if" the agent generated his subjective probabilities for the sequences by having a subjective probability distribution over the range of possible biases of the coin.

Furthermore, let the agent modify his subjective probability distribution over the sequences by conditionalizing on the observed tossings of the coin. Then, if the agent's original distribution was such that when represented as a distribution over propensities or biases it gave no propensity probability zero or one, the evolution of the agent's probability distribution over sequences can be represented as the agent having his subjective probability converge on an ever narrower range of biases, indeed converging to a propensity equal to observed relative frequency in the limit as evidence tossings go to infinity. And two such agents will then behave "as if" they started with different subjective probabilities over biases, but, learning from experience, both converged to the "real objective propensity" in the long run. For de Finetti, of course, there is no such "real" probability of heads for the coin. All that exists are the convergences to observed relative frequency, convergences themselves dependent on the particular structure of initial subjective probabilities of the agent (these being exchangeable) and on the agent's learning from experience by conditionalization.

This result of de Finetti's is generalizable in interesting ways. The key to the representation theorem is the symmetry of the agent's initial subjective probability, the probability given to a sequence being invariant over any permutation of heads and tails in the sequence, so long as the number of heads and tails remained invariant. A general result shows that such symmetries in the agent's subjective probability will lead to his acting as if he believed in an objective propensity to which observed relative frequencies would converge. Even more general results can be proven to show that agents who start in agreement about which sets of outcomes receive probability zero, and who modify their probabilities in a conditionalizing manner, will converge on identical probability distributions "in the limit." These results constitute important formal facts about subjective probabilities that are closely analogous to other results usually interpreted in an objectivist vein. We will discuss these related results in the context of the equilibrium theory of statistical mechanics in Chapter 5 when we examine the important Ergodic Theorem of classical equilibrium statistical mechanics.

An important concept, emphasized by B. Skyrms, of "resiliency" should also be noted in this context. A subjective probability is, basically, resilient if the agent would not modify the value he gives the outcome in question (or at least would not modify it much) in the light of additional evidence. Resiliency, then, is closely analogous to the kind of objectivist demand that "real" probabilities be those that are genuinely tychistic, so that no exploration of hidden variables could divide the kind of trial in question into sub-classes in which the probability of the outcome would differ from the probability in the class as a whole. Resiliency is a kind of subjectivistic "no hidden variables" for the probability in question. It can be argued that what appear to be objective probabilities, can be, at least in part, reconstrued from the subjectivist point of view as resilient subjective probabilities. Again, this subjectivistic notion is related to a kind of objectivist resiliency (metric indecomposability) that we will explore in Chapter 5.

6. Logical theories of probability

The subjectivist places the restraint of coherence on an agent's initial probability distribution. If we take conditionalization to be justified by the arguments for it, then additional constraints exist on how subjective probability distributions are constrained by rationality to change in the light of new evidence. But is there any further constraint of rationality on an agent's initial probability distribution, on his probability distribution "in the light of no evidence," or, as it is frequently designated, on his "a priori probability?" Pure subjectivists frequently answer "no," coherence being given, that one a priori probability distribution is just as rational as any other.

Others deny this. "Objective Bayesians," as they are sometimes called, maintain that there are indeed further constraints of rationality that can be imposed on an a priori probability distribution. Their claim is closely related to the claims made by those who would offer what was once thought of as an alternative interpretation of probability. Deductive logic presents us with a relationship among propositions, in particular with the idea that the truth of one proposition can assure the truth of another. A premise can entail a conclusion, so that if the premise is true, so must the conclusion be true. Could there not be a similar but weaker relationship among propositions so that the truth of one, although not guaranteeing the truth of the other, could provide some degree of assurance, less than certainty perhaps, that the other was true as well? Qualitative relationships of this kind were explored by J. Keynes and others. Later attempts – in particular, the extended program of R. Carnap – sought to develop a quantitative theory of so-called "inductive logic."

But how can the "degree of support" or "degree of confirmation" of one proposition grounded in another be determined? First, it is usually assumed that these degrees of confirmation must obey the usual formal postulates of probability theory. In his later expositions of the work, Carnap took the line that these probabilities were to be thought of, ontologically, in the subjectivist vein. Then the rationale for holding them to be constrained by the familiar laws of probability theory would be the rationales for subjective probabilities being coherent that were discussed in the last section. But these constraints were plainly insufficient to uniquely fix the degree of probability of one proposition relative to another, and further constraints were sought.

A natural way of viewing the problem is to think of propositions as classes of possible worlds, the proposition being identified with the class of possible worlds in which it is true. The degree of confirmation of proposition *h* on proposition *e* could be thought of, then, as the measure of the proportion of worlds in which *e* is true and in which *h* is true as well. We can get the results we want, a logical probability obeying the formal theory of probability, by distributing an initial probability of one over the distinct possible worlds. The naive way of doing this that suggests itself – letting each possible world have an "equal share" in the total probability – leads to an inductive logic in which we don't learn from experience. A subtler method of first dividing the probability evenly over a class of kinds of worlds, and then evenly over the worlds in the kinds (kinds that don't have equal numbers of individual possible worlds in them), gives probabilistic weighting to "orderly" worlds, and leads to an inductive logic that raises our expectation that a property will be instanced as we experience its instancing in our observed sample, a kind of inductive projection from the observed into the unobserved.

Trying to rationalize a unique confirmational measure as the only one justifiable, or to find criteria of adequacy for a probabilistic measure that will pin down the possibilities for confirmation functions to a small number of choices, and then to rationalize those criteria of adequacy, is a task that Carnap only partially achieves to his own satisfaction. He relies frequently on "intuition" even to get that far. And finding a way of extending the method originally designed for simple and finitistic languages to the richer realm of higher order languages and to worlds with an infinite number of individuals proves problematic as well. But, most importantly for our purposes, another hidden difficulty with the program exists. Notice, though, that if we had achieved the program to our satisfaction, the problem of a rational constraint on an agent's a priori probability distribution would be solved. Each proposition has its "logical probability" relative to any other proposition, and indeed to the empty set of propositions. This "probability of *h* relative to null evidence" would then be the rational probability with which to hold a proposition before any evidence came in. Probabilities after the evidence could then all be arrived at by conditionalization, using the a priori probabilities to compute, in the usual way, the conditional probabilities needed to know how to modify one's initial probability as new evidence was uncovered.

The basic rule invoked for determining the a priori probabilities – treat all symmetric propositions equally when distributing probability – is a modern instance of one of the oldest ideas in probability theory, the Principle of Indifference. The idea is that "probabilities are to be taken as equal in all similar (or symmetric, or like) cases." A priori, heads has a probability of one-half, there being only two possible symmetric outcomes. Later tossings of the coin might indicate bias and lead us to modify the probability we give to heads, but it is the symmetry of cases that provides the a priori probability where we start. But *why* should we believe, a priori, in equal probabilities for the symmetric case?

In fact, the Principle of Indifference has more problems than its lack of apparent rationalization. It is, without further constraint, incoherent. For it depends in its probability assignment upon *how* the possible outcomes are categorized or classified. A die has six faces, so the a priori outcome of a one coming up is one-sixth. But the die can either come up with a one or else with a "not-one." So there are *two* cases, and by the Principle of Indifference, the probability of a one coming up ought to be taken to be one-half. In this case, we might resort to the fact that "non-one" can be decomposed into five cases, each of which is, in some sense, indecomposable, leading us back to the natural one-sixth and away from the counter-intuitive one-half. But what if the number of possible outcomes is infinite? Here, each indecomposable outcome has probability zero, and each interesting class of these (an interval on the real line, for example) has an infinite number of elementary outcomes contained in it.

This dependence of the a priori probabilities on the way cases are categorized has been emphasized by what are frequently generically referred to as "Bertrand's Paradoxes." Imagine, for example, a container with a shape so that the surface area on the inside that is wetted varies non-linearly with the volume of the container that is filled. An a priori probability of "amount of container filled" that distributed probability uniformly over the possible volumes of fluid would give radically different results from one that distributed probability uniformly over allowable interior surface area wetted by the fluid. What principle of rationality, designed to fix the unique one of the coherent possible a priori probabilities as the rational one for an agent to adopt, will tell us which categorization of the possible outcomes to choose? Only given a principle for selecting the right categories in which to view the problem can we then apply a Principle of Indifference or symmetry over those possibilities so construed.

H. Jeffreys initiated a program of selecting from among the categorizations by examining the invariance of probabilistic conclusions under various transformations of the data. Sometimes it seems as though we would want our probabilistic results to remain invariant under certain transformations. We feel, for example, that in some cases picking a designation of one value of a quantity as zero point as opposed to another, or picking one interval as unit scale as opposed to another, should not modify our expectations. Under those circumstances, one can sometimes fix upon a single categorization, with its associated uniform probability distribution militated by the Principle of Indifference so applied. But, as we shall see in Chapter 5,III,5 and 7,III,3 when we examine attempted applications of the Principle of Indifference (or its modern day reformulation, the "Maximum Information Theoretic Entropy Principle"), such a rationalizable invariance rule to fix our a priori probabilities is only rarely available to us. And when it is, what we can obtain from it may be less than what we would like.

When investigators in the foundations of statistical mechanics allege that they are understanding probability in statistical mechanics as subjective probability, it is usually, rather, a belief on their part that there is a legitimate applicability of the Principle of Indifference to physical situations that can be applied to ground the positing of initial probabilities so essential to getting what we want in statistical mechanics that is the core of their position.

III. Probability in statistical mechanics

A great deal of Chapters 5 through 9 will be directed toward problems that, at least in part, will be explored by examining in some detail the role played by probabilistic assertions in the description and explanatory account of the world offered by statistical mechanics. It will be of use here, though, to give a preliminary survey of how probability attributions are embedded in the statistical mechanical picture of the world, and of some of the peculiarities of probability in statistical mechanics that lead to difficulties in fitting an account of its role in that theory into one or another of the philosophical views about probability that have been outlined in this chapter.

In our survey of the history of the foundational problems of statistical mechanics we saw how attempts were made to account for the equilibrium state as being the "overwhelmingly most probable" state of a system. Here, we view the system as subject to some fixed macroscopic constraints (say volume and pressure for gas) and look for some justification of the claim that of all the microscopic dynamical states of the system compatible with these macro-conditions, with overwhelmingly great probability they will correspond to the state of the gas being at or near its equilibrium value of some other macroscopic constraint (say the temperature of the gas).

In trying to understand the non-equilibrium behavior of a system, we imagine it as having been prepared out of equilibrium, again subject to some macroscopic conditions. We then try to show that the combination of a reliance upon the detailed facts about the microscopic constitution of the system, the facts about the dynamical laws governing these microscopic constituents, and some other assertions framed in the probabilistic guise can lead us to the derivation of a description of the approach to equilibrium of the system as characterized by the macroscopic hydro-dynamic equations of evolution appropriate to the system in question. In some accounts, the probabilistic assertions will be about probabilities of initial microscopic states of the system compatible with its initial macro-scopic condition. In other accounts, they will be about the probability of future interactions of the micro-components with one another. Once again, the invocation of the probabilistic claims is one that cries out for justification.

As we shall see in Chapters 5 and 7, the justification needed for the probabilistic claims will vary quite radically depending on just what the claim under consideration is. And getting clear on that will sometimes require some serious preliminary conceptual clarification. At this point I will make only a few general remarks about some kinds of conceptual puzzlement we will run into. Here I will be concentrating, in particular, on those issues that will be problematic because they introduce features of probabilistic description of the world in statistical mechanics that most directly confront the various philosophical accounts of probability we have just seen outlined. The kinds of problems noted here are not meant to be either exhaustive or exclusive of one another. Some of the puzzles noted here are closely intertwined with one another, and an acceptable understanding of the world will require a grasp of issues that does justice to answering a number of questions simultaneously.

The origin and rationalization of probability distributions. Some problems arise out of the questions concerning the physical ground on which a given correct probability attribution rests. This is intimately con-nected, of course, with the questions concerning the rationale we can offer if a claim we make that a probability distribution ought to take a certain form is challenged. The problem is complicated in statistical mechanics by the fact that probability distributions are introduced on quite different grounds, with quite different rationales, and for quite different purposes in different portions of the theory.

In equilibrium theory, or at least in the version of it that seems clearest and most defensible, the aim will ultimately be to show that a unique probability distribution can be found that satisfies a number of con-straints. One of these constraints, time invariance, comes from the very notion of equilibrium itself. The other, absolute continuity with respect to the usual phase space measure, is a constraint harder to justify, assum-ing as it does a portion of the original probability attribution we meant to justify. The rationalizing ground is sought in the constitution of the system and in the dynamical laws of evolution at the micro-level. The role played by this probability distribution in our description and ex-planatory account of the world is itself somewhat problematic.

In non-equilibrium theory, the understanding and rationalization of one probabilistic postulate of the usual theory plays a role in attempts to ground the posits that are the immediate descendants of Boltzmann's *Stosszahlansatz.* The kinetic equations of non-equilibrium theory are usually obtained by one version or another of some rerandomizing posit governing the interaction of the micro-constituents of a system. But the dynamic evolution of a system, and hence even of an ensemble of systems (or probability distribution over systems subject to a common macroscopic constraint), is fully fixed by the dynamical laws of evolution governing the micro-components of the systems. So a rationalization of the probabilistic posit here will come down to an attempt to show the *consistency* of the probabilistically formalized rerandomization posit with the accepted dynamical laws and the facts about the constitution of the systems. Here, probabilistic postulation is thought of much as a device to generate the correct result (the kinetic equation of evolution for a reduced description of the ensemble), a device to be instrumentalistically justified by reference to the real laws governing the dynamical evolution, and hence not directly a fundamentally independent component of the full description of the facts of the world we are seeking.

But another role for probability distributions in non-equilibrium theory is even more fundamental to the theory. This is the appropriate distribution to impose over the microscopic initial conditions of a system started in non-equilibrium subjected to some macroscopic constraints. In some accounts, the rationalization of rerandomization will itself depend upon this distribution. And if we want a full explanatory account of the structure of the approach to equilibrium of a system (its relaxation time, the form of its equation of evolution toward equilibrium on a macroscopic scale, and so on), it seems clear that such an assumed distribution will need to be posited. But here, much is still opaque in the theory. There is, in fact, and as we shall see, no universally accepted account of even what such distributions should be taken to be, much less a universally agreed upon account of their origin and rationalization. Microdynamical laws, constitutions of systems, places of systems in interacting external environments, modes by which systems are prepared, a priori distributions determined by principles of general inductive reasoning, and cosmological facts have all been invoked in one or another of the competing accounts of the physical basis of these essential initial probability distributions over possible micro-conditions.

Probability and tychism. We have outlined a debate between those who would take probability to be a mode of description of a world that might be underlain by a fully deterministic picture – that is, who would take non-trivial probabilities to be consistent with an underlying description of the situation that would so bring factors into account that with their full specification, the outcome would either definitely occur or not occur – and those who would hold that probability exists in the world only if there is a genuine failure of determinism – that is, only if such hidden factors fail to exist.

Quantum mechanics, with its "no hidden variables" theorems, has typically provided the context in which pure tychism and irreducible probabilistic descriptions not underpinnable by determinism at a lower level are defended as the correct picture of the physical world. Should the initial probabilities over initial conditions in non-equilibrium statistical mechanics be construed in a similar manner, or should they be viewed as merely the accounting of actual distributions over the underlying possibilities, one of which constitutes the real and fully determining condition in any particular instance of a physical system?

As we shall see in Chapters 7,III,6 and 9,III,1, this is a controversial matter. Particularly interesting in this context are attempts to argue that the probability of statistical mechanics is neither the "pure chance without hidden variables" of quantum mechanics, nor the "distribution over fully determining hidden variables" of the usual "ignorance interpretation" variety, but, rather, a third kind of probability altogether. The argument will be that the instability of the dynamic evolution of the systems with which we are concerned in statistical mechanics makes the characterization of the system as having a genuine microscopic dynamical state a "false idealization." The arguments to be offered rest upon the contention that no matter how close we get to any initial condition in systems of these kinds, we will find other initial conditions whose future evolution, as described by the familiar trajectory in phase space starting from this other condition, will diverge quickly and radically from the trajectory starting from the first initial state we considered. Under these conditions, it will be alleged, it is misleading to think of any individual system as really having a precise initial "pointlike" dynamical state and "linelike" dynamical trajectory. Rather, it will be argued, the individual systems in question ought to have their states characterized by probability distributions over phase-space points and their evolution characterized in terms of the evolution of such distributions over phase-space.

This view is like the "pure tychism" view in taking the probability distribution to be irreducibly attributable to the individual system, but unlike the "no hidden variables" view in that the grounds for denying any further deterministic specifiability of the system are quite unlike those that are used to deny hidden variables in quantum mechanics. In the latter case, we are presented with alleged proofs that the statistics in question (those predicted by quantum mechanics) cannot be represented as measures over underlying phase-space regions of points. But in the statistical mechanical case, there plainly is such a representation of the probability distribution. It is worth noting that although we will be discussing this issue in the context of classical statistical mechanics, the view takes on exactly the same aspects in the context of a statistical mechanics whose underlying dynamics is quantum mechanics. The "irreducible probability" in both cases is *sui generis*. In the quantum mechanical case, it is quite independent of the pure tychism that might be alleged to hold of the underlying dynamics.

Applying the posited probability distribution. Only occasionally are probability distributions used directly to determine observable quantities in statistical mechanics, although they sometimes do function in that way as in predictions of molecular velocity distributions and attempts to confirm these predictions by direct surveys of velocity distributions in samples of the population of molecules. More commonly, the probability distribution is used to calculate some value that is then associated with a macroscopically determinable quantity.

Care must be taken, however, to make sure that the quantity calculated can plausibly play the role demanded of it in the explanatory-descriptive account in which it is to function. Sometimes it is easy to lose sight of what one was interested in showing in the first place, and to think that a goal has been accomplished when a quantity is derived having the formal aspects sought, but whose role in the overall account is less than clear. Often a precise disentangling of the notions involved, and the application of derivable connections between quantities will go a long way to clarifying the structure of the theory. Thus, in the equilibrium theory the careful distinction must be made between time averages of quantities, computed using a probability distribution over the reference class of time states of a given system, and averages for these quantities taken over all possible micro-states of different systems compatible with given constraints and relative to some probability measure over these possible states. These latter averages are called phase averages. Also important is the distinction of the latter averages from most probable values of the quantities in question (calculated using the same set of available micro-states and the same probability distribution). The justifications of asserted relations among these values by means of Ergodic Theory and by means of results dependent upon the vast numbers of degrees of freedom of the system, play important parts in clarifying the formal and philosophical aspects of equilibrium statistical mechanics.

In some cases, it remains a matter of controversy which probabilistically generable quantity is the correct one to associate with the macroscopic feature of the world we are trying to understand. For example, in the non-equilibrium theory, most accounts of the statistical mechanical regularity to be associated with the monotonic approach to equilibrium described by thermodynamics follow the Ehrenfests in associating the curve of monotonic behavior of the macroscopic system with the concentration curve of an ensemble evolution – that is, with the curve that goes through the "overwhelmingly most probable value of entropy" at each time. And these accounts agree with the Ehrenfests that this curve should not be thought of as representing in any way a "most probable evolution of systems." But other statistical mechanical accounts do in fact think of the monotonic approach to equilibrium as being representable in statistical mechanics by just such an "overwhelmingly most probable course of evolution." Such conflicts of understanding point to deep conflicts about the fundamental structure of the theory insofar as it is designed to allow us to understand montonically irreversible thermodynamic behavior.

Probability versus proportion. In our discussion of foundational views on probability, we noted important doubts that probability could, in any simple-minded way, be identified with actual frequency or proportion in the world. Indeed, even moving to a view of proportion in possible worlds, doubts remained that one could define probability as proportion – even in the "long run."

Statistical mechanics presents us with additional puzzles concerning the relationship between probability and proportion. These are exemplified by the paradox considered by Boltzmann and discussed in Chapter 2: If equilibrium is the overwhelmingly probable state of a system, why do we find systems almost always quite far from equilibrium?

We will need to consider a variety of possible answers to questions of this sort. One might be to deny the correctness of the claim that equilibrium really is overwhelmingly probable. Another might be to deny that probability has anything to do with realized proportion, something not too implausible from some subjectivist or logical probability viewpoints. Still another group of attempts at resolving such a puzzle might be to argue that we have simply looked at too small a reference class in seeking the proportion in the world to associate with the asserted probability. Such a suggestion is that offered by Schuetz and Boltzmann to the effect that our region of the universe is only a small portion of its extent in space and time, and that if we looked far and wide enough in space and time we would indeed discover equilibrium to be dominant in proportion in the universe as a whole. Modern variants of such a "cosmological way out" will, as we shall see in Chapter 8, suggest that we would need to look far and wide indeed to find the appropriate proportion to identify with the probability of our theory. We shall also need to examine those modern variants of Boltzmann's suggestion as to why proportion differs so wildly from probability in our local region of the larger world – that is, updating his argument that such a deviation from the norm ought to be expected by us for the simple reason that in the dominant regions of equilibrium, no sentient observer could exist to experience this dominant proportion of equilibrium in his neighborhood.

Idealization and probability. In exploring how physical theories deal with the world, we are accustomed to the fact that our theories deal only with "idealized" systems. In mechanics, we talk of systems composed of point masses, or of frictionless systems, or of systems in which all inter- actions save one can be ignored. Idealization plays a prominent role in statistical mechanics, sometimes in ways that require rather more atten- tion than a mere nod to the fact that the results hold exactly only in the ideal case.

Some of the idealizations we will encounter include going to the thermodynamic limit of a system with an infinite number of degrees of freedom, or to the Boltzmann–Grad limit in which the relative size of molecule to scale of the system becomes vanishingly small, or to the limit as time "goes to infinity" invoked in the ergodic theorems of equilibrium statistical mechanics or in the mixing results of the non-equilibrium theory. In some cases, the role the idealization is playing will be clear and relatively uncontroversial. Thus, for example, the role played by the thermodynamic limit in allowing us to move from average values of some quantities to the values being identified with the overwhelmingly most probable values of these quantities will be rather straightforward. In other cases, however, the place of idealization in the overall scheme can be quite controversial. One derivation of the kinetic equations of non- equilibrium – that of Lanford, for example – makes a radical use of the Boltzmann-Grad limit to obtain results puzzlingly incompatible (at least in a conceptual way) from those obtained by the "mixing" type results. And these latter rely on the "time going to infinity" idealization for their accomplishments in a way objected to by the proponents of the Lanford scheme. Which idealization ought to be taken as "really representing how things are in the actual world" is by no means clear.

Here, I want only to note how the very concept of probability intro- duced in statistical mechanics is sometimes heavily determined by the idealizing context. Take, for example, what is sometimes called a prob- ability in equilibrium statistical mechanics viewed from the perspective of the Ergodic Theorem. Here, one starts off with systems idealized with reference to their structure, molecules interacting only by collision, and then perfectly elastically, for instance, and a system kept in perfect en- ergetic isolation from the outside world. Next, one shows that for "almost all" initial conditions the proportion of time spent by a system with its representative point in a phase-space region will be proportional to the "size of that region in a standard measure. But one shows this only in the limit as time "goes to infinity." There is of course nothing wrong with such a notion. Indeed, it fits nicely with those accounts of probability that, we saw, emphasized proportion in the limit of an infinite number of trials. But the relation of such a probability to such things as the actually observed behavior of systems in finite time intervals is only indirect. And unless one is careful to keep in mind the essential role played by the idealization, one can think that more has been shown than is actually the case.

IV. Further readings

An excellent elementary introduction to the theory of probability is Cramér (1955). Feller (1950) is a classic text on the subject. The axiomatic foundations are explained in Kolmogorov (1950).

A survey of philosophers' ideas on the nature of probability can be found in Kyburg (1970). For the dispositional theory, a good source is Mellor (1971). Kyburg (1974) surveys many variants of the dispositional account. For the view of probabilities as "theoretical quantities," see Levi (1967) and (1980) and Fine (1973). Cox's derivation of probability is in Cox (1961). See also, Shimony (1970).

Fundamental papers on the subjectivist approach to probability are in Kyburg and Smokler (1964). The relation of subjective probability to comparative believability can be found in Fine (1973).

For one version of a "logical" theory of probability, see Carnap (1950). For other versions, so-called "objective Bayesianisms," see Jeffreys, H. (1931) and (1967), Rosenkrantz (1981), and Jaynes (1983).

On Humeanism and its opponents, see Chapter V of Earman (1986). Lewis (1986), Chapter 19, is important on the relation of subjective and dispositional probability.

For an introduction to the variety of ways of characterizing "objective randomness," see Chapter VIII of Earman (1986). Fine (1973) has detailed and comprehensive treatments of the major approaches to this issue.

4

Statistical explanation

I. Philosophers on explanation

1. Causation and the Humean critique

We wish to know not only what happens in the world, but why it happens. We are interested, we say, in explanations of the phenomena of the world. We want explanations of both particular occurrences ("why did this rock dissolve when I placed it in the water?") and of general kinds of happenings ("why does salt dissolve when placed in water?"). But what is it to give an explanation of a happening or a class of happenings?

Aristotle is famous for offering a first philosophical account of the nature of explanation. It consists, he says, in giving the "causes" of the phenomenon in question. Here, he includes what is called the "efficient cause," corresponding most closely to our present idea that an event is produced by continuous earlier events that produce the phenomenon. Influenced by his experience in biology, he thinks of all events as governed by function, end, or purpose as well, so that a full account of why they occur must bring out their place in the accomplishment of some end or purpose. That is, a full explanatory account must offer the "final cause" of the phenomenon as well as its proximate efficient cause. Bringing into play his metaphysics of change, in which all change is the change of some "form" on an underlying unchanged "substance," he calls reference to the newly imposed form of the phenomenon reference to its "formal cause," and reference to the underlying unchanged substrate reference to its "material cause."

For a very long time we find little more offered to us by philosophers that goes beyond this interesting but still vague reference to causation as the core of explanation. Leibniz, like Aristotle, believed that were we but intelligent enough we could derive the full structure of the world by pure ratiocination from self-evident first principles. But, he believed, given our limitations of intellect, we must look for explanatory accounts in generalization from our experience of causation in the world. He offers as the basic postulate of his metaphysics the claim of the Principle of Sufficient Reason. This is the assertion that every happening is explicable by the sufficient causation for it which, if we but search hard enough, can be found. He uses the Principle to great effect both in founding his metaphysics and in his brilliant critique of the substantivalist view of space. Leibniz and others ponder over the relation of mind to body, held to be causal interaction by Descartes but seeming to many others to be too anomalous a relation to fit into the usual cause-effect pattern of the world. Later, Berkeley, in his critique of the representative realist's idea of matter, offers the claim that only mind could be "active" enough to serve as a cause, the notion of having causal efficacy being incompatible with the "inertness" of matter.

Finally, we get Hume's brilliant critical investigation into the notion of causation itself. Taking the basic empiricist principle that concepts obtain their meaning by serving as labels for items of sensory experience, and asking for what we actually experience when we experience the cause-effect relation, he offers his famous critique of the idea of "necessary connection" as an item in the world. We find events related to one another spatially and temporally. Causes and effects, he holds, are events that are spatiotemporally contiguous. The asymmetry of the causal order in time requires that causes always precede their effect in time. Finally, and most importantly, events are in a cause-effect relation to one another only if they are constantly conjoined, only if the cause is always accompanied by the effect and the effect always preceded by the cause, where, of course, it is now a *type* of event that is being considered as cause and a type as effect. It is this constant conjunction of kinds of events in the world that is the core of the cause-effect notion. The causal order of the world is essentially the order of *regularity* in the world, the order of things that are constantly conjoined to one another.

The naive believer in causation agrees, of course, that the regularity exists, but argues that the regularity is itself explained by the causal connection between the events. For Hume, such an "explanation" of regularities is a non-starter, for it invokes a notion of necessary connection in the world whose very intelligibility is dubious on empiricist grounds. Rather, the regularity is all there is to the physical phenomena. Hume does offer us an account of why we have been misled into thinking that there is such a thing as necessary connection in the world. The source of the error is a projection onto the phenomena themselves of something that is instead a psychological response to them on our part. Experiencing a multiplicity of constantly conjoined events of a certain kind, our mind, when presented with the idea of the cause event, leaps by a conditioned reflex beyond our will or control to the idea of the effect taking place. It is this Pavlovian response on our part that makes us leap from idea of cause to idea of effect. And it is this inevitable expectation on our part that by a mistake of projection of the subjective into the objective, we take as necessary connection in the world.

Needless to say, this Humean analysis of causation as spatio-temporal contiguity, temporal precedence, constant conjunction, and a psychological response to conditioning by observed constant conjunction leading to a custom or habit of expectation has not been without its critics. Hume's epistemology has been challenged by the assertion that we can indeed determine a cause-effect relationship to exist even from the observation of a single case, casting doubt on the origin of our determinations of causality in observed repeated regular constant conjunction. To this, the usual Humean response has been that there is always in such cases a background of constant conjunction into which the events in question fit. (You may never have seen billiard balls collide before, but you have seen many collisions of other kinds of material objects. This is the source of your immediate understanding that the impact of one billiard ball causes the other to move.)

Metaphysically, many have wished to find something in the world that can serve the pre-Humean's purpose of serving to *explain* the constant conjunction, rather than just expressing it. This is similar to the desire of the dispositional probabilists – or at least some of them – to argue that the disposition explains the revealed frequency of occurrence rather than just summarizing the fact that it or something like it actually occurs. Some seek the answer to Hume in some metaphysically primitive causal relation binding individual cause and effect event together. Others, noting the general nature of causation between *kinds* of events, argue that it is a relation of the universals, of which the particular events are instantiations, that grounds the causal relation. However, how such an invocation of realism with respect to properties is supposed to answer the Humean critique of the empirical non-presence of the causal relation is unclear.

One very interesting approach traces back to a famous remark of Hume himself, a remark hard to reconcile with his prevalent expression of his views about causation. At one point, Hume resorts to the subjunctive mode in discussing causation, speaking of the cause as that event which, had it not occurred, the effect event *would* not have occurred. Many have focused on the important role of such counter-factual discourse in trying to understand causation. Genuine laws of nature that ground causal relations are said to support counter-factuals, whereas accidental regularities do not. ("If this were copper it would conduct electricity" is true, because it is a law of nature that copper conducts electricity; but "If this were in my office it would weigh less than 300 pounds" is false, said of an elephant, even though it is true that all the objects in my office do in fact weigh less than 300 pounds.) Can we understand the causal relation in terms of the counter-factual relation events can have to one another, the cause being, essentially, as Hume suggested, that which, had it not occurred, the effect would not have occurred either?

Much suggestive work has been done in this direction, but puzzles abound. Some counter-factual dependence is non-causal, requiring us to delimit in the class of counter-factual dependencies those that are causal dependencies. Problems of multiple causation and of preemption – where the actual cause, had it not occurred, would not lead to the non-occurrence of the effect because if the cause hadn't occurred, some other preempted by it would have – require sophisticated handling. Most importantly, questions arise regarding the epistemology, metaphysics, and semantic status of the counter-factual claims themselves. Invocation of possible worlds and a relation of "nearness" among them provides an interesting formal semantics for counter-factual locutions, but we remain, with Hume, puzzled about what in the actual world grounds the "modal" force of causation (the necessary connection of the events, the "wouldness" of the "would not have occurred" in the counter-factual analysis).

We will not be probing too deeply into any of these issues, but will, instead, look at them from a slightly different perspective. Our primary concern will be with the notion of statistical explanation. And within the discussion of that notion, our focus will be on looking for some frameworks useful for later asking what kind of explanations of the phenomena statistical mechanics can be held to provide us. But, as we shall see, in the approach to those issues some of the puzzles about causation and its relation to "mere constant conjunction" will reappear in a slightly different guise.

2. Explanation as subsumption under generality

As noted briefly in Chapter 2, the early history of the kinetic theory of gases was afflicted by a persistent critique at the hands of such distinguished physicist-philosophers as Mach and Duhem of the whole point of the kinetic theory project. Their objection to the theoretical program was partly motivated by a phenomenalistic objection to the positing of unobservable entities and properties as explanatory devices in science. The molecules of the gas, being revealed only indirectly by the observational effects at the macroscopic level, were objected to on grounds that, ultimately, go back to Berkeley's attack on the representative realist's "matter." If an entity (or feature) is immune to direct observational awareness by us, there is no good reason to posit it. More strongly, the very idea that one can meaningfully talk about it is rejected by the empiricist account of meaningfulness for our terms. For Mach it was finally the immediate contents of our perceptual awareness, the sense-data and "red patches here now" of the phenomenalist that constituted the domain of scientific inquiry. The aim of science was to seek the regularity and order that governed the world of our direct experience. And the positing of the "something I know not what" unobservables to explain the observables was a methodological howler. There is much confusion here. Although many would object to the phenomenalist account of theory on a number of grounds, at least one thing seems clear: Even if you are a phenomenalist, there is nothing methodologically abhorrent in positing the molecular constitution of matter. Of course, this positing must itself be phenomenalistically understood if one remains a phenomenalist, but there is nothing more absurd in this than in understanding macroscopic material objects from a phenomenalist point of view.

But an empiricist-phenomenalist distaste for the unobservable was not the only general consideration motivating the "energeticist" opponents of the molecular theory of matter and of heat. They viewed the kinetic theory program as an unjustified continuation of what they took to be an illegitimate demand that all physical theory conform to the model of mechanism as characterized by Newtonian mechanics. They believed that the new realms of physical phenomena being dealt with by physics – in particular, the subject matters of electromagnetic theory and of thermodynamics – were not appropriately dealt with by a search for some underlying "mechanism" that could be described in the Newtonian mechanical terms and the operation of which would explain the lawlike regularities among the physical phenomena described by the newly discovered laws. These laws were the laws of electromagnetism and the thermodynamic laws of heat, energy, and their transformations. To demand such a Newtonian mechanism as the basis of any physical account, they maintained, was to believe that the appropriate concepts for describing one part of the world had to be, of necessity, applicable to all. But this, they argued, was a mistake. The aim of physical science was to find the general, deep regularities governing a particular class of physical phenomena. Once these laws were found, physical science had done all that could be expected of it. To demand in every case that the laws found be underpinned by some mechanical model was to promote Newtonian mechanism out of its proper sphere as the correct description of some of the physical phenomena of the world, and to demand, irrationally, that all physics be subordinated to it.

As things turned out, of course, the warning against seeking mechanical models of phenomena proved to be healthy in the case of electromagnetism, where Maxwell's models of electromagnetic wave motion as a form of motion of components of a mechanical aether remain mere curiosities of the history of physics. But in the case of the kinetic theory, and more generally of the theory of heat as a "quantity of motion" of not directly observable micro-constituents, the anti-mechanistic program offered the wrong advice. Up to a point, the theory of heat and its transformations *is* the theory of the mechanical behavior of the micro-constituents that form a "mechanism" underlying the macroscopic thermal phenomena.

Nevertheless, the discovery of a wide diversity of physical phenomena, each realm subject to its appropriate lawlike regularities, when combined with the highly influential Humean approach to causation as constant conjunction, did lead to a general account of what it is to explain phenomena in scientific terms. That is, it led to an account of what it is to answer a "why" question in the only way science was qualified to answer such questions. From this point of view, to explain a particular event is to relate that event to other events by reference to some general regularity under which their co-occurrence falls. To explain why *this* occurred is to refer to *that*, and to the general fact that this and that always (or, sometimes only in a statistically regular way) co-occur. And what is it to explain why the regularities are true? Not, according to this picture, to proffer some metaphysical cause that "explains" the regularity, but rather merely to subsume the regularity in question under some more general regularity of nature.

The point of explanation from this perspective was made clear by indicating the relation between explaining and other things we want to do. We wish to be able to *predict* what will happen in the world. And we wish to be able to *control* what will happen. From this perspective, the search for explanations is the search for just those resources we would need in order to be able to predict and control the phenomena of the world. If we know that the occurrence of e_1 is lawlike regularly followed by the occurrence of e_2, then e_1's occurring will lead us to predict the future occurrence of e_2. And we can sometimes bring it about that e_2 will occur by bringing about the occurrence of e_1. The broader and more general – the deeper – the regularities we know obtain, the richer the resources we will have available to us for prediction and control. To explain is not to predict. But the point of looking for scientific explanations is that once we find them, we will have automatically obtained the resources we need for prediction and control.

This picture of explanation, for the case where the regularities under which the event was to be subsumed were without exception, was given a formal codification in the deductive nomological account of explanation of C. Hempel and P. Oppenheim. Here, an event was taken to be explained when a description of it could be deduced from a list of descriptions of other events and a list of statements of lawlike regularities. The law statements were required to be both empirically confirmable and, for a successful explanation, true. Additional constraints were needed to avoid certain trivializing cases of the model. For a regularity, a scientific law, to be explained was for the statement of the law to be deducible from a set of other, distinct, law statements.

One set of difficulties confronting the model were cases (like those of historical explanations) where we feel that an explanation has been given although no laws are invoked and where no deduction is possible. Other objections focused on the explanation of rational action where the regularities invoked were alleged to fail to have the contingent nature demanded by the model. Still other objections pointed to the "pragmatic" aspects of explanation. Which description of an event was appropriate in an explanatory context, which other events (and under which descriptions) were appropriate to invoke in the explanatory context, which laws could serve as part of the explanation – all were alleged to depend upon the situation of the agent who was asking the "why" question. What that agent knew and assumed, what his epistemic lacks consisted in, all determined what would be an appropriate answer to his "why?"

More important for our purposes was the objection that a putative explanation could meet all the constraints of the deductive nomological model and yet not be explanatory at all from an intuitive point of view. Although it seems appropriate to explain the motion of a billiard ball by the collision with it of another, the two motions linked by the laws of dynamics, it seems inappropriate to think of the lawlike determination of the past motion by the later motion as explanatory of the earlier motion. Although, to use a favorite example, the height of a building, the position of the sun, and geometric optics can explain the length of a shadow, it seems absurd to say that one could explain the height of the building by reference to the length of the shadow, the position of the sun, and geometrical optics, even though the derivation works that way as well. Here, the idea seems to be that to explain is to give the *cause* or the *mechanism* by which the event is brought into being. Mere lawlike correlation of event to be explained to any other event constantly conjoined with it is insufficient for explanation. The explaining event must be part of the "cause of the event to be explained" or part of the "mechanism by which the event to be explained is brought about." We will return to these issues in sub-section 3 following.

Those who take explanation to be subsumption under generality do not want to restrict the basic features of this account to those explanations that invoke strict unexceptionless regularities. Rather, it is claimed, a similar picture will do as well for our understanding of explanations of events that invoke statistical regularities. Here the idea is, once again, that a particular event is explained when its connection to other events with which it is correlated by means of a "lawlike" statistical regularity has been displayed. Of course there must be differences in the accounts.

If we can connect an event with other events by an unexceptionless law, then we can say that the event in question, at least relative to the explanatory events having occurred, *had* to happen, or could not have not happened. But if we connect an event, which in fact does occur, with other events by means of a merely statistical regularity connecting the events, then the explanatory argument carries no such strict modal force. Although the event to be explained did happen, and although its happening is connected by a statistical regularity of science to other events that happened, intuitively it "could have not occurred" even given the occurrence of the explanatory events.

Sometimes it is claimed that the structure of a statistical explanation is to associate an appropriate probability with the occurrence of the explained events by means of its connection to the other events by the statistical regularity. Thus, if the statistical regularity tells us that given the explanatory events having occurred, the explained event has a probability of occurrence p, then we can take the explanatory "argument" as telling us that on the premises that the statistical regularity in question holds, we can "infer with probability p" the occurrence of the explained event. Here, probability in the statistical sense, as expressed in the statistical regularity, is associated by means of the statistical explanation to a probability of a proposition, some kind of measure of justified "degree of belief" in the truth of the proposition describing the explained event. Here again, we have a kind of rationale for the search for statistical explanations similar to that offered for the search for lawlike explanations. The rationale works by tying the search for the resources needed to provide explanations with the usefulness of these same resources in situations where prediction and control are our aims. Knowing that an event has occurred, it is a little misleading to speak of its probability as inferred by some statistical explanation as offering us the degree to which we should believe in the occurrence of the event in question. After all, we know that it has occurred, and our degree of belief in its occurrence is, if anything, one. Rather, it is that the resources brought to bear in a statistical explanation could, in similar circumstances where we don't know what has occurred or where we wish to bring about or prevent some occurrence, provide us with the resources needed to make probabilistic estimates of the likelihood of an event of the kind in question. These probabilities could then be introduced into our scheme for rational belief or action (say, by combining them with desires in a "maximize expected utility" scheme for rational behavior and belief). This would give us the resources requisite for prediction and control in what are traditionally called situations of risk.

But what kind of probability for the explained event must the subsumption of it relative to other explaining events by means of a statistical generalization generate in order that we think of the event in question as *explained* by the other events and the statistical generalization? One natural demand to make is that the event to be explained is explained only if it is shown to follow with "sufficiently high probability" from the explaining events. Wouldn't this be the natural extension of the showing of the inevitability of the event that is a feature of deductive-nomological explanations into the statistical realm? Naturally there will be some indeterminateness in how high probability must be in order that we take the event to be explained, but such a looseness in the concept of statistical explanation would be hardly surprising.

But there are many cases where we seem to think an event has been statistically explained but where its occurrence has been derived from the occurrence of other events and statistical generalizations with quite *low* probability. We have explained the bursting into fire of material by spontaneous combustion when we point out that such things at least occasionally happen in the circumstances obtaining, even if they happen only very rarely. Here, the notion of the explaining event *raising* the probability of the event explained is sometimes invoked. The reason we feel that an event is explanatory in certain circumstances is not because it made the explained event probable, but because it made that event more probable than it would have been had not the explaining circumstance obtained. Here, reference is frequently made to two senses in inductive contexts in which evidence can confirm an hypothesis. One sense is that in which the evidence makes the hypothesis probable, but the other sense is that in which the evidence raises the probability of the hypothesis over that which it had on the background information.

Consideration of other examples has led some to drop even the demand that the explaining facts raise the probability of that which is explained. Consider the case of the patient who dies because of the unfortunate (but fortunately, rare) effect of a drug. Doesn't the taking of the drug and the statistical probability of its leading to death explain the patient's demise, even if, had the patient not taken the drug, the probability of death would have then been higher? Taking the drug lowers the probability of death, but it was the taking of the drug with its statistically correlated concomitant of fatality that does, in this case, explain the death of the patient.

Sometimes it is argued that for a circumstance to be statistically explanatory it must at least be statistically relevant, modifying the probability of the outcome to be explained from what it would have been had the explanatory circumstance not obtained. But even this seems, to others, to demand too much. Suppose the probability of outcome *e* was one-half in the circumstances of some causal situation not having obtained. But suppose that circumstance did obtain and was the statistically causal generator of what occurred. Isn't that circumstance statistically explanatory of the occurrence of the event in question, even if, in that circumstance, the probability of the event occurring was still one-half?

A most liberal version of the account of statistical explanation that holds to it as being subsumption of the explained event under a statistical generalization relative to other events, so generating a probability of its occurrence, would allow for any value of probability, and any change of probability from background, positive, negative, or zero, to be permitted, and the systematization still counted as explanatory. After all, the information generated to produce the systematization will have its appropriate uses in circumstances of prediction and control. And isn't that what we demand of the resources brought to bear in explanation from this point of view? None of this will, of course, be to deny that high probabilities or raised probabilities are of interest in the explanatory or prediction-control situations.

One problem that has received attention in the context of the analysis of statistical explanations being understood as subsumptions under statistical regularities is the problem of the "specificity of the reference class." Once again, reference to the related situation of probabilistic or inductive inference is useful here. If we are to decide how much credence to give to a hypothesis, we might consult its "probability relative to the evidence," assuming we believe in such things and have some alleged method of computing them. But surely our actions should be guided not by some partial sub-set of the evidence available to us, but by our total (relevant) evidence. The natural constraint to impose is to use the probability generated by the placing of the situation in the "narrowest" reference class generating reliable probabilities. If our evidence consists of disparate bodies of background that fail to imply one another in any way, and that generate unlike probabilities for the hypothesis, we may simply have no ground for attributing any definitive probability to that hypothesis on that evidence.

Related considerations suggest to Hempel that we ought to take an event as statistically explained not when it merely has sufficiently high probability, that being his model, relative to any class in which the situation can be placed, but only when the probability is generated by a placing of the situation in the narrowest reference class for which the explainer has reliable statistics. Thus, knowing that x is an A and that 99% of A's are C's doesn't explain x's being a C to us if we also know that x is a B and that only 1% of all things both A and B are C's. From this perspective, statistical explanation has an "epistemic relativity" built into it that is absent in the case of full deductive-nomological explanation. If being an A guarantees being a C, then it doesn't matter what else is true, or what we know about what else is true. Knowing that x is an A assures us that it is a C. But in statistical explanation, from this point of view, new knowledge can result in what was an explanatory fact ceasing to be explanatory.

Others would avoid this relativity by insisting that the only appropriate reference class relative to which probabilities can serve as statistical explanations is some appropriately objectively most narrow possible class. Here, the idea would be that only knowing that something was in a reference class A that generated C's with a purely tychistic probability p – that is, which was such that C resulted from being an A with that probability and with there being no additional "hidden variables" that could further specify the event more narrowly and thereby lead to a different probability – would allow x's being an A to be explanatory in a legitimate statistical explanation. Other classifications of the event with their associated probabilities might have a kind of weak legitimacy insofar as they constituted incomplete sketches of explanations, sketches that when filled out would bring to the surface the real reference class to which the event belonged and that was the proper one for statistical explanation.

There are, I believe, a number of distinct issues entangled in the debate over what constitutes a legitimate "reference class" in which to place an event and relative to which to calculate its probability for statistical explanatory purposes.

One dispute centers around the problems introduced by the fact that an event can have one probability relative to a given reference class and quite a different probability relative to others. Here are the two extreme positions. The "conservative" position takes only some maximally specific reference class as legitimately referred to in explanatory contexts. This class might be construed either as a matter relative to the epistemic state of the explainer or in the "objective" sense as being maximally specific "all possible information being in." The "liberal" position allows us to put an event in any reference class in which it is known to fall and legitimately generate an "explanatory probability." The former position would bolster its case by pointing out how in the inductive inference case we would consider ourselves irrational to rely upon probabilities generated by reference classes known to us not to be the maximally specific ones for which we had good statistics. It would ask us to note, also in the inductive case, that we would withhold inference were we to know the event in question to fall into two reference classes, neither of which contained the other and that generated different probabilities for the outcome in question. The latter position would emphasize the fact that bringing to light the probability of an event relative to any reference class legitimate on other grounds would provide us with resources that could be applied in appropriate contexts of prediction and control. The "liberals" would argue that it is the elucidation of such structural facts about the world that is the goal of the search for scientific explanations.

A second set of issues centers, rather, around the appropriateness of reference classes for statistical explanatory purposes where it is the nature of the classifying kind, rather than its epistemic or objective maximal specificity, that is in question. Suppose we know that an event has (hasn't) taken place. Then surely we can find ways of characterizing reference classes into which the event falls that make its "probability" relative to those classifications as close to one (zero) as we like. The trick is simply to use very specially and narrowly constructed classes that will contain only the specific event in question and as few other events as we like. But we feel that such artificially constructed reference classes would be illegitimate for explanatory purposes. Only "natural" classifications of events into kinds are genuine generators of "real probabilities" and only placing events in such natural classifications provides the resources for genuine statistical explanations.

But what are the legitimate "natural kinds" and what gives them their special legitimacy? Here, the expected "Humean" reaction is to seek the distinction between legitimate "natural" classifications and illegitimate "non-natural" kinds in the place played in our general scheme for de-scribing the regularities, lawlike and statistical, that we discover in the world. The overall idea is that a classification is a legitimate one for deriving the probabilities usable in statistical explanation only if it is a kind that fits at some level into the broad scheme of generalizations that, ordered hierarchically, provide the inductively established scientific set of regularities describing the world. From this perspective, the artificial and too narrow classifications that pick out individual events by more or less definite descriptions of them (or of them and a few individual fellow events) or that peculiarly link together categories – "is a cow or a type II supernova of brilliance greater than that of all but 10% of such super-novas," for example – cannot function legitimately in statistical explana-tions, because the kinds in question do not function in the overall scientific scheme of generalizations that fit into the hierarchy of scientific laws and statistical regularities. Only scientific classifications and their associated probabilities have the importance that is sufficient for us to grant that their elucidation provides an "explanation" of the event in question. This issue concerning legitimacy of classificatory kinds is one we will touch on again in the next section, where issues of causation and mechanism are discussed.

Finally, there is another debate between the "tychists" and the "proportionists" The former are those who would take a statistical sys-tematization as explanatory only when the classification relative to which the probability is generated is the objectively maximally specific such classification, so that it is not partitionable into sub-classifications, at least in any "natural" way, that would generate probabilities differing from the original classification. The latter are those who would allow that any otherwise legitimate reference class can be brought into play in a statistical explanation, ultimateness not being required. From the "proportionist" point of view, a probabilistic schematization of an event can be statistically explanatory of it even if the reference class in question can objectively be partitioned into smaller classes in a natural way and which are such that some of them generate probabilities for the event in question differing from those generated by the reference class in the proffered explanation.

In Section III of this chapter we will explore in a preliminary way how some of these issues come into focus in the particular context of the search for statistical explanations in statistical mechanics.

3. Subsumption, causation and mechanism, and explanation

In order to situate the conceptual problems of explanation in statistical mechanics, it will be useful to probe here just a little further into some of the general philosophical disputes on the nature of explanation. In particular, we should look at the claim that notions of causation and mechanism are essential ingredients in any analysis of what it is to offer a scientific explanation of phenomena. And we should look at the related claim that these notions cannot be cashed-out in terms of subsumption under generality in any straightforward way. These claims play an important part in current debates over the nature of scientific explanation. We will be only able to explore the issues here in a superficial way, but some summary survey of the main issues of contention will be helpful.

The debate starts with the observation, noted earlier, that in many cases regular connection is insufficient for explanation. Even in the strictly lawlike deterministic cases, merely showing that the event to be explained is constantly, even as a matter of law, conjoined to some other event is insufficient to credit the other event with explanatory force. The most obvious cases are those where the putatively explanatory event is really an effect, rather than cause, of the event to be explained (the length of the shadow not explaining the height of the building, but vice versa). Other important cases are those where two events are lawlike correlated because they are both causally generated by some third event, in which case neither of the first two events explains the other, despite the lawlike correlation between them.

Similar, if a bit more complicated, considerations come into play where it is the probabilistic explanation of events that is in question. If we demand that an explanatory event make the event explained highly probable, we are presented with cases where event 1 makes event 2 highly probable, but doesn't explain the latter event, because event 1 doesn't bear the appropriate relation to event 2 of cause to effect. Similar cases where one event raises the probability of another but fails to causally explain the other are also easy to construct. Indeed, for any probabilistic relation among events one may construct, it is possible with enough ingenuity to find cases where the probabilistic relation is satisfied but where, intuitively, explanation is not obtained due to a lack of the appropriate causal structure among the events. One can go further in the probabilistic case. Not only do various probabilistic relations seem insufficient to establish the appropriate explanatory connection, none of them seem necessary either. Let event 1 cause event 2 (intuitively). It is easy to find cases where event 2 has low probability on event 1, and cases where event 2 has its probability lowered by event 1 occurring. Spontaneous combustion is an example of the former kind. Examples of the latter kind include such cases as where a remedy, while lowering the probability of an illness, causes the illness as a rare "paradoxical" effect of its being applied, a not uncommon case in medicine.

Naturally there have been numerous attempts to find a sufficiently subtle nexus of regularities involving high or raised probabilities that would allow us to reinstate the idea of causation as being fully analyzable in terms of relations of high or raised probability among events. But intuitive counter-examples have been quickly constructed to each such proposal. Or, in other cases, the proposals have been found to involve, explicitly or implicitly, the use of otherwise unexplained distinctions between causal and non-causal relations among events, vitiating the proposal as a fully reductive analysis of causation in terms of regularity.

Isn't it the case, it is argued, that what makes one event explain another the fact that the first event caused the other, or constituted part of the mechanism by which the other was made to occur? Mustn't we take some notion of causation or mechanism, then, as the core ingredient in explanation, and not the notion of the subsumption of events under patterns of lawlike or statistical regularity?

But what is the causal relation? To even begin to explore this problem area would be to enter into some of the most puzzling and intractable areas of philosophy. Metaphysical issues abound. What are the relata of the causal relationship: events, states of affairs, facts, or universals? And what are these? Are events, for example, to be taken as primitive, irreducible components of our ontology? Or should they be viewed, rather, as composites – for example, as ordered structures of things, properties, and space-time locations? Is the causal relation an ordinary "extensional" relation – like "being next to," for example – so that if one object of the related pair bears the relation to the other object, this is true no matter how either object is redescribed? Or, instead, is there an "intensional"

aspect to causation, so that in the manner alleged to hold in belief contexts and by some in modal contexts, the truth of the statement attributing the causal relation to the relata can change upon mere substitution
of one way of denoting one of the relata for another? Alleged intensionality
in belief is illustrated by the fact that I can believe a bank robber to be
wicked but not believe my brother-in-law to be wicked, even if the
brother-in-law is, in fact, identical to the bank robber. In the causal case,
it is sometimes said, driving home drunkenly can cause an accident not
caused by driving home, even though the drunken driving home is identical to the driving home.

Additional problems abound concerning the connection of causation
with a variety of modal notions. Can any sense be made of a "necessary
connection" between cause and effect event, or at least any "realistic"
notion over and above such projectivist accounts as Hume's? Can a notion
of causation grounded in a counter-factual basis – the cause being that
without which the effect would not have occurred – be made plausible
against the particular difficulties such an account faces? In the case of
probabilistic causation, where we feel that it is wrong to say that the
cause in any way necessitated the effect, and where even if the cause
had not occurred we sometimes will grant that the effect might have, all
the connections among events being merely probabilistic, can the modal
notions still be made to fit properly with the causal? And even if our
modal and causal intuitions can be systematically reconciled, can we
understand the modal notions in such a way as to counter the claims that
such facts, dependent on what happens in "other possible worlds," seem,
at first glance, epistemically objectionable. Do they not cut our notion of
causation off from an adequate evidential basis in actual experience?
Again, can we respond to the critic who will accept such modal notions
only if they can be metaphysically grounded in the actual – that is, made,
in some way or other – facts "determined by" or "supervenient upon" the
actual facts of the actual world?

One natural program is to simply take the causal relationship to be a
primitive, irreducible to any other notion and a fundamental component
of the full description of the world. To some, the causal relation will be
one that holds between individual events, with the support of general
regularity by causal connection, a fact to be explained as a kind of truth
about the fact that event pairs showing certain other similarities turn out
to show the similarity of being causally connected as well. To others, the
generality is built in to the causal notion, because causation is taken to
be an irreducible relation among universals, properties, or kinds. Pairs
of individual events are then causally related because they instance the
appropriate general properties that bear the causal connection, in its
most basic guise, between them.

All such approaches suffer, however, from Hume's famous objection: What constitutes the evidential access we have to such a primitive, irreducible kind of causality? If the causal relation holds between individual events, why is it that we cannot determine this relation by some sort of "direct inspection" of the events, the way we could determine without further ado that one object was indeed next to another? Why is it that reference to repeated trials of events of similar and differing kinds is what we require to determine causality? If causality is a primitive irreducible relation among universals, why is it that we can't discover this relation by a mere process of inspection either, in this case inspection of the universals and the relation between them? Instead, we require a process of varying the features of pairs of events in order to determine just which of the infinite pairs of properties the events instance are the universals connected by causality. This doesn't constitute a final and irremediable refutation of any positing of causality as some additional, primitive, and irreducible aspect of the world. But the absence from phenomenal experience of "causation itself," as opposed to constant conjunction, spatio-temporal contiguity, temporal precedence, and so on – the absence that Hume so clearly noted – does present a group of puzzles the causal primitivist must resolve. He must tell us how we come to understand the concept of causation, and must provide for us an adequate account of just how it is that we come evidentially to determine the presence of causality in the world, as opposed to the presence of the regularity, and so on. For the primitivist, regularity is not constitutive of causality but only evidential of its presence and, in part, explainable by its presence.

A quite different approach to pinning down causation in the world looks toward the facts about causal relations as they appear in our most fundamental physical theories and in our general presuppositions about causation in the world. The idea seems to be to describe the structure of causation in the world, as we actually take it to be, without the invocation of any primitive causal vocabulary. W. Salmon, for example, following up ideas in B. Russell and H. Reichenbach, proposes such an approach to causation. Causation, according to our best available physical theories, is a spatio-temporally continuous process, causal influence being propagated from event to event along continuous space-time paths. Indeed, within the relativistic picture the possible paths of causal influence propagation are limited to a proper sub-set of all one-dimensional continuous paths in space-time (the timelike and null paths). This idea of spatio-temporal continuity of causal influence even holds in our most advanced quantum picture of the world, quantum field theory, appearing there in the guise of a posit of no causal influence outside of relativistic constraints (for example, vanishing of commutators of field quantities at

spacelike separations). There are curious "non-causal correlations" derivable from quantum mechanics, as illustrated in the cases used to discuss
the Einstein-Rosen-Podolsky thought experiments and the related Bell
Theorems to the effect that the correlations described in those thought
experiments can have no explanation in terms of an antecedent common
cause spatio-temporally continuously connected to the correlated and
separated events. But they are thought of as "non-causal" in nature – and
remain one of the greatest conceptual mysteries of the interpretation of
quantum theory. So one thing we ought to say about causation is that it
has this spatio-temporal and relativistically constrained aspect to it.

Going further, it is argued that one need only think of causal influence
propagating along the appropriate allowed continuous space-time paths
as being a continuity of feature of the events along the path. Thus the
continuity of the height of a water wave along a path in space and time
is the causal propagation along the path, and no primitive notion of the
wave height at one point "causing" the height at the next space-time
location need be invoked.

It isn't that easy, though, for at this point problems concerning "pseudo-
causal processes" arise. These are the result of causal structures that
generate a process along a continuous space-time path that looks causal
but is not. The propagation of the effect along the path is due, intuitively,
to a remote common cause that generates each event along the path
of pseudo-causal propagation in turn. For example, consider the spot of
light generated on a distant wall by a searchlight as it moves along a
continuous path. Each light image in turn is the result of the continuous
propagation of causal influence from searchlight to wall, not an influence
from spot to spot propagated along the path followed by the moving
patch of light. Salmon tries to exclude these continuous changes mas-
querading as causal propagation by using Reichenbach's notion of
"marking." Put a red filter in front of the searchlight and all along the
continuous path from it the light becomes red. But marking one of the
light spots as red, say by inserting the filter into the beam generating that
spot just before it is generated, doesn't color the other spots along the
path on the wall red. It is, however, far from clear that this proposal for
delimiting genuine from spurious causal processes doesn't itself invoke
a primitive notion of causation. Why say, for example, that holding the
filter in front of the searchlight, and thereby making all the spots red,
isn't a case of "marking" the path of light spots at the first spot and seeing
that mark propagate down the path of spots, unless one already knew
that the causation was in the holding of the filter and the propagation
of redness along the searchlight beam? In any case, Salmon attempts to
characterize causation as the combination of features propagating along
relativistically allowed space-time paths, and the interactions of such

paths that generate changes in the otherwise continuously propagated features.

Along with considerations of spatio-temporal continuity of the propagation of a feature as an essential component of causality there is, of course, also the fact that we take it in general that causes always are precedent to their effects in time. As we shall see, this assumption, either explicitly or implicitly, appears again and again in attempts to explain the asymmetry in time of systems that is a central problem of statistical mechanics. And we shall later in Chapter 10,III,1 and 10,III,2 explore the question of whether, in some way, this intuitive asymmetry of causation can be grounded in, or reduced to, the asymmetry of the process of the world that thermodynamics and statistical mechanics attempt to explain. Suffice it to say here that the place of this time asymmetry in the physical theory that seems to be most intimately associated with the intuitive aspect of causation in question is much less clear than the place of the spatio-temporal continuity of causation as it fits into our general space-time physics. But even that latter placing of an intuitive notion into contemporary physical theory is itself not as open and shut a matter as it is sometimes made to appear.

A general methodological point about approaches of the kind we have just been discussing is important. One wants to be very wary of tying such a general and programmatic notion as that of causation too closely to specific features of how causation acts in the world as current science describes it. The basic intuition with which we started was that to explain was to reveal causes, and that, intuitively, the mere subsumption of events under regularities, lawlike or statistical, was not, in general, sufficient to provide an explanation. Furthermore, it seemed in the statistical case that no provision of a specific regularity condition was in general necessary to provide an explanation. But tying explanation to causation, and causation to some specific intuitive or scientific accounts of how causal influence actually behaves in the world, may tend to put us in the disturbing position of tending to assert that when the orthodox view of the world we have been holding is subject to radical shifts, we must forgo looking for explanations of phenomena at all.

To consider an example, think of the intimate "constitutive" association of spatio-temporal continuity with causation described in the viewpoint just outlined. If to explain is to reveal causes, and if causes are, in their very nature of definition spatio-temporally continuous in their action, then in a world in which action-at-a-distance governs the interactions of matter, no genuine explanation at all will be possible.

Now, to be sure, action-at-a-distance has sometimes seemed so disturbing in its nature as to lead people to just the conclusion that should it hold, the world is inexplicable. Or, at least, to the conclusion that regularities invoking action-at-a-distance can't serve as explanations. Newton, formulating gravity mathematically as just such an action-at-a-distance force, is well known for remarks that can be construed as denying that this postulate could constitute a physical explanation of the gravitational interaction, reliable as it was in its predictive aspects. But most would now think action-at-a distance theories possible, and, if true, possible constituents of a genuinely explanatory theory. Further, the difficulties of current spatio-temporally continuous theories of causal interaction among particles – that is, the difficulties of field theories such as divergent self-interactions of particle and emitted field, have led physicists to try to construct relativistically respectable action-at-a-distance theories as at least viable explanatory alternatives worth exploring.

Although the energeticists who criticized atomism in the nineteenth century were wrong as far as thermal phenomena went, their general claim – that one ought not to insist that all explanations be framed in the form of Newtonian mechanistic explanations – was surely correct. And the generalization of this – that it is a mistake to constrain explanation to any specific facts about the behavior of the world, beyond such general facts as the existence of distinct systems and their states and the possibility of some form of regularity among them – is surely correct. It may be reasonable to demand that explanation requires causation and mechanism. But if one goes on to fill this out with a demand that causation and mechanism have specific forms – say, of spatio-temporal continuity in their action – one may very well find oneself in the embarrassing position of insisting that what later science takes to be perfectly good explanations of the phenomena don't count as explanations at all on this methodological stance.

An alternative immediately suggests itself – to argue, in a sense, that causal relations and mechanisms are just regularities, but that not all regularities are equal as far as explanation goes. The distinctive character of the regularities we count as revealing causes and mechanisms, it might be argued, is their broad generality and their fundamental place in the hierarchy of our best available science. It would be claimed, from this point of view, that insofar as notions of causal necessity, connection with counter-factual inference, and so on, are to be associated with causal relations, these are notions that in turn are to have their conceptual grounding in an understanding of the positing of a causal relation as being the positing of regularity, lawlike or statistical, framed in terms of characterizations of events by means of the concepts that play the most general and foundational role in our scientific theories. And the regularities themselves are the basic regularities of that science, or are related to those basic regularities as some simple regularity derivative from them.

Here, the connection of causality with spatio-temporal continuously propagated effects would then be grounded in the claim that it is indeed true of the most fundamental regularities of our current best physical theories (relativistically compatible field theories) that they are framed in such spatio-temporally continuous ways. But we would be prepared for a possibility of a scientific revision to a novel theory in which the fundamental regularities were quite different in their nature if, say, action-at-a-distance theories became fundamental once again. In that case, we would still believe in causation and mechanism as newly construed, but it would be a form of causation and mechanism quite different from that we previously had in mind.

The intuition that the invocation of regularities is sufficient for explanation only when the regularities are those revealing causation and mechanism would make at least some sense from this point of view. The argument would be that given our understanding of science as constituted by a body of discovered regularities hierarchically ordered in terms of fundamentality and generality, the regularities that play a fundamental role in our science are distinguished from other isolated or very limited regularities. Invoking the fundamental regularities in an explanatory context brings to view generalizations whose importance for prediction and control is clearly distinguishable from those regularities invoked that do not so fit into the scientific scheme. Hence we honor the subsumptions under generality that invoke the fundamental generalities with the title of being explanations and honor the regularities so invoked by saying that they describe causes and mechanisms.

Needless to say, this approach will hardly satisfy many, especially those who find in the notion of causation and its alleged necessary connection something disparaged by Hume but, according to them, actually an indispensable component of our intuitive picture of the world. One aspect of approach to explanation through causation – the search for the scientific ground of the intuition that causation is something asymmetric in time, always going from past to future – will receive further attention from us, because the science invoked to account for this asymmetry is usually statistical mechanics, and because the alleged asymmetry of causation is itself so frequently invoked (often unconsciously) in attempts at explaining the important temporal asymmetries of thermodynamics and statistical mechanics.

But aside from this, I will forgo trying to offer any more persuasive case for the proposal that the necessarily causal nature of explanation has its origin in the structure I have suggested. I will turn, rather, to a preliminary outline of the problem that will be central to later considerations. The basic problem is that in the context of statistical mechanics, we presuppose an understood level of causal explanation of the behavior of systems. Yet statistical mechanics imposes on top of that level of description and explanation another, invoking probabilistic notions of description and explanation. But just how does this latter level of comprehension "fit" over the presupposed grounding causal level?

II. Statistical explanation in statistical mechanics

Statistical mechanics is a theory constructed on the presupposition that systems can legitimately be described by an underlying causal picture of the world. Here, the constitution of the system as a structure built up of micro-constituents, and the lawlike behavior of that structure that follows from the laws governing the micro-constituents and their interaction, constitutes the ground of the supposition that the system as a whole has its behavior governed by the causal structure generated for it out of its constitution. Essentially the same picture holds whether we are dealing with presupposed classical laws, as in this book, or with the more refined picture in which the state of the system is constituted by its quantum state, and the dynamical laws are those of the evolution of quantum systems. Most of the fundamental conceptual problems of statistical mechanics arise out of the problem as to how to fit this theory, with its probability distributions, features generated out of these probability distributions like average or most probable values, and the dynamics of these features, onto the underlying causal understanding of the systems, their nature, and their dynamics.

The causal behavior of the system is determined by its structure and its governing dynamical laws. In one sense, there are "givens" over which the additional elements of statistical mechanics must be superimposed. But in another sense, these two components are not unquestioned features that we must take for granted, but are themselves subject to critical investigation in the pursuit of the foundations of statistical mechanics.

Take structure first. Although the constitution of a system out of micro-constituents and the fundamental response of those constituents to external and interactional forces is given to statistical mechanics from the outside, other structural features are subject to questioning and open for consideration in a foundational investigation of statistical mechanics itself. This comes about from the fact that the statistical mechanical results are always obtained from systems that are characterized by idealizations of an appropriate kind. Although some of the idealizations are harmless and easy to justify and understand, others have a more fundamental place in the theory and are sometimes open to critical objection.

Many accounts of the approach to equilibrium of a system, for example, treat the system as energetically isolated from the external world. But, as we shall see in Chapter 7,III,2, other accounts insist that the irreducible interaction of system and external world is a fundamental component of any adequate explanation of the approach to equilibrium. Many accounts of behavior of systems in statistical mechanics treat the systems as having an unlimited number of degrees of freedom, allowing the theoretician to work in the so-called thermodynamic limit. Although the vast number of degrees of freedom (vast number of micro-constituents) of realistic systems certainly make such an idealization seem reasonable, the specific role of the idealization in generating crucial results must be examined with care in each case. Other results are obtained by idealizing the inter-actions among systems as having "singular" natures, treating molecules, for example, as "hard spheres" that have no interaction when separated but that are absolutely impenetrable to one another upon collision. Here, again, the question of the extent to which the results obtained in statistical mechanics are dependent on the idealization is a critical issue. Sometimes, trying to put together the account appropriate for the highly idealized system with what can be shown to be true about systems not so idealized becomes an issue. Again, it is sometimes important to treat systems as being of low density and as having their micro-constituents vanishingly small relative to the size of the macroscopic system. This idealization is sometimes used to generate a model of approach to equilibrium radically at variance with the conceptualization usually used in some crucial ways.

In cases like the last, a careful examination of how the idealization is being used and a comparison with the systematic use of the alternative idealizations and their uses is important. As we shall see time and again, it is often an unresolved issue as to which idealization of structure and which accompanying systematic view of the correct application of prob-ability and constructs from it is the appropriate one to take as represent-ing in our theory the observed, actual behavior of systems in the world.

The dynamical laws governing the micro-constituents, and derivatively the macroscopic system constructed out of them, are, again, usually taken as a given from the general dynamical theory. However, the persistently intractable difficulties encountered in trying to fit irreversible thermo-dynamics onto a reversible underlying dynamics has often led to the suggestion that the source of irreversibility must be found in some modification of the familiar dynamical laws, or in some understanding of them that introduces irreversibility at the level of micro-dynamics. This leads to non-orthodox versions of the foundation of the theory that we shall at least have to touch upon.

In any case, it is on top of an underlying causal picture of the system that the statistical or probabilistic account of phenomena, essential to the statistical mechanical attempt to account for the thermodynamic features of the world, is superimposed. This causal picture is framed in terms of the characterization of the structure of the system derived from our theoretical account of the structure of matter, perhaps supplemented by

the choice of idealization noted earlier, and in terms of the characterization of dynamical laws derived from our fundamental chosen dynamical
theory, classical or quantal. Various probabilities are assigned. Relative to
them, quantities such as averages and most probable values are calculated. Associations are made between these statistical quantities and
quantities representing observed features of macroscopic matter. From
these assignments, calculations, and associations, it is hoped that the
familiar thermodynamic regularities – specific like the equations of state
or the hydrodynamical equations of approach to equilibrium, or general
and all-encompassing like the Second Law of Thermodynamics – can be
"derived."

The probabilities are assigned as measures over a diversity of fields
of basic events. Sometimes it is frequency in time of micro-states of an
individual system we have in mind. Sometimes it is proportion of systems in an imagined infinite collection of systems having micro-states of
a specified kind that is what is meant. Sometimes it is the proportion of
initial conditions of a system satisfying some feature that is taken to be
probability. The general warning given by the Ehrenfests must always
be kept in mind. Unless one is careful to the point of obsession in
differentiating these various concepts, it is easy to think that one has
solved a puzzle when the hard work of solving it mostly remains to be
done. One of the recurrent conceptual problems we will face is trying to
figure out how some quantity calculated on the basis of some probability
assignment or other is to be connected with the macroscopic feature the
behavior of which we were trying to account for in the first place. And,
as the Ehrenfests told us, we must always be sure that we haven't hidden
some of the most difficult components of the explanatory problem in an
ambiguous use of probability and items constructed from it.

Statistical mechanics is usually divided into a theory of equilibrium and
a theory of non-equilibrium. The former theory is certainly in a more
completed and less controversial state than the latter. But, as we shall see
in Chapter 7, the notion of "explanation" invoked in the latter, more
difficult and controversial, portion of statistical mechanics looks more
familiar to a philosopher, accustomed as he is to the standard notions
of explanation as either subsumption of pairs of states of systems under
regularities or the giving of a causal account of one state by reference to
another.

In equilibrium theory, the basic aim is to account for the regularities
among macroscopic quantities characterizing a system at equilibrium (the
equations of state), and to understand such basic features of equilibrium
as the equi-partition of energy among all available degrees of freedom.
Differing ways of looking at the results of the theory quantities may be
associated with proportions of time spent by a system over infinite time

in various kinds of micro-states, or, instead, with proportions of an infinite collection of systems whose micro-states obey some constraint. In both cases, the "probability" comes in as a measure of those proportions, and the fundamental problem of the foundations of the theory is rationalizing or justifying the probability distribution chosen. Other foundational issues center around the problem of making clear just why a given quantity calculated given a certain probability is to be associated with a particular macroscopic observable.

As we shall see in Chapter 5, the source of the appropriate probability measure is usually traced to the structure of the system and the dynamical laws of the micro-components, the crucial point being that the notion of a probability distribution over initial conditions plays no crucial role. Actually, this is not quite true, for the results obtained will be conditioned on accepting the posit that sets of initial conditions that are given zero probability in the measure rationalized are not to be expected to characterize actual sets of systems in the world. As we shall see in Chapter 5, the sense in which equilibrium features are "explained" in the equilibrium theory is a curious one. The existence of equilibrium is, rather, presupposed and not accounted for. And no explanation is really offered of the facts about the approach of systems to equilibrium in the world, these matters being left to the non-equilibrium theory. The "explanations" obtained in equilibrium theory don't look at all, in fact, like the generative or causal explanations of how a state "comes into being" that are so familiar from our ordinary explanatory context or from the bulk of our understanding of what we are looking for when we look for explanations of phenomena in physics. What we obtain in equilibrium theory is rather a demonstration that equilibrium, assumed to exist and to be connected with certain theoretical states must, if certain plausible assumptions are made, have certain features. These follow from the structure of the system, the laws governing the micro-dynamics, and the very notion of what an equilibrium state is supposed to be.

When we move to the non-equilibrium case, the aim and structure of explanation takes on its more familiar shape. Here, in the manner we are so accustomed to from other contexts, we deal with systems initially in one condition that move in a specifiable way to some other condition. In particular, systems in non-equilibrium move to equilibrium and stay there. What we want to understand is *why* this dynamic process exists and why it has the features it does. We want a general account of the approach to equilibrium, indeed a general account of why the Second Law of Thermodynamics is true. For specific systems, we want an understanding of why their approach to equilibrium has the time scale it does – that is, an understanding of the magnitude of "relaxation times." And we want an understanding of why the route taken to equilibrium of systems of particular kinds started in non-equilibrium conditions of specific sorts has the detailed nature that it has, a nature characterized by the hydro-dynamical equations of evolution toward equilibrium.

In non-equilibrium theory, the presupposed structure of the system and the presupposed micro-dynamical laws will of course be essential components of the explanatory scheme. Choice of appropriate idealization of structure will become an essential and controversial matter. But now the postulation of a probability distribution over the micro-states compatible with the initial state of the system characterized in macroscopic terms will also be an essential component of the explanatory schemes. In at least some of these schemes, far richer assumptions about this probability than the reasonableness of "neglecting sets of measure zero" will come into play.

As usual, there will be problems involved in trying to determine the correct relation between some probabilistic construct of the statistical mechanical basis with some observable quantity whose behavior is described at the thermodynamic level. In the non-equilibrium context, these "reductive" associations become more problematic and controversial than they are in the simpler equilibrium case. Most particularly, which statistical quantity to associate with the thermal quantity of entropy becomes a question fraught with difficulty, because rather different perspectives on how to understand the approach to equilibrium conceptually are tied intimately to different opinions as to what to construe as the entropy of a system.

Another problem comes into focus when one asks for the right general probabilistic model to use to understand the overall approach to equilibrium. This is the question as to whether to understand the standard equations describing that approach as descriptive of the "overwhelmingly most probable course of evolution of a system" or, instead, to understand them, in the manner made so clear by the Ehrenfests, as describing the "concentration curve" that runs through the "overwhelmingly most probable" states of the collection of systems at each time. As we shall see in Chapter 7,II,1 and 7,III,7, despite the Ehrenfests' critique of the former position, and apparently incompatible as it is with the recurrence theorem, some contemporary models of non-equilibrium posit it as the correct way of understanding the approach to equilibrium at the statistical mechanical level.

An additional problematic area concerns the interpretation of the probability distribution over the initial conditions of a system, and the connection of this divergence of interpretation of probability with a divergence of interpretation of the kind of statistical explanation of the approach to equilibrium we should be offering in statistical mechanics.

One can think of two ways in which a "probability distribution over initial conditions" can be fit into a causal structure of explanation. In one way, one thinks of the distribution as specifying, in some sense, proportions in which specific initial conditions, each of which initiates a deterministic evolution from itself, are realized in nature. In the other way, one thinks of the probability distribution as specifying, rather, a "dispositional probability" of individual "tychistic" initial states. From the latter point of view, each individual system has as its initial state a state containing in itself a manifold of causal possibilities for the states that it will causally determine to succeed it. The probability is then a measure of this "causal propensity" of this genuinely tychistic initial state to generate its followers.

Most versions of non-equilibrium theory opt for the former interpretation of these initial probabilities. The model of the statistical explanation of evolution they offer is that each individual system has a deterministic causal evolution from its initial state, and statistical mechanics provides us with a way of characterizing how these individual evolutions in the world will be found to behave when examined as a collection or an ensemble. Nonetheless, despite the similarity of explanatory structure up to this point, accounts of this kind may differ in fundamental ways from one another when the question of the nature of the statistical explanation they offer of the phenomena of the world is pushed further. For it is at this point that the question of explaining *why* the probability distribution over initial conditions has the form it does arises.

As we shall see in Chapters 7 and 8, a number of quite different answers are offered to this further "why" question. Some, adopting what has sometimes been called the "real ensemble" and sometimes the "de facto" approach, will simply posit this distribution over initial conditions as a contingent, otherwise inexplicable, fact about the world. Others will seek the origin of this probability distribution in a form of causal explanation, attempting to trace this distribution over initial conditions of a vast manifold of systems to the particular features of the initial condition of one system of a very special kind, the universe itself. This second approach will give rise to many important methodological problems of its own, not the least of which is the puzzle over what kind of further explanation, if any, we could offer for the special nature of this ultimate initial condition that generates causally all of the others. Others will find the origin of the probability distribution in the notion of the preparation of a system and on alleged impossibilities of preparing initial collections of any kind except those characterized in the manner needed for the statistical mechanical results. Each mode of trying to account for the essential initial probability distribution will lead to a rather different conception of what sort of explanation of the phenomena of the world statistical mechanics can be expected to offer us.

The other fundamental approach – treating the initial probability distribution as descriptive of the inner probabilistic nature of an initial tychistic-propensity state – is far less orthodox than the approaches noted. But, as we shall see in Chapters 7,III,6 and 9,III,1, it has its proponents, and their arguments are not easily dismissed. From this perspective, the statistical explanation of the behavior of systems becomes quite different. It is not an ensemble of systems, each evolving deterministically, that statistical mechanics is describing. Rather, it is the causal evolution of tychistic state to tychistic state of individual systems that we are attempting to model in our theory. From this point of view, statistical mechanics does not simply presuppose the underlying causal structure of the microdynamics. Rather, the theory forces upon us a radical revision of the very notion of the state of a system and a radically different understanding of its causally determined evolution. Here the argument will be that the very success of statistical mechanics, and an understanding of the grounds of that success, shows us that the standard causal picture at the microlevel is a false idealization.

There are some other alternatives to consider as well. Some ideas, such as those that account for evolution toward equilibrium in terms of the non-isolation of the system from the external world and that world's intervention into the system's dynamics, fit easily into causal ideas of explanation. Others that try to understand the probabilities of statistical mechanics in a "subjectivist" or "logical" manner offer quite a different conception of the kind of explanatory work we can expect from a statistical theory of this kind.

Finally, all of these approaches will have to confront that touchstone of any adequate explanatory account of the world that is to do justice to the phenomena described by thermodynamic theories: Where in the explanatory account does the irreversible nature of the phenomena we macroscopically observe become accounted for by the scheme? The very *kind* of explanation we should be looking for here is a matter of continuing controversy.

III. Further readings

The seminal work on causation as subsumption under regularity is that of Hume. See Hume (1988) and (1955).

A thorough study of explanation as subsumption under a regularity in the lawlike case is Hempel and Oppenheim (1948). For further exploration, replies to criticism, and so on, see Hempel (1965).

For some recent discussion of causation, see Lewis (1986), Chapter 21, and Salmon (1984).

Statistical explanation as subsumption under a statistical regularity is explored in Hempel (1965).

Salmon (1984) is a wide-ranging treatment of statistical explanation as seen by philosophers, and very useful for the history of several approaches to the problem. Another extended treatment of the notion of probabilistic causality is Eels (1991).

Proposals that view statistical explanation as a variety of causal relation involving objective dispositional probabilities can be found in Railton (1978), and are worked out at length in Humphreys (1989).

5

◁▭▭▭▭▭▭▭▭▭▭▭▭▭▭▭▭▭▭▭▭▭▭▭▭▭▭▭▭▭▭▭▷

Equilibrium theory

I. Autonomous equilibrium theory and its rationalization

1. From Maxwell's equilibrium distribution to the generalized micro-canonical ensemble

The existence of an equilibrium state, describable by a few macroscopic parameters whose values and interrelations do not change with time, is the most fundamental thermodynamic fact for which statistical mechanics must account. A full understanding of the equilibrium situation would require a demonstration within the context of the dynamical theory of non-equilibrium that the equilibrium state exists as the "attractor" to which the dynamics of non-equilibrium drives systems, and it is that approach to the problem we shall examine in Chapters 6 and 7.

Beginning with Maxwell's first derivation of the equilibrium velocity distribution for molecules of a gas, though, there have been approaches to deriving the equilibrium features of a system that at least minimize considerations of the underlying micro-dynamics and of the time evolution toward equilibrium that one hopes to show is driven by it. As we saw, Maxwell's first derivation of the equilibrium distribution relied solely on the assumption that the components of molecular velocity in three perpendicular directions were probabilistically independent, an assumption clearly recognized by Maxwell as "precarious."

Whereas Boltzmann's first essay on equilibrium treated it from the general dynamical perspective, using the kinetic equation and H-theorem to attempt to show that the equilibrium distribution was the unique stationary distribution, he later developed, in the course of his probabilistic response to the earlier criticisms of the kinetic equation and H-theorem, his new method applicable to the case of the ideal gas. Here, one counts distributions of molecules among "coarse-grained" boxes in μ-space and identifies the equilibrium distribution as the one obtained by the overwhelmingly largest number of permutations of molecules among boxes.

Once again, dynamical considerations are not totally lacking here, because the choice of position and *momentum* as the appropriate dimensions for the μ-space was clearly guided by the dynamical considerations with which Boltzmann had become familiar.

Next in this evolutionary development of autonomous equilibrium theory was Maxwell's introduction of the system phase-space approach. Here, the notion of an ensemble of systems – a probability distribution over possible micro-states for the systems – is introduced. The standard micro-canonical distribution as at least one distribution invariant under dynamical evolution, and the identification of equilibrium values with phase averages of micro-quantities relative to this probability distribution, make their appearance. At the same time, Maxwell introduces what becomes the classic problem of equilibrium theory, the problem of demonstrating that the standard probability distribution is *uniquely* stationary in time.

For our purposes at this point, we can view the major contribution of Gibbs to be his extension of the notion of the "equilibrium ensemble" beyond Maxwell's consideration of an ensemble of systems energetically isolated from the world to systems in thermal contact with a (vast) heat reservoir and to systems that can also interchange numbers of molecules with their external environment. These latter cases are to be described by using the canonical and grand-canonical ensembles respectively. But our attention will be focused on the micro-canonical ensemble. This is the imagined collection of systems, each energetically isolated from the surrounding environment, having among them all possible micro-states compatible with the macroscopic external constraints placed upon the system, and having these micro-states distributed according to the standard, time invariant, micro-canonical distribution function.

The "recipe" for equilibrium theory by means of the micro-canonical ensemble is clear. Represent the possible micro-states of the system in question by points in an appropriate system phase space whose dimensions are the position and momentum coordinates of the micro-constituents. Given the fixed energy of the system, confine your attention to the appropriate "energy hyper-surface" in the phase space. Pick a probability distribution over this now restricted phase space by constructing from the dynamically invariant volume measure of the phase space as a whole an "area" measure over the energy surface. This new measure is itself dynamically invariant in time. This is done according to the simple procedure of Maxwell, Boltzmann, and Gibbs. Distribute probability uniformly with respect to this new measure.

Given this probability distribution, pick the appropriate microscopic phase function for any given macroscopic thermodynamic quantity.

Identify the macroscopic quantity as the average value of the phase function over the allowable phase points, calculating the average by using the constructed time-invariant probability distribution as one's weighting function. The usual thermodynamic equilibrium relations then follow, as do the general results on equi-partition of energy among degrees of freedom, detailed balance of transitions from one momentum range to another of molecular motions, and so on.

We ought to note here, however, one useful generalization beyond the micro-canonical ensemble. In the usual case, the only known constant, macroscopic, dynamic parameter of the system is its energy. A gas confined to a box, for example, cannot be expected to have a constant angular momentum due to the interaction by collision of molecules with the box. But astronomical cases and others may lead to a situation where we know a fixed angular or linear momentum for the system in question. Can the micro-canonical ensemble approach be generalized to allow us to construct a dynamically invariant ensemble where the existence of constants of motion for the system over and above its total energy allows us to fix more closely the allowed region of phase space a point representing the micro-state of the system must occupy?

An affirmative answer is given by the work of R. Lewis, who showed that one could generalize the micro-canonical ensemble in just this way, resulting in panta-micro-canonical ensembles. These are sub-spaces of the original phase space of lower dimension than the energy hyper-surface. They are determined by the imposition of additional known constants of the microscopic motion and they are equipped with a probability distribution over them provably invariant in time under the dynamical evolution of the system.

Although this recipe for the solution of the equilibrium problem is clear, it is the *rationalization* of it that is our primary concern. Clearly, the rationale will not ignore the results of dynamic evolution entirely. The equilibrium state has many important roles. One is its role as the "end state" toward which a system tends evolving from an initial non-equilibrium starting point. But any hope of showing that the statistical mechanical treatment of equilibrium is appropriate for equilibrium so construed must rest in the understanding of the full non-equilibrium theory. Short of this, what considerations can be adduced to convince us that the micro-canonical recipe for the computation of equilibrium features is indeed the right way to calculate equilibrium features of a system subject to appropriate macroscopic constraints and built up of micro-constituents in a particular way? And if we are already convinced that the recipe is reliable on the basis of its success in practice, what considerations can we adduce to allow us to understand why the recipe works as well as it does? It is to these questions that we now turn.

2. The Ergodic Hypothesis and its critique

What are some of the questions we must answer in order to be fully satisfied that we understand why the micro-canonical recipe works?

(1) Surely it is reasonable that the probability distribution we use to calculate equilibrium quantities would be one that is invariant over time. Equilibrium is, after all, a state that is temporally unchanging, and we would expect the statistical surrogate by which we calculate equilibrium values to share this temporal invariance property. But just *exactly* how does the dynamical invariance of the micro-canonical probability distribution function to rationalize, justify, or explain its use?

(2) As Boltzmann clearly realized, we will not be satisfied merely with a demonstration that the micro-canonical distribution is dynamically invariant. Couldn't there be other, inequivalent, probability distributions that were invariant in time as well? Is there any way in which we can demonstrate that the micro-canonical probability distribution is the *unique* distribution invariant over time?

(3) The imaginative picture of an innumerably vast number of systems, all subject to the same macroscopic constraint and taking on the infinite variety of possible microscopic states compatible with those constraints, provides a picturesque model to explicate the formalism of the statistical mechanical procedure of averaging by means of a probability distribution over the system phase space. But what we observe and measure is the behavior of individual systems in their equilibrium state. Just how are we to understand the introduction of the imagined ensemble of systems, and how to connect the quantities we calculate relating to this ensemble to what we observe and measure about an individual system?

(4) A particular puzzle, related to this last question is generated by the fact that it is *average* quantities that we calculate using the statistical mechanical approach. But an average quantity is especially difficult to relate to the behavior of an individual system. After all, a population can have an average value for a quantity even if no individual in the population has the feature in question at that average value. A population of equal numbers of five foot tall and seven foot tall persons has an average height of six feet. What is the connection between average values calculated over an ensemble and the magnitude we observe of some quantity attributable to an individual system?

A rich and fascinating, if often puzzling and frustrating, route to find the answers to these questions has its beginnings in the suggestions of Maxwell and Boltzmann that the evolution of an individual system followed over all time would be such that its path would go through every possible micro-state compatible with the macroscopic constraints on the system. Although Boltzmann's remarks sometimes seem to indicate that he held this would be true even if the system remained in perfect energetic isolation from the outside environment, Maxwell invokes the small but continuous impingement of disturbance from the outside as responsible for this so-called "ergodic" behavior. In this chapter, we shall, for the most part, follow out the Boltzmannian way of viewing the situation. Later, in the discussion of non-equilibrium, we will try to show why one ought to be quite cautious in invoking imperfect isolation of the system as a rationale for one's probabilistic assumptions in statistical mechanics. But we will also see why such a resort to the imperfect isolation of the system has proved to be appealing to many theorists.

Suppose we make the bold conjecture that an isolated system is such that over the course of all time it will, no matter what its initial micro-state, eventually go through every micro-state possible for the system in question given its macroscopic constraints. What would follow from this so called Ergodic Hypothesis?

The trenchant argument of the Ehrenfests shows us that given the Ergodic Hypothesis – if $f(q,p)$ is any function of the phase (micro-state) of the system – the infinite time average of f will equal its phase (or ensemble) average computed by means of the micro-canonical probability distribution. Once more, remember how the argument goes. Because the probability distribution is temporally invariant, the phase average must equal the time average of the phase average. By reversibility of averaging processes, the latter is equal to the phase average of the time average of f for a system started in each initial micro-state. But if the Ergodic Hypothesis is true, the time average of f is independent of the initial micro-state, because there is only one trajectory followed by every system no matter what micro-state it starts in. So the phase average of the time average is just the unique time average common to all initial micro-states. So the phase average of f, computed by means of the micro-canonical probability distribution is just the average of f over infinite time for a system whose motion starts in any micro-state whatever.

Even more follows in a fairly trivial way. Consider any fixed region of the phase space. What fraction of its life history, in the infinite time limit, does a system spend with its micro-state in the given region? If the Ergodic Hypothesis is true, the answer is simple indeed. Consider the function $g(q,p)$, which has value 1 for points in the region and value 0 for all other points. By the earlier result, the infinite time average of g must be equal to its phase average. But its phase average is just the *size* of the region in the standard measure, for, by the uniformity of the micro-canonical distribution relative to this standard way of measuring size of regions in phase space, the phase average of g is just the *size* of the designated region. So, over infinite time, a system will spend a portion of its life history with its phase point in a given region that is proportional to the size of that region in the standard measure.

Yet more follows in a straightforward way. Consider *any* time in-variant probability distribution on the allowed region in phase space. Consider any region of non-zero size. The probability assigned to this region by all such invariant measures must be the same, for each must assign to the $g(q,p)$, which has value 1 for points in the region and value 0 elsewhere, the *same* phase average, that which is equal to the time average of g over infinite time – that is, given the Ergodic Hypothesis, the size of the region. So if the Ergodic Hypothesis is true, there is only one invariant probability measure to consider, at least if we insist that regions of measure zero receive zero probability. So the problem of the uniqueness of the invariant measure has been substantially resolved.

But can we reasonably believe the Ergodic Hypothesis to be true? It suffers from one technical problem in that it presupposes the existence of the infinite time limits of phase functions, something that must in fact be demonstrated. But its real difficulty is much more severe. In its strict Boltzmannian form, it must be false.

In 1913, A. Rosenthal, using newly developed work in the topological theory of dimensionality, showed the impossibility of the one-dimensional trajectory of a system bi-continuously filling the multi-dimensional allowed phase space. A distinct argument by M. Plancherel, appearing in the same year, and utilizing a subtle argument from measure theory, proved the same result. The Ergodic Hypothesis in its Boltzmann version cannot be true.

For some time, attention was directed to the so-called Quasi-Ergodic Hypothesis. Here it was conjectured not that a single trajectory went through every point in the phase space over infinite time, but rather that a trajectory, started at any point would, in the fullness of time, come arbitrarily *close* to every point in the allowed phase space. In the mathematician's language, the conjecture was that each trajectory in phase space was *dense* in the space.

But quasi-ergodicity proved to be a blind alley. First, there was the difficulty of proving of a given system that it was quasi-ergodic, a prob-lem that proved intractable. Second, it later became apparent, in the light of results obtained through the ultimately successful approach to ergodic theory, that quasi-ergodicity even if provable would be insufficient to gain the results one wanted. It could be shown, for example, that there were quasi-ergodic systems that failed to demonstrate the equality of time average with phase average that followed from the Ergodic Hypothesis and that grounded the other desirable results for which the Hypothesis was first conjectured.

The difficulties encountered led to skepticism in general about the ergodic approach. Typical was R. Tolman's text of 1938, which took the standard probability distribution, the uniform distribution over the standard measure – the micro-canonical probability distribution – as a basic postulate of the theory of equilibrium statistical mechanics. The correctness of the postulate was to be justified epistemically be its success. Ergodic theory was dismissed as both something whose aims could not be accomplished, and that, even if successfully obtained, would fail to serve any useful foundational role in the theory.

3. Khinchin's contribution

A program of research carried out by A. Khinchin, while not circumventing the need to fine a replacement for the discredited Ergodic Hypothesis, did initiate a series of results that in the longer run, played an important part in the understanding of the place of the micro-canonical probability distribution from the general ergodic point of view.

Consider a certain probability distribution over an allowed region of phase space. Let $f(q,p)$ be a function on the phase points. Suppose $f(q,p)$ has only a small probability of differing from the phase average of $f(q,p)$ by a specified amount using the given probability distribution. It can then be shown that the probability is at least as low that the time average of $f(q,p)$ will differ from the phase average of the function by that amount.

But how can we be assured that $f(q,p)$ differs from the phase average of $f(q,p)$ only with low probability? Here, Khinchin relies on the special nature of the f's we must consider in statistical mechanics, on the particular constitution of the system out of its microscopic components, and on the large number of microscopic components making up the macroscopic system.

First, he notes that the only functions $f(q,p)$ we need be concerned with are those whose averages we identify with some thermodynamic quantity. They all turn out to be so-called sum functions – functions of the total phase of the system that are sums of functions of the phases of the components where each function in the sum has the same form. Then he assumes that the system is "ideal" in the sense that its total energy is the sum of the energy of its component subsystems. With these restrictions and assumptions in place he is able to show that given the micro-canonical probability distribution, the probability of $f(q,p)$ differing from its phase average by a given amount does indeed rapidly become small as the number of components in the system increases.

So in the thermodynamic limit of a vast number of degrees of freedom, we can be assured, if the system is indeed ideal and if we restrict our attention to suitably restricted phase functions, that given the micro-canonical probability distribution, the probability of the time average of a function $f(q,p)$ differing by much from the phase average of $f(q,p)$ will become vanishingly small.

Our full appreciation of what we can and cannot infer from these results of Khinchin will have to wait until Chapter 5,III,I, when we explore the inferences we can and cannot legitimately infer from other core aspects of ergodic theory. But for now, two observations can be made.

First, there is the restriction of Khinchin's results to the case of the ideal system whose total energy is merely the sum of the energies of its constituent components. This is a restriction we would like to get around, because it is the realistic case of systems with intercomponent interactions contributing to the total energy that is the real concern of statistical mechanics. Here, the more recent results on the thermodynamic limit obtained in rigorous statistical mechanics have gone a long way to weakening Khinchin's restriction in ways that allow us to apply results that generalize his to cases of systems with realistic intercomponent energetic interaction.

But there is a much more disappointing aspect to the Khinchin results. The point of the Ergodic Hypothesis was to allow us to identify the phase average of a function of phase with its infinite time average. The Khinchin result restricts us to only certain functions of phase, but this will do no harm if all of the phase functions we are interested in – all of those whose average value is associated with a macroscopic quantity – obey the necessary restriction. But the Khinchin theorem only tells us that there is a "high probability asymptotically going to one" that a system in a given micro-state will have its infinite time average of the appropriate phase function equal the phase average of that function.

Isn't "probability one" good enough? No. Here, the "probabilities" are themselves being computed by means of the micro-canonical probability distribution whose rationale was in question in the first place. Suppose there is only "zero probability" according to the micro-canonical probability distribution that the time average of a phase function will differ significantly from its phase average. Can we then assume that there is "no chance" that we will encounter such a divergence of averages in nature? Certainly not. Suppose there is some global constant of motion of the system above and beyond its constant energy that we have ignored (such as the constant angular momentum in the astronomical case). The system in question will have its motion confined to a sub-space of phase-space one dimension lower than that of the energy hyper-surface. And it may very well be the case – in fact, usually will be the case – that the infinite time average of nearly all systems confined to this sub-space will differ substantially from the phase average computed over the full energy hyper-surface. In such a case, using the micro-canonical probability to compute phase averages and identifying these with equilibrium thermodynamic quantities will give us the *wrong* results. But the collection of systems that are confined to this smaller sub-space is a collection of "probability zero" in the micro-canonical probability distribution. We cannot conclude, then, that having such "zero probability" is grounds for being considered irrelevant in our physical expectation.

Nonetheless, as we shall see in Chapter 5,III,I, aspects of the results of Khinchin will function prominently in some of the ways in which the results of a deeper ergodic theory are understood. They will help reveal the conceptual relevance of the deeper mathematical results.

II. The Development of contemporary ergodic theory

1. The results of von Neumann and Birkhoff

Results obtained by J. von Neumann and G. Birkhoff initiated the course of research leading to contemporary ergodic theory. The initial suggestion of von Neumann's was to look for sufficient conditions on the dynamics of a system to prove directly the aimed-for end result of the Ergodic Hypothesis – the identity of infinite time average and phase average for any function of the microscopic state of a system. Here, the point was to avoid the demonstrably false Ergodic Hypothesis and the quasi-ergodic substitutes suggested for it and to try to prove directly, from some appropriate condition on the dynamical structure of the system, the sought-for identity of averages. First, results were obtained by von Neumann, and a strengthened theorem was soon obtained by Birkhoff. Because the strengthened theorem is both easier to understand and also more directly usable for the intended purpose of ergodic theory, I shall confine my attention to it.

Birkhoff's first result fills a gap we noticed earlier in the Boltzmannian approach. We noted that the Boltzmann argument, as formulated by the Ehrenfests, assumed that the time average of a phase function existed. Birkhoff proves a result that has this consequence: If we use an invariant probability measure on the allowed phase-space, then except for a set of micro-states that has probability zero, the infinite time average of any (integrable) phase function, $f(q,p)$, along the trajectory started from a given point, will exist. Further, the value of this limiting time average will be independent of the starting point we choose along the trajectory.

But will this time average equal the phase average of $f(q,p)$ over the allowed phase space using the invariant probability distribution to compute this phase average? Suppose the allowed phase-space can be broken into pieces such that (1) there is more than one piece, (2) each piece has non-zero measure – that is, the probability that a point is in the piece is non-zero, and (3) the trajectory started from any point in one of the pieces remains forever contained in that piece. If this is so, the

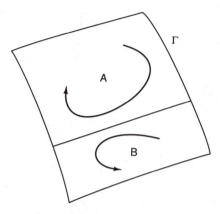

Figure 5-1. Metric indecomposability. Suppose the phase space region for a system Γ can be split into two (or more) regions A, B, . . ., that are of non-zero measure but are invariant. That is, they are such that a system whose initial micro-state is represented by a point in one of the regions always evolves in such a way as to have its representative point remain in the region. Then Γ is said to be metrically decomposable. In such a case, it is clear that one can find a quantity that is a function of the micro-state of the system whose time average in the infinite limit will not equal the phase average of that quantity over all of Γ. Just assign each system in A the invariant quantity 1 and each system in B the invariant quantity 0. If the measure of A is p and that of B is $1 - p$, the phase average of the quantity will be p. But for A region systems, the time average of the quantity will be 1, and for B region systems it will be 0. If no such decomposition into invariant sets of non-zero measure exists, the phase-space for the system is metrically indecomposable.

allowed region of phase-space is called metrically decomposable. If such a decomposition of the phase-space cannot be made, the phase-space region is metrically indecomposable. Then it is easy to show this: If and only if the phase space is metrically indecomposable, then, except possibly for a set of points of zero probability, the phase average of an integrable phase function over the allowed phase-space, using the invariant probability to compute this average, will equal the infinite time average of the function along the trajectory starting from an initial point in the phase space. (See Figure 5-1.)

What is the import of metric indecomposability? We can gain some insight by considering the notion of a global constant of motion of a system. Suppose there is some phase function $g(q,p)$ such that starting from any given point, $g(q,p)$ remains constant along the trajectory from that point, no matter how far into the future we take the trajectory. Then $g(q,p)$ defines a global constant of motion for the system. Suppose we could break the allowed phase space into two invariant pieces of non-zero probability. Then we could define a function $g(q,p)$ that has value zero for all the points in one piece and value 1 for all the points in the other. This would define a new global constant of motion, one we neglected to fix in delimiting the allowed phase-space for determining our uniform probability distribution and computing our average values. On the other hand, suppose there is a $g(q,p)$ that takes continuous values in a range $[a,b]$ and that is a global constant of motion. Then we can define the region of points for which $g(q,p)$ is $\geq c$, for c in the range $[a,b]$ and those for which $g(q,p) < c$. These two sets will then metrically decompose the region. So the condition of metric indecomposability is essentially equivalent to the condition of there not being any neglected global constant of motion that differs in its value on sets of trajectories of non-zero probability.

Notice also the way in which the Birkhoff result (sometimes called the Pointwise Ergodic Theorem and that we will call, henceforth, the Ergodic Theorem) avoids the necessity for anything like the Ergodic or Quasi-Ergodic Hypothesis. It is now metric indecomposability that is the necessary and sufficient condition for the equality "almost everywhere" (that is, except possibly for a set of points of probability zero) of the infinite time average and the phase average. There is a condition equivalent to metric indecomposability, though, that replaces the old Ergodic and Quasi-Ergodic Hypotheses, for metric indecomposability is equivalent to the condition that given any set of positive measure in the phase space, the trajectories from all but perhaps a set of measure zero of phase points intersect that set. (See Figure 5-2.)

If a system is ergodic (metrically indecomposable) on a specified region of phase space, it then satisfies what can be thought of as a generalization of the Law of Large Numbers from probability theory. In standard probability theory, if we consider an infinite sequence of trials of a process with a variety of outcomes, and if each outcome has a constant probability in each trial, and if the trials are independent – that is, if the probability of an outcome in a trial conditional on the outcomes of other trials is the same as its unconditional probability on a single trial – then one can prove that in the limit as the number of trials goes to infinity, the relative frequency of an outcome will, with probability one, converge to its probability on a single trial.

If a system is metrically indecomposable, we can show, by means of an argument that parallels exactly the argument given earlier when we looked at consequences of the Ergodic Hypothesis, that the fraction of time a given system spends in a region will converge to the probability assigned to that region by the invariant probability measure, except, possibly, for trajectories initiated on a set of probability zero of points. So we get a "with probability one frequency in the long run equals

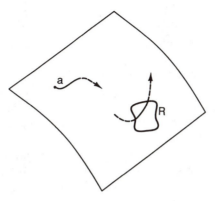

Figure 5-2. The ergodic theorem. Let a system be started in some micro-state a. Let R be any region of micro-states possible for the system given the system's constraints. Let R have a definite, non-zero, size in the phase space. Then, when a system is ergodic, it will be the case that except possibly for a set of initial micro-states of size zero, the trajectory from the initial micro-state a will eventually pass through the region R.

probability" result, even though in this case the "trials" are hardly independent. Indeed, we are assuming a strict determinism that fully fixes the point occupied by the system at a later time given its micro-state at an earlier time.

An additional consequence of metric indecomposability needs to be mentioned here, again paralleling a result that would follow if the Ergodic Hypothesis were true. Suppose the allowed region of phase space is metrically indecomposable. Then, any probability measure that is invariant in time, and that is such that it assigns measure zero to all those sets assigned probability zero by our initial invariant probability measure, will agree with that initial probability measure on all sets. Suppose, for example, that the indecomposable region is the energy hyper-surface and our invariant probability measure is the familiar micro-canonical probability distribution. Then it follows that the micro-canonical distribution is the *only* invariant probability distribution that assigns probability zero to regions of zero size in the standard measure on the energy hyper-surface.

2. Sufficient conditions for ergodicity

The mathematically profound results of the von Neumann–Birkhoff theorems are, alas, not very useful by themselves to a program of rationalizing the probabilistic assumptions of statistical mechanics. Although they provide a clearly sufficient condition for a system to be ergodic – that is, to have the infinite time average of any phase function equal its phase average "almost everywhere" – the condition, metric indecomposability, must itself be demonstrated to hold of a specified dynamical system if we are to show that system to be ergodic. Hence, there begins the program, a program that required thirty years to be brought to its successful conclusion, of demonstrating for some ideal dynamical system that bears a reasonable resemblance to the systems of real interest in statistical mechanics, that the allowed phase space of the system is in fact metrically indecomposable. The aim of the project is to begin with the laws governing the microscopic dynamics of the system and with the structure of the system – in particular, its macroscopic confinement and the governing interaction among the microscopic constituents – and to end with a proof of metric indecomposability for the allowed phase space for the system.

The core idea is that of the instability of the trajectories in phase-space of the system. Here, instead of trying to prove some conjecture about single orbits, such as the Ergodic or Quasi-Ergodic Hypothesis, one studies groups of orbits that neighbor a point in the phase space. If one can show that such orbits diverge from one another with sufficient rapidity, both in the future and past directions of time, then it will be possible to prove metric indecomposability. That is, one will be able to show that not more than one collection of phase points exists that is of more than zero measure, and that is such that the orbits started from a point in it remain for all time (past and future) within that collection. The one collection that is of non-zero probability and closed under time is then, of course, the entire allowed phase space except, perhaps, for a set of points of measure zero.

Because we will need to review the nature of this instability or divergence condition in the trajectories when we examine the rationalization of non-equilibrium theory in Chapter 7, I will not pursue it in detail here, except to remark that the instability condition used to prove ergodicity in the important standard cases proves to be rich enough to prove more powerful results about the "randomness" of the dynamical motion as well. These more powerful measure theoretic results prove to be vital in going beyond the rationalization of the equilibrium theory to gaining ground in the project of understanding the success of the non-equilibrium theory.

The problem of demonstrating ergodicity has, however, now been pushed back one more step. For now, it is necessary to demonstrate, from the dynamical laws and structure of the system, the appropriately strong divergence of nearby trajectories. A lengthy development of important results at the hands of E. Hopf, N. Krylov, A. Kolmogorov, D. Anosov, and others finally resulted in a demonstration of ergodicity for

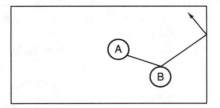

Figure 5-3. Trajectory instabilities of hard spheres in a box. Because of the extreme sensitivity on the exact location of the point of impact of the path A will follow when it scatters off B, even the system composed of two "hard spheres" in a box, one moving and one a fixed scattering center, will show the kind of radical instability of trajectories in phase space needed to prove results such as ergodicity.

an interesting dynamical model by Ya. Sinai. The work begins with the study of the geodesics (curves of minimum curvature and length on a curved surface) on surfaces of constant negative curvature. These have a strong divergence property and are related to dynamics in the sense that motion along a geodesic in a curved space can be thought of as "free" motion in that space. One can prove that such "geodesic flows on surfaces of constant negative curvature" are indeed ergodic.

Sinai was able to show that the motion of spherical particles confined to a parallelpiped box – spheres that move freely except when in collision with one another or with the walls of the box, at which point they suffer a perfectly elastic collision – was also ergodic. Here, the vital point is that particles impinging at nearby points on a given spherical particle are, because of their divergent reflections, soon traveling trajectories quite distant from one another. As a result, trajectories that are quite close at one point are soon wildly divergent from one another. (See Figure 5-3.)

Because the model of hard spheres in a box is the familiar Boltzmann model for an ideal gas whose molecules interact only upon collision and are otherwise moving freely and independently of one another, the Sinai result finally has demonstrated ergodicity, in our mathematically refined sense, for a "realistic" model for a system of interest in statistical mechanics.

3. The KAM Theorem and the limits of ergodicity

Although it has proven possible to show that "hard spheres in a box" constitute an ergodic system, it has also proven possible to show that other dynamical systems have features exactly contrary to those required for ergodicity.

In statistical mechanics, we would like to show a dynamical system radically unstable. It is only the strong divergence of nearby trajectories that allows us to demonstrate ergodicity or any of the other more power-ful "randomness" features on an ensemble we will discuss in Chapter 7. In celestial mechanics – the theory of planetary motion – we would like to prove *stability* of trajectories rather than instability. If we neglect the gravitational interaction of the planets with each other, then they will follow stable orbits around their center of mass with the sun. But the planets do interact gravitationally. What assurance can we find that the mutual interaction of the planets with one another will not, in the full-ness of time, throw one or more of the planets out of their stable elliptical orbits, sending them either crashing into the sun or "off to infinity?"

This was a classic problem of Newtonian dynamics as early as the eighteenth century. A solution of the equations of motion that take the mutual interactions of the planets into account in closed, simple func-tions is impossible. Instead, one resorts to solving the equations by means of series approximations. Early "proofs" of the stability of the solar system suffered, however, from a severe defect. It proved impossible to show that the terms of the series being used to approximate the solution to the dynamical equation would not, at some point in time, diverge – that is, become unboundedly large.

The fundamental difficulty is the problem of resonances, or "the prob-lem of small denominators." The terms in the series expansion will be fractions, fractions whose numerators will decrease in size as one goes out in the series. But the denominators will typically contain a factor of the form $nw_1 - nw_2$, where n and m are integers and the w's are the orbital angular frequencies of the two interacting planets. There will be terms where n and m are such that this factor will be close to zero (indeed, exactly zero in the highly improbable case where w_1/w_2 is itself an integer). These terms in the series will "blow up" because of their small denominators, and in the usual series expansion one has no guarantee that the numerators will decrease fast enough to keep terms far out in the series from diverging.

The physical reason behind these terms is intuitively not hard to under-stand. The planets will exert the maximum disturbing effect on one another when they are nearest. These positions of near attraction will occur periodically. It is when n traversals of the orbit of one planet equal m traversals by the other planet of its orbit that these closings will recur. The situation is the familiar one of resonance. A swing pushed at random by a slight push will suffer only minor changes in its motion. A swing pushed by slight pushes timed appropriately to the "natural frequency" of the swing will result in large changes of motion of the swing. A small push of one planet on another may result in a large effect if the interactions are timed appropriately to the natural orbital period of the effected planet.

The program that provided the first solutions to the stability problem

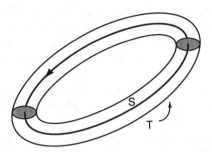

Figure 5-4. The KAM Theorem. The closed curve S represents a system that passes from a given initial state through a series of intermediate states and then returns to its exact initial state. It then repeats the process ad infinitum. An example is a planet that, unperturbed, repeats a closed orbit forever. The KAM theorem states that for systems satisfying certain conditions, a small enough perturbation of the system (say, of the planet by the gravitational tug of another planet) will result in an orbit that although it will not in general any longer be closed, will be confined to a finite region in phase-space, indicated by the tube T, surrounding the initial curve S. Such a system cannot then be ergodic and wander over the available phase space.

and avoided the difficulties discussed was initiated by H. Poincaré and carried to completion by A. Kolmogorov, V. Arnold, and J. Moser. The major result of the research of the latter three researchers is known as the KAM Theorem.

The fundamental trick is to so rearrange the terms of the series approximations to the equations of motion under perturbation as to obtain series whose convergence could be guaranteed. Basically what the KAM Theorem tells us is this: Suppose a system has been obtained by introducing a perturbation into a system that when unperturbed, followed a multiply periodic trajectory in the phase space. Such a trajectory would be like the one we would obtain if we kept track of the position and momenta of a group of planets simultaneously, each undergoing its periodic motion around the center of mass but not, of course, sharing a common periodicity. If the perturbing influence on the motion – say, from mutual gravitational interaction or interaction of the components of the system by means of some other interaction potential – satisfies a rigorously specifiable set of conditions, then in the perturbed system there will be a set of initial conditions that will define trajectories such that all of these trajectories will be confined to a region of phase space that contains the original unperturbed multiply periodic trajectory, but that is not the entire allowed phase-space. This region of trajectories that show a stability property will have non-zero size, in the usual measure for size of regions in phase space. (See Figure 5-4.)

Actually, demonstrating that the conditions sufficient for the regions of KAM stability to exist can only be done for simple cases. But there is strong reason to suspect that the case of a gas of molecules interacting by typical intermolecular potential forces will meet the conditions for the KAM results to hold. The case of the Boltzmann model of hard spheres in a box escapes the KAM Theorem results because of the singular nature of the intermolecular interaction, in which the forces are zero until the molecules meet each other and then become infinite. The existence of regions of stable trajectories that are of non-zero size, regions proven to exist by the KAM Theorem, is plainly incompatible with the ergodicity of the system. For ergodicity to hold, all but a set of measure zero of initial conditions must give rise to trajectories that wander all over the available phase space, rather than remaining confined to a proper part of it. So there is plausible theoretical reason to believe that more realistic models of the typical systems discussed in statistical mechanics will fail to be ergodic.

Further support for this claim comes from the ingenious computer simulations of systems of many degrees of freedom that have been carried out. Here, one picks a particular structure for a system and sets up on the computer a vast number of initial conditions for the system. Then one follows out the trajectory generated by each initial condition. The hope is to find by this method those regions of initial conditions that give rise to stable trajectories and those that give rise to the wildly wandering trajectories characteristic of ergodicity.

The computer simulation method leads to highly intuitively graspable results when a method, again due to a technique of Poincaré, is used. A single two-dimensional surface in the phase-space is chosen, and the points of intersection of a trajectory started from a point with that surface are calculated. If the trajectory is a simple periodic one, it shows up as a single point on the surface where that trajectory repeatedly intersects the surface at the same phase-space location. A multiply or conditionally periodic trajectory shows up on the surface as a closed curve, the inter-section of toroidal surfaces in the phase-space with the plane surface. Ergodic-like trajectories show up as an apparently random scattering of points over a region of the surface. (See Figure 5-5.)

The expression "ergodic-like" is used here deliberately. In simulations, one frequently sees regions of stability that one identifies with the re-gions whose existence is assured by the KAM Theorem. Outside these regions there are regions where the motion seems to be random with individual trajectories wandering all over the region. But are these really ergodic regions? It isn't completely clear. In the case of those systems that we can prove metrically indecomposable, the full results of ergodicity

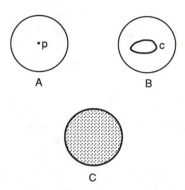

Figure 5-5. Computer models of stable and chaotic trajectories. If the path in phase space followed by the point representing the micro-state of a system is a closed curve – that is, if the system is periodic, the intersection of that trajectory with a Poincaré section will be one (or more) simple points, as in A. If the system is conditionally periodic, with its trajectory confined to the surface of a torus in the phase space, the intersection with the Poincaré section will be a curved line in the section, as in B. If the trajectory is chaotic, this shows up in the computer model by a distribution of points of intersection of the trajectory with a Poincaré section that seems to fill a region on the section as the number of intersections increase, as in C.

follow. But in the cases where the KAM Theorem gives a region of stable trajectories it is not yet possible to prove that the regions outside the regions of stability are genuinely ergodic. The appearance of randomness of trajectories in the computer simulations could be compatible with there still being metric decompositions of the "random" region. The existence of additional constants of motion that are, however, wildly non-analytic or "unsmooth" in the phase parameters could give us such a random appearance without the full measure-theoretic notion of indecomposability being satisfied. The program of trying to get a theoretical grip on the appropriate notion of randomness or chaos to apply to the regions outside the regions of KAM stability is one that only now is getting under way and in which little progress had been made.

The problem of trying to understand what physical situation causes the transition from the regions of stable orbits to the regions of unstable ones is also now under active investigation. Some results seem to show that the instability is the result of a multiplicity of resonances in the system. Each resonance will effect a strong disturbance of the trajectories in some regions of the phase space. It seems that the overlapping of regions affected by distinct resonances is responsible for the chaos observed outside of the regions of stability. It is very difficult, though, to get a theoretical grip on the structure of the stable and unstable regions for even the simplest cases. A pattern of regions of stability containing within them regions of instability that contain within them islands of stability, and so on through an infinite number of levels of fineness, can be found in even very simple cases.

For our purposes, the crucial question is the degree to which we can have confidence, given the KAM Theorem, that a reasonable model of a system of the kind treated in statistical mechanics, over and above the special cases like that of the hard spheres in a box that are provably ergodic, can be expected to be an ergodic system.

It seems clear that at this point resort will be made to the vast number of component micro-systems in our macro-system. As the number of interacting components of a dynamical system increases, we can expect that the multiplicity of interactions among them will result in ever smaller regions of stable orbits and an ever greater fraction of the allowed phase space being occupied by orbits of the apparently ergodic sort. Computer simulations do indeed suggest just this sort of diminution of the stable and increasing dominance of the unstable.

An ideal result would be a demonstration that in the thermodynamic limit of an infinite number of interacting systems, the regions of stability shrink to zero size, in the familiar measure, and the entire remaining phase-space consists of genuinely ergodic trajectories. Even this result would leave us with less than we might desire, because the systems with which we deal in the real world we are trying to describe are of course composed of only a finite, if very large, number of micro-components.

But we cannot even obtain this result of "ergodicity in the thermodynamic limit" for realistic interactions, at least at the present state of mathematical development. Some theoreticians have in fact expressed doubts that the result is even true. The best we can hope for at present would seem to be persuasive arguments to the effect that for systems of a vast number of components, interacting by means of reasonable interactions, the regions of stable trajectories will be quite small (in the usual measure relative to the size of the allowed phase space, and the overwhelmingly largest part of that phase space will be of the sort that in computer simulation, shows up as apparently ergodic. Even if the apparently ergodic region is a region of genuine ergodicity, then – something we cannot yet demonstrate – we can only show that a typical system is "very probably ergodic." And because our notion of probability here presupposes the use of the standard measure of probability, even this result requires that at least that amount of probabilistic thinking has to be introduced into our rationale of the standard probability measure in an a priori manner.

III. Ergodicity and the rationalization of equilibrium statistical mechanics

Summarizing the mathematical-physical results of the last few sections, we can say the following:

(1) For certain special cases, such as hard spheres in a box, metric indecomposability is provable. In such cases, ergodicity follows. For any mathematically well-behaved – integrable – function of the phase point, each of the following will then be true: (a) For all except a collection of points of probability zero in the standard probability measure, the infinite time average of the phase function over the trajectory starting from that point will exist and will be independent of the starting point along the trajectory. (b) The time average of the phase function will equal its phase average for all points except possibly a set of them of probability zero. (c) The proportion of time a system spends in a region of phase-space as one traverses a trajectory will, again except for a set of initial points of probability zero, in the limit as time goes to infinity, be the proportion of the phase-space occupied by that region where sizes are computed according to the standard measure. (d) The only invariant probability distribution that assigns zero probability to the same regions as those assigned zero by the standard probability distribution is the standard probability distribution. Further, none of these results depends in any way on the number of micro-components in the system.

(2) For more realistic systems with non-singular interactions among the micro-components, there is good reason to believe that the KAM Theorem will hold. In this case, for any finite such system there will be regions of initial conditions of non-zero size (in the standard measure) that are such that the trajectories starting from them will not be ergodic. Outside of these regions there will be a region that seems on the basis of computer simulations to be ergodic-like, but for which genuine ergodicity cannot be established. Although it is conceivable that in the thermodynamic limit the systems will once more be ergodic, the regions of stability shrinking to measure zero and the remaining region being genuinely ergodic, it is not clear that this will in fact be so. There is good reason to think, however, that large finite systems will have small regions of stable trajectories, and that the overwhelmingly largest part of the available phase space, once more with all sizes being measured in the standard way, will be at least ergodic-like.

In the next sections I will assume that we are dealing with a system that is genuinely ergodic. Here, we will be concerned primarily with the following question: Given that a system is ergodic, how can we legitimately use ergodicity to answer any of the standard questions we noted about the rationale for invoking the standard probability measure in the statistical mechanics of equilibrium systems? As we shall see, the answers to that question are not simple or transparent.

1. Ensemble probabilities, time probabilities, and measured quantities

One proposal for utilizing the results of ergodicity to account for our experience with equilibrium systems has the advantage of being remarkably straightforward. This approach focuses on the fact that our measurements of macroscopic thermodynamic quantities take time. The measurement of the pressure of a gas by a manometer, for example, is a process that takes a very long time if we measure the passage of time in such microscopically appropriate units as the mean time between molecular collisions, for example. Because this is so, can we not argue that a macroscopic quantity should be identified with the average over an infinite time of its appropriate microscopic phase quantity? In the case of pressure, for example, ought we not to identify the pressure of a gas on the walls of its container with the average over an infinite time scale of that microscopic phase function that keeps track of the transfer of momentum to the walls of the container by molecular impact? Then, by the ergodic theorem, we can, for almost all microscopic states of the gas, identify the infinite time average of the phase quantity with its phase average, thereby explaining why phase averages calculated by means of the usual ensemble approach work to give us the familiar equilibrium results.

This argument, although straightforward, is, alas, not very plausible. To be sure, macroscopic measurements must be time averages in microscopic quantities over "long" periods of time. But "long" is not forever. And it isn't merely a matter of adding a reasonable additional hypothesis that "long" is long enough to take the infinite time average as a reasonable idealization. For we have quite good reason to believe this is false.

Indeed, patently false. If our macroscopic measurements could all be legitimately construed as infinite time averages, then every macroscopic measurement would have to result in the equilibrium value for the quantity in question. We could have no macroscopic determinations of the values of thermodynamic quantities for systems not in equilibrium, nor a macroscopic ability to track the approach to equilibrium by following the variation in these quantities as they approached their final equilibrium values. But of course we can macroscopically determine the existence of non-equilibrium states, their features, and the laws that govern their approach to equilibrium. Macroscopic measurements may indeed be averages over times that are long on various microscopic time scales, but they cannot be treated, even in idealization, as infinite time averages.

There is, however, a subtler way of utilizing the results of ergodic theory – in particular, its ability to identify ensemble quantities with temporal quantities of an individual system – to understand the success of equilibrium statistical mechanics. But the form this understanding takes must itself be carefully understood.

Let us return to the picture of a system isolated energetically over a time from past infinity to future infinity as proposed by Boltzmann in response to some of the criticisms of his earlier versions of the *H*-Theorem. In this picture, a system over infinite time moves from micro-state to micro-state. For the overwhelmingly largest part of the time, the system is in micro-states near the equilibrium micro-state – that is, the micro-state described by the Maxwell–Boltzmann distribution. Excursions from this "most probable" state do occur, but the greater the excursion of the system away from this dominating most-probable state, the less frequently it occurs. The picture is, remember, completely time-symmetric, transitions from non-equilibrium to closer to equilibrium states being matched by transitions in the opposite direction.

If the system is ergodic, we can, at least to a degree, back up this Boltzmannian picture with an argument that the picture is in some sense a correct representation of the facts about the system. First, we must note that the ergodic results will tell us only that the Boltzmann picture holds "with probability one in the usual probability measure." Ergodicity tells us that non-ergodic behavior is possible only for a set of systems started in micro-states from a set of probability zero in our usual probability picture. But being a member of a set of measure zero is not being an impossible state. For the time being, however, let us focus on the "highly probable" systems, leaving the qualification to be attended to later.

By ergodicity, we know that over the infinite time interval the proportion of time spent by a system in a region of phase-space is proportionate to the size of that region relative to the total size of the accessible phase space in the standard measure. But what assures us that the size of a region of points that represent systems close to equilibrium will, in fact, be large in the accessible phase space as a whole? Not ergodicity. Rather, we must resort here to the results that had their origin in Khinchin's program and have continued in the rigorous study of the thermodynamic limit. Here, we restrict our attention to those phase functions that obey the suitable structural conditions demanded by the Khinchine program, the hope being that we can show that all of those phase functions that we relate to thermodynamic macroscopic quantities are of the appropriate sort. Next, we invoke the special features of the interaction of the components of the system needed to prove the results that hold in the thermodynamic limit. Combining these we obtain the result we sought – that the regions of phase space that are those corresponding to systems that are close to Maxwell–Boltzmann in their distributions are overwhelmingly large in the phase space, and that regions of systems deviating from the equilibrium condition have their sizes diminish rapidly with increasing deviation from the dominant distribution.

Although the exact results hold only in the thermodynamic limit, it is useful to note that the rate of convergence of quantities to the limit quantities is something one can compute in the theory of the thermodynamic limit. And assurance that normally macroscopic systems are large enough to appropriately idealize them by the thermodynamic limit can be had from the theory. So, ergodicity combined with the results on the way in which large systems appropriate phase functions have their distribution converge to one sharply and symmetrically distributed around the overwhelmingly dominate most probable value, which is then identifiable with the mean or average value, gives us a clear underpinning to the Boltzmann picture.

But we still must proceed with caution if we are to understand how these results are to be connected to what we find experimentally in the world. Ergodicity, where it can be demonstrated, does certainly provide us with something of deep conceptual interest – that is, a proof that for almost all systems, one formalistically probabilistic notion (size of region of phase space in the standard measure) is provably equal to another formally probabilistic notion (proportion of time spent by a system in a region of phase space in the infinite time limit). But what questions about the nature or origin of equilibrium states does that allow us to answer?

It does not seem plausible to expect ergodicity to play a role as a component of a causal explanation of the equilibrium condition of a system. Presumably, that account will have to deal with whatever initial micro-states led to the dynamical movement toward a micro-state of the equilibrium sort. Statistical propositions about micro-states will come into play when we deal with the probabilities of initial condition micro-states having differing dynamical consequences. As we shall see in Chapter 7,II,3 when we discuss the rationalizing of non-equilibrium theory, a set of results stronger than the result of ergodicity but related to it in their derivation will indeed function in that statistical-causal picture. Even there, though, we shall see that hopes of a complete statistical-causal explanation of the origin of the approach to equilibrium will not be fully gratified. On the other hand, the derivation of the ergodic results from the pure dynamics of the micro-components does give us, in some sense, a "causal" origin for a probabilistic result.

Nor does ergodicity function in any simple straightforward way to explain equilibrium by subsuming it under a statistical generalization. If we experimentally found states by randomly selecting them from a system maintained in isolation over an infinite time, then we could, presumably, subsumptively account for an observed state being of a certain macroscopic nature by making reference to the time proportion of the system spent in such kinds of states. And we could offer a high-probability subsumptive account of the observation being of a micro-state of the near to equilibrium sort. But of course we really don't experimentally fix on equilibrium as that state in which a system isolated over infinite time spends nearly all of its time. Rather, it is that state quickly approached by non-equilibrium processes and that maintains itself over what are reasonably long periods of experimental time.

Perhaps the most we can say is something like this: Proportion of time spent in a region of micro-states over an infinite time is a formally probabilistic notion. It is, in the case of ergodicity, a well-defined dispositional notion even from the actualist point of view, because it is derivable from the underlying actual lawlike propositions governing the microscopic dynamics. Ergodicity does allow us to bring the formally probabilistic notion of size of region in the standard probabilistic measure closer to "what we physically determine," in the sense of allowing us to see that the measured equilibrium quantities are not just those overwhelmingly probable in an a priori presented probability measure. They are also overwhelmingly probable in the sense of being the time-dominant states in the idealized case of a single system maintained forever in isolation.

Here we see a case both of probability as legitimized disposition and of statistical explanation as showing the association between what occurs and some idealization. So long as we keep ourselves aware of what such an "explanation" of the nature of equilibrium isn't – not a causal explanation and not a simple subsumptive explanation either – there will be no harm in saying that an explanation of some kind of why equilibrium is the way it is has been provided. Ergodicity does serve to satisfy Boltzmann's demand that probabilities as proportions in imagined ensembles of identically prepared systems be brought down to earth by being shown to be legitimated representatives of probabilities as proportional durations of isolated individual systems over infinite time.

2. The uniqueness of the invariant probability measure

In the last section, we utilized the consequence of ergodicity that stated that for almost all points, in the limit of infinite time, the proportion of time spent in a measurable region of phase space was the same as the size of that region in the standard measure. Here, we focus on another rationalizing role ergodicity can play in equilibrium theory. This rests upon the simple corollary of the theorem to the effect that any probability measure that assigns measure zero by the standard measure and that is invariant in time under the dynamic evolution of the systems in the ensemble must agree with the standard measure on the probability to be assigned to all measurable sets.

How can this result be utilized? Consider our notion of the equilibrium state of a system. Its empirical origin is in terms of a system whose macroscopic thermodynamic variables remain the same over observable periods of time, idealized to invariance over an infinite time in our theoretical picture. But, of course, in our statistical mechanical picture of the world we realize that spontaneous fluctuations must be allowed for. As we have seen in Chapter 2, and as we shall explore in much greater detail in Chapters 6 and 7, the probabilistic picture that takes the place of the thermodynamic lawlike approach to equilibrium of a system is that of an ensemble of systems, originally prepared by fixing certain macroscopic constraints, evolving to a new ensemble when these macroscopic constraints are changed, thereby leaving the ensemble in a non-equilibrium condition. Here, the notions of being in non-equilibrium or being in equilibrium are predicates not of the individual systems, but of the ensemble. Of course, the ensemble, from most people's perspective, is just a picturesque way of representing a probability distribution over possible micro-states of a system compatible with specified macroscopic constraints. It is this probability distribution whose dynamic evolution we follow in statistical mechanics, and it is to it that the terms "non-equilibrium" and "equilibrium" should appropriately be applied.

What probability distribution should be described as the equilibrium probability distribution? Surely it is the one that under the dynamic evolution that determines the time trajectory beginning with each phase point, continues to assign the same probability at each time to each given region of points in the phase space. The picture is of systems in the ensemble having their micro-state move out of a region of micro-states due to their dynamic evolution, and of other systems whose micro-state was initially outside the region having their micro-state enter the region because of their dynamic evolution. But the proportion of systems in the ensemble having their micro-states in the given region remains constant over time.

It is easy to show, using the definition of the standard probability distribution and the general law of dynamic evolution of a system, that the standard probability measure has this invariance feature (Liouville's Theorem). But, as Boltzmann clearly pointed out, it does not follow that it is the only such invariant measure. The existence of a partition of the phase points into two or more pieces of non-zero measure closed under

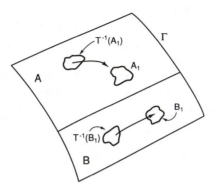

Figure 5-6. Non-uniqueness of invariant measure if metric decomposability holds. Suppose the phase space for the system Γ is decomposable into two invariant subspaces of non-zero measure, A and B. We can allocate any portion of the probability we like – say p, to A, and the remainder, $1 - p$, to B. Then we can spread those probabilities uniformly over their respective invariant subspaces A and B, using the standard measure. The resulting probability assignment will be invariant, because the probabilities assigned to a region in A, say A_1, and the probability assigned to $T^{-1}(A_1)$ will be the same, and the probabilities assigned to a region in B, say B_1, and that assigned to $T^{-1}(B_1)$ will be the same. But it will not in general be the standard probability measure, for A_1 and B_1 need not get the same probability even if their size in the standard measure is the same.

time in their evolution would allow us to find other such measures. We could, for example, put all the probability into one such piece using the standard measure renormalized there and give probability zero to all the others. The regions of the piece in which the probability had been concentrated would have their probability remain invariant as before, and the subregions of the other pieces would have invariant measure zero. (See Figure 5-6.)

Ergodicity makes this last situation impossible. If the phase space of the ensemble is metrically indecomposable, we can then be assured that there is only one probability measure that is invariant in time, at least as long as we confine our attention only to probability measures that assign measure zero to every set of points assigned measure zero by the standard measure. In this sense, then, we can justify our choice of the equilibrium probability measure by its uniqueness relative to the conditions of dynamic invariance and "absolute continuity" with respect to the standard measure.

At this point, we can then reinvoke the thermodynamic limit results used before to assure ourselves that having chosen the standard probability distribution as the equilibrium probability distribution, the set of points that are such that the phase functions in them used to define macroscopic quantities have values close to their values that give rise to the standard equilibrium values for the macroscopic quantities will be overwhelmingly dominant in the available region of phase space. Having picked as the equilibrium probability distribution the standard probability distribution, we can be sure that it is "overwhelmingly probable" that a system will have as its macroscopic values the standard values familiar from thermodynamics. This further argument utilizes, of course, the vast number of micro-components in the system, the special nature of the phase functions used to reduce the macroscopic quantities, and the special structural features of the system required to prove the thermodynamic limit results.

3. The set of measure zero problem

At this point, we ought to deal with a problem that has repeatedly surfaced in our treatment of the use of ergodicity to rationalize equilibrium statistical mechanics. If a system is ergodic, then phase averages of phase functions will equal their infinite time averages, *except*, possibly, for a set of measure zero of initial trajectory points. Standard probabilities of regions will equal, in the infinite time limit, proportions of duration the system spends in the region, *except*, possibly, for a set of measure zero of initial phase points. The standard probability measure will be the unique invariant measure, *except*, possibly, for other, non-equivalent measures that fail to assign probability zero to sets assigned probability zero by the standard measure. But by what right can we assume that sets of probability zero in the standard probability assignment can be ignored?

That a set has probability zero in the standard measure hardly means that the world won't be found to have its total situation represented by a point in that set. After all, every phase point is the member of an infinity of sets of measure zero, such as the set of that point by itself. We could of course simply adopt as a postulate that it is reasonable to assume that situations that have zero probability according to the standard probability measure can be taken to not be the case. So we would take it as simply a priori for an ergodic system that we can be assured that we will not find, for a particular system, that it is in such a micro-state that its infinite time average of a phase quantity as it evolves from that state on will not equal the phase average of that quantity. Such an assumption, a priori as it may be, is certainly a weaker, purely statistical, addition to the underlying dynamics than was the original assumption of the full standard probability measure as correctly representing all probabilities. But can we do without even this weaker assumption?

Attempts at explaining why we should give "real" probability zero to sets of measure zero in the standard measure – that is, why we should assume that there is no likelihood above zero of a state of the relevant kind being encountered in the world – usually rely upon ways of connecting measure zero in the standard measure with other features of the set.

One topological feature of these measure zero sets is that they have no "interior" points. That is, for each point of these sets, and for any neighborhood of that point (that is, open set containing that point), there are points *not* in the set contained in that neighborhood. This suggests the idea that any system of that kind would be instantly driven out of the relevant set by even the slightest interference from the outside. Therefore, it is argued, we ought to take the likelihood of a system's having its state be of that kind as non-existent.

But as D. Malament and S. Zabell point out, one must be careful with such arguments. Some sets that fail to have interior points are such that we certainly do not want to attribute zero "real" probability to them. The set of points having at least one coordinate with an irrational value is also such an "unstable" set, but we would expect with probability one that a system would have its phase point in that set! And indeed the set gets measure one in the standard measure.

Malament and Zabell suggest an alternative property associated with having measure zero in the standard measure to focus on. Suppose we demand that probability have a continuity property – that is, if one set of points in phase space can be obtained from another by a small displacement – then the probability of a point being in the second set should be close to the probability of its being in the first. The idea seems to be that a method of preparing an ensemble – that is, of fixing constraints so that one guarantees that a system will be in a given set – ought to have this continuity property. The condition implies that sets of measure zero in the standard measure have probability zero.

But we must still be very cautious. As Malament and Zabell clearly realize, we can sometimes so prepare systems as to guarantee that they will be in a set of measure zero. By fixing the energy of a system, we guarantee that its phase point will be on the energy hyper-surface, a set that has zero measure in the volume measure of the full phase space. In general, fixing the value of any global constant of motion will confine the system to a lesser dimensional sub-space of the phase space than one started with, and such lesser dimensional sub-spaces are always of measure zero.

Malament and Zabell respond to this by pointing out that one could then simply restrict one's attention to the phase space defined by fixing all the global constants of motion and use the appropriate panta-micro-canonical measure for that limited phase space. Their continuity argument would then guarantee that one ought to assign probability zero to the sets of measure zero in that refined standard measure.

One can go even further, because we are working in a context in which ergodicity is being assumed. As we saw earlier, that will guarantee that there are no global constants of motion that vary continuously over some interval and that we have neglected. For if there were such constants of motion, the region of phase space would be metrically decomposable and the system not ergodic.

Yet we must still be cautious. What exactly do constraints on our ability to delimit an ensemble have to do with real probabilities in the world? Even if it were the case that we could not control systems in such a way as to select out an ensemble of only those systems in a designated set of measure zero, would that fully justify our claim that in nature we have grounds, over and above purely postulational ones, for assuming that we have a right to take as zero the real probability that a system in the world could be in the deviant set? I think that this question is more easily discussed in the context of the rationalization of non-equilibrium theory, for it is in that context that the restrictions in our ability to pre-pare systems in certain ways and the role these restrictions play in grounding the postulates of statistical mechanics are most fruitfully treated.

Let me mention here, though, in anticipation of the fuller discussion in Chapter 7, a few salient facts. We shall see that our ability to confine the state of a system to a particular region of phase space by purely macroscopic means is rather greater than one might at first think. For this reason, we will be a bit hesitant about accepting as uncontrovertible even such plausible principles as the continuity principle invoked by Malament and Zabell to rationalize attributing "real" probability zero to sets of measure zero in the standard measure. We shall also see that claims to the effect that certain classes of points in phase space are so unstable with regard to even infinitesimal interferences from the outside that we ought not ever to consider such classes as relevant kinds of systems that we could view as a prepared ensemble ought also to be accepted only with caution. In Chapter 7,II,1, we will examine some cases where interference from the outside can be demonstrably reduced to such a degree that modes of preparation we might have thought made impossible by such disturbance of the system from the outside become quite possible.

The suggestions of Malament and Zabell discussed, and numerous other suggestions made in the context of trying to rationalize probabilistic assumptions in statistical mechanics, make use of important topological features of the space of points representing the micro-states of systems. Here, I will try to put this discussion in a somewhat broader context.

Topological features of a space are those dealing with issues of continuity. Without going into the technicalities of how the panoply of topological concepts are formally defined, we can simply note that such notions as the continuity of a curve, the "neighborhood" of a point, the "openness" or "closedness" of a set (whether it does not or does contain all of those points that are the "limit points" of "converging" sequences of points in the set) are the standard topological notions. In a space in which a distance function can be defined as so-called metric space, one can define a topology out of the metric in a simple way. The most common standard formulation of topology defines all other topological notions from the notion of open set, taken as a primitive. The open sets can be specified as being all those generated out of a proper sub-set of them called a basis. Given a metric, a basis of open sets can be specified by taking as an open set all points less than a specified distance from a given point. To say, then, that a point has points of a specified kind in any neighborhood of it becomes the claim that points of the specified kind can be found as close as one likes to the given pont.

Just as measure theory allows for the specification of a set that is "small" in the measure – that is, of measure zero – topology allows for sets that are "topologically small," the so-called sets of first-category. Formally, these are all sets that can be generated as countable unions of sets that are "nowhere dense" in the space in question. That is, they are countable unions of sets whose complements contain a dense subspace. The sets not of first-category are said to be of second category. The definition is framed in such a way as to make the rational numbers "sparse" – of first-category – in the reals, just as the rationals are a set of measure zero in the reals in standard measure theory.

Many notions applicable in measure theory and the study of spaces under transformations from a measure theoretic point of view, the notions we have explored in talking about ergodic theory, reappear transformed in a "dual" topological theory. Thus, metric indecomposability is paralleled by a property of a space under a transformation called minimality. There is, in fact, an important "duality" theorem to the effect that whole classes of theorems in measure theory are converted into topological theorems when the appropriate substitutions are made of topological notions for measure-theoretic notions, such as substituting sets of first-category for sets of measure zero.

To what extent could equilibrium statistical mechanics (and perhaps aspects of the non-equilibrium theory as well) be grounded in topological facts about the behavior of the trajectories mapped out in phase space as systems evolve according to the laws of the micro-dynamics? Some explorations in this question have been made. There are a number of topological results one might desire from such an alternative foundationalist perspective, but there are the usual questions to be asked.

Are these results true? Can we show that for appropriately idealized systems they are true? And even if true, could they be used in the justificatory way one intended without debilitating criticisms being raised against them?

One might hope to show, for example, that for a class of states all of whose members are close to equilibrium, a topologically "dominant" set of trajectories – a set of trajectories of the second category – are such that systems following those trajectories have their states in the set of near equilibrium states almost all of the time – that is, with time average one in the limit as time goes to infinity. Alas, proving such results for reasonably idealized physical systems is generally a near impossible task. Again, proving the topological equivalent to ergodicity, called transitivity, and which says that if a set is invariant under the dynamic transformation it is either of the first-category or has its complement of first-category, is also usually an impossible task.

In answer to the third question noted earlier, it is sometimes the case that we can be pretty sure that the plausible topological result we have in mind just won't be enough to give us the rationale we want for statistical mechanics anyway. The problem presented to us at the beginning of this section was to explain why we ought to believe that sets of measure zero in the standard measure could be ignored. Is it a basic posit of statistical mechanics that such a procedure is legitimate, or can this posit be grounded on something seemingly less arbitrary? One suggestion might be that sets of states of topological sparseness, of first-category, ought to be ignorable, the topology of the phase space being (seemingly) less open to our "choice" than choosing a measure absolutely continuous with the standard measure. But, alas, there is a result to the effect that the real line can be decomposed into two sets, one of measure zero and the other of first-category. It would seem, then, that being of the first-category is hardly reason to assume that a set of states can then be ignored, because such a set can contain "almost all" the points in a set from a measure-theoretic or probabilistic point of view.

For another example, we need only refer back to the notion of quasi-ergodicity mentioned in the historical survey of Chapter 2 and earlier in Chapter 2,I,2. There we noted that the original hypothesis that each trajectory went through every accessible point of the phase space could be rejected on topological grounds. And, we noted, the suggestion had been made that this original Ergodic Hypothesis might be replaced by one to the effect that each trajectory came arbitrarily close to each point of the phase space, the Quasi-Ergodic Hypothesis. Here is a posit framed in topological terms rather than measure-theoretic terms.

But, as we noted earlier, this hypothesis suffered from two defects. First, it could not be shown to be true for reasonably idealized systems.

And, more importantly for our discussion, it could be shown that being quasi-ergodic was not sufficient for infinite time averages for each trajectory to equal the phase average of a quantity over the accessible region of phase-space. And it was this latter claim that the ergodic posit had been introduced to justify in the first place. Here we see a topological result proving insufficient to provide the grounding of a statistical mechanical assumption that was desired.

Other, weaker, results can be obtained, however. Given a topology, the open sets are sets that have interior points – points surrounded by neighborhoods, all of whose points are in the given set. It is plausible to think of such sets as "larger than zero in size" and to demand of a measure on the space that it give all open sets non-zero measure. Sometimes this demand that a measure be "regular," as it is called, can eliminate some possible measures in the abstract sense, and may even pin one down to the standard measure as that one measure uniquely regular relative to the chosen topology. Here again is a suggestion for alleviating some of the arbitrariness in choice of measure that afflicts the usual foundational approaches to equilibrium theory.

But, on reflection, more than one question arises. *Why* should we demand regularity of a measure, given our intention to think of measures as some kind of indicator of the "real proportion" of cases in which micro-states occur in the world? And how non-arbitrary is our choice of a topology, and the associated metric distance function in the phase space, in the first place? The standard use to which the arguments are put that attempt to ground the basic posits of statistical mechanics on topological facts about the flow of points along trajectories in phase-space all rest upon implicit assumptions that relate topological facts to facts about the possibilities for "preparing systems" or "selecting a group of systems to constitute an ensemble." The arguments are of the sort that alleges that if each micro-state of a system can be approached arbitrarily closely by a micro-state that leads to a trajectory of a particular sort (say, of a system that spends almost all its time near equilibrium), then we are right to say that any system, no matter how prepared or selected, will justifiably be thought to have that feature. But such arguments depend upon implicit assumptions that the metric used, usually the familiar standard metric of the phase space, will be appropriate for making such assertions. And this assumption depends upon the correctness of the assumption that "arbitrarily close" systems in the given metric cannot be distinguished by any physical process within our control.

But, as we noted in the comments on the arguments of Malament and Zabell, questions of just what we can and cannot do to select out systems from one another by the deft manipulation of their initial micro-states are by no means trivially answered. As we noted there, some of these questions will be taken up in more detail in the treatment of non-equilibrium theory in Chapter 7. At this point, we need only remark that any attempt to get around issues of the "arbitrariness" of the imposed measure used in the usual foundational formulation of statistical mechanics that relies upon some deep association of the standard measure with important topological features of the phase space and its flows in the standard topology must be supplemented with careful arguments. First, we must show how the topological feature invoked is to be presumed relevant to the foundational issue in question. Then we must show that the choice of topology and associated concepts, such as "arbitrarily close" for phase space points, can be backed up with a demonstration that this topology is the appropriate one given the physics of our abilities to control systems and sort them into ensembles.

4. Ergodicity and equilibrium theory in the broader non-equilibrium context

The results we have been examining in these last few sections have utilized a number of crucial resources. The basic laws of dynamics of the micro-components of a system, the particular interactions governing them and the large size of the system, in the sense of its having a large number of micro-components, have all been invoked in obtaining the ergodic results.

The results obtained have been significant, but we ought to reflect on some of the matters that have not been dealt with. We have seen that for a very limited class of systems, full ergodicity can be demonstrated. In this case, we have, neglecting the problem of a possible class of exceptional micro-states of measure zero, the result that the standard probability measure is uniquely invariant and hence uniquely suited to represent an invariant equilibrium situation. Again, neglecting the measure-zero problem, we are able to show that for an idealized system kept isolated for doubly infinite time, the proportion of time spent by the system in a region of micro-states is proportional to the size of that region in the standard measure. Invoking the vast size of the system and the special nature of the phase functions used to compute thermodynamic quantities, we can also show, for reasonable interactions of the micro-components, that the dominant region of phase space in the standard measure is that corresponding to systems whose thermodynamic values are close to the values predicted by standard equilibrium thermodynamics.

For systems outside the limited class for which genuine ergodicity can be proven, we have much less. For these systems, we have only the hope that by utilizing their large size, a demonstration will someday be obtained that the computer simulations that show us that the apparently ergodic-like region outside the regions of stable trajectories become overwhelmingly dominant as the size of the system increases are hinting at some kind of genuinely rigorous ergodic-like result in the thermodynamic limit.

In this limited context, we cannot hope to gain any insight into the route by which equilibrium is obtained. By restricting our attention to the standard equilibrium ensemble and the standard technique of calculating equilibrium values by means of phase averages, we have stayed away entirely from the problems of characterizing non-equilibrium states and their dynamic evolutions. For this reason, we have made no attempt at all to characterize one of the most crucial features of equilibrium, its nature as an "attractor" state to which other states of systems converge. Nor have we gained any insight, of course, into the details of the process whereby systems evolve into equilibrium, such as their standard rates of approach to that state. Most crucially, we have made no attempt at all to understand the peculiar time-asymmetry of the process summarized by the Second Law.

Not only does the approach we have been taking exclude certain topics from being investigated, it also limits the kind of account we can give of what we are investigating. We can, in the equilibrium approach, offer some degree of "transcendental" rationale of the equilibrium probability distribution, by showing it uniquely invariant. And we can show that equilibrium states are, in an interesting sense, temporally dominant for idealized ever isolated systems. But we cannot offer anything like a causal or statistical-causal account of the origin of equilibrium, because this, perforce, requires reference to the states out of which equilibrium evolves and statistical generalizations over their likelihood of occurring.

Nor can we even offer an explanation of equilibrium of the simple subsumptive kind. A genuine account of the conditions under which we are likely to various degrees in the actual world to encounter equilibrium states will require fitting equilibrium into the much broader context of all states of systems, equilibrium and non-equilibrium, and their dynamic interrelations. The reference class of all states over doubly infinite time of an ever isolated system is an interesting idealization, but of much greater importance are the reference classes fixed by our selection and preparation of systems in the world.

Because our general treatment of selecting or preparing systems will require a much broader context for investigation than the one to which we have so far limited ourselves, the problem of the rationalizing or ignoring sets of measure zero will also be immune to its full treatment until we broaden our perspective beyond the pure equilibrium theory. Only in the investigation of the attempts to rationalize the full treatment of non-equilibrium will we be able to get further insight into this issue, although even there definitive results will be hard to come by.

5. The Objective Bayesian approach to equilibrium theory

There is an interesting approach to the rationalization of equilibrium statistical mechanics, primarily due to E. Jaynes, that it would be useful to explore at this point. Jaynes suggests that equilibrium statistical mechanics can be viewed as a special case of the general program of systematic inductive reasoning. From this point of view, the probability distributions introduced into statistical mechanics have their basis not so much in an empirical investigation of occurrent frequencies in the world, although as we shall see such empirical results are to be factored into the basis of probabilities at some point, but instead in a general procedure for determining appropriate a priori subjective probabilities in a systematic way. Jaynes has argued that these results can be extended to non-equilibrium cases as well, a matter we shall explore in Chapter 7,III,3 in the appropriate context.

In the general theory of statistical inference, the word Bayesian is used to designate someone who argues that the inductive process ought to be viewed as the updating of subjective probabilities we hold prior to an experiment by utilizing the results of the experiment. The updating is done by the use of Bayes' Theorem that allows us, by conditionalization, to revise our subjective probability assignment in the light of the new data.

But where do our initial subjective probability assignments come from? Many Bayesians would hold that they are simply reflective of our propensities to believe. Insofar as they are *coherent* probabilities – that is, insofar as they obey the formal axioms of probability theory – they are rational. Failure to obey these axioms will result in betting behavior in certain situations in which a loss is guaranteed (or, at least, in which a loss is possible but no gain is), and this is sufficient to demand that the probabilities obey the formal principles. But no stronger constraints of rationality can be demanded.

The "Objective Bayesians" demand more. For some of them, only a specific class of initial a priori probabilities are admitted to be rational. Almost invariably, the rule for delimiting the class of rational initial probabilities is a generalization of the famous Principle of Indifference. That rule tells us to assign equal probabilities to all symmetrically equivalent outcomes. An example is the assignment of equal probability 1/6 to the outcome of any specific face of a tossed die ending up topmost. The generalized principle is to apply also to cases where there is a continuum of possible outcomes.

An elegant version of the rule can be constructed by generalizing from Gibbs' construction of the standard statistical mechanical probability distributions. They, remember, were the distributions that maximized the (fine-grained) ensemble entropy, as Gibbs demonstrated. The suggestion now is to do this: Let there be some appropriate measure over the continuum of values, where the range of possible values is delimited by some specified constraints. Choose a probability assignment over the values that maximizes the ensemble entropy calculated in the Gibbsian manner. This assignment will be the generalization of the probabilities determined by the Principle of Indifference in the case of a finite number of basic outcomes, and it will specify one's rational choice of a priori probabilities.

At this point, there is usually reference to the important work of C. Shannon on the foundations of information theory. Consider any probability distribution over a finite number of possible outcomes. Shannon shows that the entropy of that probability distribution, $-\Sigma_i p_i \log p_i$, is the unique function that is additive for independent distributions – that is such that if we have two sets of possible outcomes with their respective probability distributions, the value the function assigns to them jointly is the sum of the values the functions assign to them individually, and that is maximized by the uniform probability distribution over the outcomes. Adding appropriate continuity assumptions, we can generalize this to the continuous case. So this entropy is a plausible candidate for measuring the amount of "uncertainty" we have about a situation. It is maximal when our uncertainty of the actual states of affairs is maximal – that is, when we have no grounds for expecting one outcome over another. And that is the case when the probability distribution we assign to the outcomes is uniform.

We now have what looks like a justification for the standard probability distribution invoked in equilibrium statistical mechanics. It is, after all, the ensemble distribution that maximizes the Gibbs' fine-grained entropy. May we not then justify this choice not on some grounds of a special postulate about the frequencies of occurrence of micro-states of systems subject to macroscopic constraints but on the grounds instead that this is just the choice of a priori probabilities always justified on general inductive principles?

But the situation is not that simple, and for a very fundamental reason. The injunction to "spread a priori probability uniformly" over a continuum of possible values is, alas, vacuous. For "uniformly" will vary depending upon the measure we impose on the set of values. If the values are in the range [0, 1] we will get quite distinct a priori probability assignments if we consider the size of a sub-interval to be its usual value or instead fix some other measure – say, letting the interval [0,9/10] be half of the distance and the interval [9/10,1] be the remaining half – and spread a priori probability uniformly relative to the new, unusual measure.

A means of resolving this very familiar ambiguity was suggested by H. Jeffreys. Here, one looks at the experiment relative to which the a priori probability is to be assigned. On the basis of the nature of that experiment, one seeks an invariance principle, a defined set of transformations of the space of the unknown parameters and experimental values, relative to which the probabilistic consequences of the experiment must remain invariant. Suppose, for example, that we are trying to guess a parameter that is a single real number. Our guess should remain the same no matter where we arbitrarily fix the zero point of measured and guessed values. So the probability assignment we get by conditionalizing on our a priori probability distribution should remain invariant on linear transformations. This fixes the unique "uniform" a priori probability as that uniform in the usual linear measure from minus infinity to plus infinity. Actually this is a so-called improper prior because it really isn't a normalizable probability measure. Every "well posed problem," in Jaynes' terminology, should suggest an appropriate invariance principle to allow us to uniquely specify a uniform of "maximum entropy" a priori probability distribution.

As a general rule of inductive inference, this "objective Bayesian" approach has its problems. These include difficulties in reconciling the rule for a priori probability with conditionalization in the case of multiple experiments, the problem of uniqueness of results given possibilities of redescribing experiments, and problems encountered when constraints relative to which probabilities are assigned are themselves given probabilities. We shall not explore these here, but instead focus on the application of the program to statistical mechanics, and in particular at this point, to the equilibrium case.

Suppose we have a system in equilibrium, and we wish to guess at its micro-state. We impose a probability distribution over the micro-states compatible with the fact that the system has a fixed value of some controllable constants of motion and fixed macroscopic constraints. Because the system is in equilibrium as time goes on, nothing macroscopic changes. Hence our "ignorance" of the exact micro-state remains the same over time. So the probability distribution we attribute to micro-states should remain temporally invariant. The standard panta-micro-canonical distribution is such a temporally invariant probability distribution, the uniform one with respect to the invariant panta-micro-canonical measure. Hence it is "rationalized" by the principles of objective Bayesianism.

Clearly the argument here is hard to distinguish from the argument, familiar since Boltzmann, that equilibrium, being an invariant state over time by definition, its statistical surrogate must be temporally invariant as well.

But of course Boltzmann felt that a proof that the standard distribution was the *uniquely* temporally invariant measure was required to fully justify choosing it. This was one of the motivations behind the Ergodic Hypothesis, of course. What does the objective Bayesian respond to this demand?

For the most part, Jaynes dismisses ergodic results as irrelevant. His reason for this is the quite correct argument noted earlier that identifying infinite averages with measured quantities using the time duration of measurement processes as justification is fallacious. As we have seen, however, ergodicity, when it is demonstrable, does give us a demonstration of the uniqueness of the invariant probability distribution, subject to our restricting our attention to probability distributions "absolutely continuous" in the standard measure. Although this still necessitates an a priori assumption, or at least an assumption that, as we have seen, requires resort to non-equilibrium theory for its justification, it still reduces what we must accept a priori in our probability assumptions, and should be of interest even from the objective Bayesian approach.

When the system isn't ergodic, there might be controllable constants of motion we have ignored. In those cases, making the probability assumptions deemed appropriate by objective Bayesianism relative to the constants we are aware of will lead to the wrong predictions about equilibrium values. Now Jaynes points out that this will in fact suggest to us that there are remaining constants of motion to be found. Indeed it is true that false predictions obtained with the standard probability assignments immediately do lead us to seek the constants of motion we have ignored. And we have always found them. But several things need to be said. The first is that if these constants do exist, our initial probability assignment was *wrong*, suggesting that there is an objective rightness and wrongness about a priori probability distributions. But from a subjectivist point of view, even of the objective Bayesian sort, our initial probability assignment was *right*, being the uniform probability assignment having the appropriate invariance characteristics relative to the knowledge that we had.

Even if there are no *controllable* macroscopic features of the system we have ignored, either because the system is ergodic or because the regions of non-zero measure in which phase averages don't equal time averages are so non-analytic in the phase space that we are unable to "fix" the system in such regions by macroscopic operations, there still remains the *explanatory* question: Why do our probabilistic assumptions work so well in giving us equilibrium values? After all, couldn't the world consist solely of systems all of which were members of the sets of ill-behaved systems, even where those sets are small or even of size zero in our standard probability measure?

As we have seen even in the best possible case, where ergodicity is provable, we still cannot give a full answer to this question within the purely equilibrium context. Neither a causal-statistical explanation nor even a subsumptive-statistical account is fully realizable. On the other hand, the ergodic results do at least suggest ways in which the question might be answered in the fuller context.

A. Hobson takes up the question "why does statistical mechanics work?" from the objective Bayesian point of view of Jaynes. His suggestion is that statistical mechanics is just inductive reasoning, and "inductive reasoning usually works." He continues, "To push the question one step further by asking 'why does inductive reasoning work?' would lead us beyond science and into philosophy. In science, one ordinarily assumes the validity of inductive reasoning, since if inductive reasoning *didn't* work science wouldn't be possible in the first place."

But this is not very persuasive. To shunt the question out of science into "philosophy" where, presumably, it can be dismissed by scientists, seems a peculiar move for an exponent of a view that assimilates an important branch of theoretical physics to "pure inductive logic." If inductive logic is "philosophy" and if statistical mechanics is inductive logic, answering the question "why does statistical mechanics – that is, inductive reasoning, work?" is precisely a demand for a scientific explanation. Even Hobson admits that when "inductive reasoning" doesn't work it is a *scientific* explanation that we seek – that is, "what controllable constant of motion have we missed?"

Admittedly in the confines of purely equilibrium theory, the best the ergodic approach can do is to offer us a partial "transcendental" rationale for the standard probability distribution. And, admittedly, even this is unavailable to us in the case of those realistic systems for which we can, by plausible extrapolation of the KAM Theorem, reasonably infer that the system is not ergodic in the full sense.

In Chapter 7,III,3, we will explore the objective Bayesian approach to the construction of probability distributions for the non-equilibrium case. And we shall use the suggestion made in that context as rationale for the approach. We shall also once more contrast this attempt to view statistical mechanics as a branch of general inductive reasoning with the more orthodox attempts to see it as a physical theory acceptable only on empirical grounds and crying out for an explanation of its success.

IV. Further readings

For the classic treatment of equilibrium ensembles, see Gibbs (1960). Balescu (1975) and Munster (1969) are typical modern treatments.

For the generalized panta-micro-canonical ensembles, see Truesdell (1961).

The classic discussion of the Ergodic Hypothesis is Ehrenfest and Ehrenfest (1959). For the refutations of the Ergodic Hypothesis, see Rosenthal (1913) and Plancherel (1913). The discussion in Farquhar (1964), pp. 76–77, is illuminating.

For the Quasi-Ergodic Hypothesis, see Farquhar (1964), pp. 77–78, and the references cited there.

Tolman's "brute posit" approach is in Tolman (1938).

For Khinchin's use of the large numbers of degrees of freedom and the laws of large numbers to evade the need for ergodic posits, see Khinchin (1949). For comments, see Truesdell (1961).

Farquhar (1964) is an excellent work on the older ergodic theory (that is, prior to the work of Sinai et al.). Part I of Jancel (1968) is also very clear and thorough. A briefer sketch is Truesdell (1961).

On contemporary ergodic theory, two very brief sketches are the appendices to Jancel (1969) and Balescu (1975). Arnold and Avez (1968) is an elegant and mathematically sophisticated precis of the basics. Sinai (1976) is a good introduction for those with some mathematical sophistication. Cornfeld, Fomin, and Sinai (1982) is a comprehensive treatise on the mathematics with physical applications in mind.

A brief sketch of the KAM theorem and its importance is in Arnold and Avez (1968), Chapter 4. A more extensive discussion including the application of the results to problems in dynamical astronomy is Moser (1973).

For an elegant sketch of the overall problem area, including the limitations on ergodicity as an ultimate rationalizer for the equilibrium problem for realistic systems, see Wightman (1985).

For the observations of some philosophers on ergodic theory and its explanatory role, see Sklar (1973) and Malament and Zabell (1980). Other philosophical pieces that should be consulted are Quay (1978), Friedman (1976), and Lavis (1977).

For the mathematical connections of topology and measure theory, see Oxtoby (1971). A thoughtful philosophical examination of topology as a foundation for equilibrium theory is in Guttman (1991).

On "objective Bayesianism" as the basis of equilibrium theory, see Jaynes (1983), especially the pieces called "Information Theory and Statistical Mechanics." Hobson (1971) explains the Jaynes approach. Another sympathetic book is Katz (1967). For a criticism alleging an incoherence in Jaynes' approach, see Friedman and Shimony (1971), Dias and Shimony (1981), and Shimony (1985). For other criticisms and Jaynes' response, see Jaynes (1983), especially the section titled "Where Do We Stand on Maximum Entropy?"

For Shannon's original definition of information-theoretic entropy, see Shannon (1949).

6

Describing non-equilibrium

I. The aims of the non-equilibrium theory

What results might we want to obtain from a statistical mechanical theory of non-equilibrium? Even a brief itemization of our ability to describe the non-equilibrium situation in macroscopic terms will be sufficient to indicate the breadth and depth of the results we would like to underpin with a statistical, microscopic theory.

It is an experimental fact that many non-equilibrium states of matter (and radiation) can be characterized in terms of a small number of field parameters – that is, assignments of values of a physical quantity by means of a function from locations in the system to numerical values. In general, these parameters are derived from those we use to characterize the equilibrium situation by a kind of generalization. Thus, such kinematic and dynamical quantities as energy, pressure, and volume are carried over, except that an intensive quantity like pressure now becomes local pressure at a point. And such purely thermodynamic quantities as temperature and entropy are generalized to local temperature and local entropy production. So the macroscopic situations with which we expect to be able to deal, at least most straightforwardly, in statistical mechanics are those of systems that are either close to equilibrium or that, although far from equilibrium, are such that they can be described in terms borrowed from equilibrium theory in small enough spatio-temporal regions of them. That is, although the system as a whole is far from an equilibrium state, a small enough part of it can be considered, for a short enough time, as if it were in a temporary, local equilibrium condition.

The states so described are in dynamic transition. And we discover experimentally that we can frequently find laws governing the dynamic evolution of the systems. These range from the simple heat transport equation relating heat flow to temperature gradient, to the hierarchy of hydrothermodynamic equations that describe fluid flows under the influence both of mechanical and thermal variation.

Typically entering into these equations governing dynamical evolution are a number of important constants, the so-called transport coefficients.

196

Typical examples are thermal conductivity, viscosity, and the coefficients of material diffusion appearing in diffusion equations. These have definite values fixed by the constitution of the system.

In the situation where a near equilibrium system is evolving to equilibrium, we frequently discover important "reciprocity" relations among various flows. Crudely, the effect of thermal gradients on momentum transfer, for example, will be matched by a reciprocal influence of velocity gradients on heat flow.

Next, there are the dramatic experimental facts revealing the existence, in systems far from equilibrium, of steady-state flows. Such phenomena as hexagonal flow cells in fluids maintained under thermal gradients, chemical oscillations in mixtures of appropriate chemicals, and so on fall under the descriptive capacity of macroscopic non-equilibrium thermodynamics. There is also the fact that these steady-state systems can exist in multiple phases, with transitions between them at critical values of the thermodynamic quantities. This interesting parallel to equilibrium phases and phase-transitions also requires explanation in our statistical, micro-dynamical account.

Finally, there is the general feature of the time-asymmetry or irreversibility of non-equilibrium processes. This is the most important qualitative fact of all about the non-equilibrium situation, the temporally one-sided drive of non-equilibrium systems to "goal" equilibrium states. There is reason to believe that we can extend the role of entropy, which appears in thermostatics as the marker of which equilibrium systems can evolve upon change of constraints into which others, into a field-like notion. Here, we talk of the entropy in a region of a system at a time, and try to characterize the unidirectional change of the system by a principle of local entropic increase or production as summarized by something on the order of the Clausius–Duhem inequality. Under special circumstance we seem to be able to characterize the dynamics of evolution by such further general principles as that the local entropic production will be minimal for the actually observed dynamic flow.

Given this rich and fruitful ability to macroscopically characterize the non-equilibrium situation, what can we hope to do in the way of accounting for its existence and success by underpinning this macroscopic account with a statistical mechanics of the microscopic dynamics?

To begin with, what resources will be available to us in our microscopic account? Certainly the underlying dynamics of the microscopic constituents, either classical or quantal. Certainly also the constitution of the system out of its microscopic constitutions including the specific laws of interaction between them – for example, the specific form of the intermolecular potential in the case of the fluid constituted out of molecules.

But we also expect that we will need to invoke something beyond these two fundamental components of theory. We will need to be able to specify the macroscopic constraints imposed upon a system, such as the volume to which a gas is constrained both initially and as it evolves. And, we expect, we will need to invoke some principles for determining the "probability" that a system subject to a specific macroscopic constraint will be found to have one of the infinite possibilities of a particular micro-state compatible with that macroscopic constraint. We also expect, and we shall see how true this is, that many of the most fundamental issues in grounding the non-equilibrium theory will rest upon the rules for assigning such a probability distribution and the explanatory grounds on which such an assignment can be justified.

What would we like to be able to derive from these basic components of the statistical micro-theory? We would like to be able to specify in the grounding theory's terms just what conditions are necessary and sufficient for there to be a description of the system in a small number of macroscopic parameters. And we would like to be able to determine just what these parameters will be when such a "reduced description" exists. Further, we would like an explanatory account of just why systems of the requisite kind are encompassible in the macroscopic-thermodynamic mode.

We would like a characterization, in the grounding theory's terms, of just when a dynamical evolution equation is possible and a derivation from the underlying theory of the structure of the macroscopic evolutionary equation. Insofar as the evolutionary equation is formulated in terms of constants such as the transfer coefficients, we would like to be able to derive the values of these coefficients from the grounding theory. Where reciprocity relations among transport phenomena exist, we would like to be able to specify the necessary and sufficient conditions for their existence in the statistical-microscopic theory's terms.

Given a statistical-microscopic theory of dynamical evolution we would ideally like to be able to derive a general solution of the dynamical equation of evolution. And we would like to be able to relate this general solution both to the nature of the macroscopic evolution equations and to their most general solutions. Insofar as the dynamics results in multiple phases of stable steady-state flow systems far from equilibrium, we would like to be able to derive this complex phase picture from the fundamental theory as well.

To the degree that these derivations of the equations of evolution, the values of their parameters, or the nature of their solutions require the positing of principles over and above the basic laws of micro-dynamics and the constitutive structure of the system, such as principles governing the distribution of probability over initial conditions compatible with macroscopic constraints, we would like, once again, an explanation of why we can with success make such posits.

Finally, we would like a characterization in terms of the statistical-microscopic theory of the all-important time-asymmetry of macroscopic phenomena. Insofar as this can be characterized in terms of entropic non-decrease of final over initial equilibrium state, we will want a characterization of equilibrium entropy in the grounding theory, and a statement of the conditions necessary and sufficient to guarantee its non-decrease. Insofar as we can go beyond this to a non-equilibrium entropy, perhaps of the field variety, and to a principle of entropic non-decrease for this generalized entropy, perhaps of the Clausius–Duhem inequality form, we will want a grounding of this generalized notion of entropy as well.

Here, once again, it will not be enough to state the conditions necessary to guarantee the existence of an entropy and to guarantee its asymmetric behavior in time. We would like, in addition, an explanatory account of the reason why such posits as are necessary to gain these results are true, at least to the degree that the posits outrun our presupposed laws of microdynamics and constitutive facts about the system.

II. General features of the ensemble approach

1. Non-equilibrium theory as the dynamics of ensembles

The general format our statistical-microscopic theory ought to take seems clear. Macroscopically, a system is prepared in a state that is unstable and that evolves into distinct states in a progression toward a state that is macroscopically unchanging, the equilibrium state. But the development of kinetic theory through the responses to early critiques makes it clear that we must understand this experimental situation in a probabilistic vein. The initial preparation of the system is compatible with many microscopic initial states of the system. The evolution of the system is determined by its initial microscopic state and the underlying dynamical laws of evolution. The macroscopically discerned approach to equilibrium must be understood in terms of some "most probable" evolution or evolution of "most probable" states of the system in question.

The basic formal schema to represent this is also clear. We use the Γ-space, the phase space in which points represent the entire microscopic dynamical state of the system. The initial macroscopic preparation restricts our attention to a limited region of this phase space. We impose a probability distribution over the set of micro-states compatible with this initial constraint. Then we look to see how such an initial "ensemble" evolves as each point of it follows the trajectory determined by the dynamical equations. Notice that this evolution is fixed once the initial ensemble is chosen, at least if we are working in the standard picture of deterministic dynamical laws of evolution and genuinely isolated systems free of intervention from outside the system as time goes on. We will reflect upon alternatives to the standard approach in Chapter 7,III,2.

From the ensemble at any time we will calculate the quantity that we associate with some measured macroscopic feature of the system. The procedure will involve, of course, some appropriate averaging done with the probability distribution over the ensemble. The ultimate aim will be to derive some equation in the macroscopic quantities that described the monotonic approach to equilibrium through time of the idealized system of the macroscopic approach. This derivation will often run through a generalization of the Boltzmann equation, which we will call a kinetic equation, and which extracts from the ensemble evolution some feature of it that has the appropriate time-asymmetric, evolutionary quality.

Here, it is worthwhile making a couple of general observations that hold irrespective of the details of the theoretical construction and independently even of the choice of foundational grounding for the theory we adopt. One observation concerns the recurrence problem. In both classical and quantum statistical mechanics, the systems investigated are usually such that the Poincaré Recurrence Theorem holds. Thus we can expect any individual system in the ensemble kept isolated for a sufficiently long time to recur to within any small degree of approximation we choose to the initial micro-state in which it started. The exceptions to this rule will constitute at most a set of measure zero of systems in our usual probability measure. But, at least in the case where our underlying dynamics is classical mechanics, the recurrence of each individual system in the ensemble is perfectly compatible with the ensemble as a whole showing a monotonic behavior without any semblance of recurrence. This is due to the "incoherence" among the recurrences of the individual system. Unless they were synchronized to some degree as to the times at which they occur, their individual "almost return" to a state at a given initial time is compatible with the ensemble never showing such an "almost return" to its initial configuration.

It is interesting that this argument fails in a statistical mechanics based on underlying quantum mechanics. In that case, the individual recurrence is matched by an ensemble recurrence, at least if the system in question has only a finite number of degrees of freedom, which, for particle systems, means a finite number of microscopic components. In the quantum case, one can only avoid recurrence for the ensemble by making the further idealization to a system with an infinite number of component parts – by going to the thermodynamic limit. Although the use of the thermodynamic limit is essential in classical statistical mechanics for other reasons, it is interesting that it must be invoked at a deeper level in the quantum context.

Next, we must be careful in the way we understand the use of the ensemble evolution to generate a time-asymmetric kinetic equation. Here we must remember the careful explication of the probabilistic interpretation of the Boltzmann equation given by the Ehrenfests. The solution to the kinetic equation with its asymmetric, monotonic behavior in time is usually not supposed to represent the "most probable" pattern of evolution of an individual system. We know that cannot be the case because all except exceptional systems will show non-monotonic recurrence behavior. Rather, the usual picture is this: We start at a time with the ensemble of systems in their initial state. At any specific later time, we look at the state achieved by each system in the ensemble. We identify the state achieved at that time by the overwhelmingly greatest number of systems. We plot the succession of such dominant states. It is this "concentration curve" of the ensemble evolution that is supposed to be represented by the solution to the kinetic equation. The monotonic, time-asymmetric behavior of this "concentration curve" is perfectly compatible, once again, with each individual system in the ensemble showing recurrence behavior, so long as the recurrences of the individual systems in the ensemble are properly uncoordinated with one another. But, we should note, some programs for rationalizing the non-equilibrium theory will in fact propose the alternative picture of the statistical theory as describing "most probable" evolutions of systems.

We should also note the role played by the large number of component micro-systems of a given system, even in the classical case. To complete our account of why systems show apparently lawlike evolutionary behavior of their macroscopic parameters, we would like to show that the probability of these parameters having at any time the value they do have for an individual system is highly concentrated around whatever value for that parameter we calculate for that time using our ensemble and the appropriate means of going from ensemble structure at the time to parameter value. Here, of course, it will be arguments of the Khinchin variety that are either implicitly or explicitly brought to bear. Again, there will be alternative rationales that utilize the large numbers of degrees of freedom of the system in rather different ways.

The Khinchin argument rests upon the specific way in which the parameter value is calculated from the ensemble. For example, there is the assumption that the quantities in question will be appropriate "sum functions" of the values of the dynamical state of the microscopic components, or some appropriately well-behaved generalization of a sum function. Here, one must rely upon the specific nature of the inter-component interaction and the results obtainable through the study of the thermodynamic limit in the rigorous theory. As in the equilibrium case, one would have to give a proof to the effect that if the intercomponent interaction is sufficiently well behaved, then, in the thermodynamic limit of the number of components and size of the system going to infinity but with density held constant, the use of law of large number theorems will allow us to prove that for the types of functions derived from the ensemble that we use to calculate observed macroscopic parameter values, the probability distribution for the parameter value calculated for an individual system in the ensemble will in fact cluster overwhelmingly around the mean value that we calculate from the ensemble. It is this argument, not usually rigorously carried through but, instead, presupposed, that will justify our assumption that the calculated evolution of the parameter will in fact represent the "overwhelmingly probable evolution" of the parameter, in the sense of "probable evolution" derived from the Ehrenfests and just outlined.

2. Initial ensembles and dynamical laws

The methodology for non-equilibrium statistical mechanics that has been outlined then comes down to this general procedure: First define an initial ensemble, that is an initial region of points in the Γ-space phase space for the system along with a probability distribution over the points in that region. Follow the evolution of that ensemble as determined by the trajectories followed by each point in the initial ensemble determined by the dynamic laws of evolution of the mechanics of the micro-components and their specific interactions. Derive from this general ensemble evolution the evolution of some function computable from the ensemble at any time, and show that this function evolves according to an equation that has the proper time-asymmetry and monotonicity to represent approach to equilibrium. That is, derive a kinetic equation. Use the kinetic equation and the appropriate association of ensemble defined quantities with macroscopic observables to derive from the kinetic equation and its solutions the appropriate macroscopic equations of evolution that represent the experimentally observed non-equilibrium behavior.

But, alas, carrying out this general procedure is very far from straightforward. The difficulties are basically two-fold. First, there is the difficulty of coming up with a general theory of initial ensembles. Second, there is the important fact that the actual derivation of kinetic equations is done in practice not by inferring them from the determined evolution of an initial ensemble, but instead by invoking some principle of evolutionary "chaos," or "rerandomization" which is the generalization of the Hypothesis of Molecular Chaos, whose role in deriving the Boltzmann equation was so clearly delineated by the Ehrenfests.

In order to characterize general initial ensembles, the most common tack has been to seek generalizations of the familiar micro-canonical and canonical ensembles of equilibrium theory. The generalization of the micro-canonical ensemble is presumably the one to apply in cases of isolated systems and the generalized canonical in the non-equilibrium situations that generalize a system in thermal contact with a constant temperature heat bath. Actually, the generalization of the canonical is often used for its computational simplicity, with a "thermodynamic limit" type argument invoked to rationalize this because the results of microcanonical and canonical will agree in that limit.

To characterize an ensemble, we need to pick a set of macroscopic parameters first, and then, relative to them, assign an appropriate probability distribution over the points in phase space compatible with the assigned values (or probability distributions) over the selected macroscopic quantities. But how are the macroscopic quantities to be determined? There are basically two approaches. One relies upon the observational-experimental facts we have about non-equilibrium systems. The other relies upon an anticipation of how, in statistical mechanical theory, macroscopic quantities are to be associated with statistical quantities.

The first approach requires some insight into what macroscopic features of a system can be experimentally determined. Here, it is not some general theory of this that is being used, but rather our experience that such quantities as local density, local pressure, local temperature, system total energy, and so on are determinable by us in a wide range of cases of interest. Another closely related "empirical" approach relies upon our observational experience that tells us that in many cases we can determine, at the macroscopic level, dynamical equations of evolution that describe the lawlike behavior of many systems. The macroscopic parameters that we ought to use to characterize the relevant initial non-equilibrium ensembles, then, are those that enter into these laws of evolution on the macroscopic scale.

The alternative approach to characterization of appropriate macroscopic parameters for initial ensembles relies on the observation of the way in which, in typical cases, information from the ensemble is extracted to define a macroscopic parameter in the standard procedure that associates statistically defined quantities with thermodynamic quantities. The function $f_n(q_1 \ldots q_n, p_1 \ldots p_n)$ is the function that gives, for the ensemble at a time, the full set of all conditional probabilities about locations of particles in position and momentum ranges. That is, it will contain such information as "the probability of finding a particle in the range $q + \Delta q$, and the range $p + \Delta p$ given the totality of information about the locations of all other $n - 1$ particles." This function is mathematically equivalent to the probability density function over the points in the allowed phase-space. The function f_1, on the other hand, is just the function that allows us to compute the overall probability of finding a particle in a region. It is obtained by "summing over" (actually, integrating over) all the probabilities that the particle is in that region conditional on each possible location for the other particles. It contains much less information than f_n, just as an average of a list of numbers contains much less information than the list. Similarly, f_2 contains only probabilities conditional on the location of one other particle; etc. Each f_{k+1} contains more information than f_k but less than f_{k+2}.

The point here is that in general, macroscopic parameters will be definable by using only f_1, or at most the f's, up to some small subscript number. Local density, for example, is clearly only dependent on f_1. The suggestion is, then, that we take as macroscopic parameters for characterizing initial ensembles only those quantities that are to be associated in our statistical theory with quantities definable by f_i, where i is less than n, the total number of particles. If one goes to the infinite system limit idealization, then all f's with finite subscripts would presumably be admissible. Of course, this gives a very abstract characterization of macroscopic parameter that may be useful for a general theoretical study of initial ensembles. In practice, to study systems we must resort to our empirical experience as noted earlier in deciding which macroscopic parameters, with their particular association with an f_i, are to be used. We shall return to these issues and others of a similar sort only touched on briefly here when we discuss the "reducibility" of thermodynamics to statistical mechanics in Chapter 9.

Suppose the macroscopic parameters have been chosen, and their values, or ranges over values (or a probability distribution over values as in the generalization of the canonical ensemble), have also been chosen. How, relative to these choices, should we then distribute probability over the allowed phase points? Here, the usual tack is to generalize from the equilibrium case and distribute probability "as uniformly as possible." In the generalization of the micro-canonical ensemble this means picking the probability distribution function to be uniform with respect to the appropriate phase-space measure. In the case of the generalized canonical distribution it is the appropriate generalization of Gibbs's canonical distribution. In both cases, an appropriate entropy function is maximized, and one can characterize the probability distribution as one with "minimal correlation" above the specification of the macroscopic parameters. In other words, although the ensemble is not, in one sense, uniform over the allowed phase space, once the constraints have been established that allow it to move toward its eventual equilibrium, we take the ensemble to be as uniform as it can be subject to our initial constraint.

We can see how this works in a very typical case, a system initially in equilibrium that becomes non-equilibrium because of the instantaneous change of a constraint. For example, consider a gas initially confined to the left-hand side of a box just after the partition has been removed. Our assumption that the system was in equilibrium when the partition was removed tells us that our initial ensemble should have probability distributed uniformly over that part of the phase space available before the partition was removed. Of course, after the partition is removed it is not uniform over the newly enlarged available phase space.

In practice, two kinds of initial ensemble are considered. One kind is that described, where a system that was in equilibrium no longer is so because of constraint change. The other is the local equilibrium ensemble. This is typically used for cases such as a system with an initial temperature distribution, density distribution, and so on. Here, the uniformity assumption is utilized in the assumption that the appropriate initial ensemble is the "local equilibrium" ensemble. This ensemble is the ensemble generated by taking the distribution of probability in regions of phase space corresponding to small enough portions of the system to conform to the equilibrium distribution appropriate to the values of the field quantities (local density, local temperature, and so on) associated with that region of the system. Many non-equilibrium systems are empirically discovered to be such that even if they are not local equilibrium systems to begin with – that is, they are in initial stages where even local specification of thermodynamic quantities is impossible – they very quickly "relax" into non-equilibrium systems where at least local equilibrium is a justifiable assumption. We then expect the local equilibrium version of the initial ensemble to do justice to describing their evolution from that point on.

The justification for making such a choice of "minimal correlation," and the explanation as to why it is so successful in those cases where it is, are of course going to be fundamental questions requiring foundational investigation as we proceed.

Now it would seem that having specified our initial ensemble our work is done. For the evolution of the ensemble is completely fixed by the evolution of each of the systems in the collection. And their evolution is completely determined by the micro-dynamics. At least this is so if we stick to the standard approach that introduces no intervention from outside the system and no purely stochastic underlying dynamical laws.

But the way in which kinetic equations are usually derived is not by a study of the effects of dynamical evolution on the initial ensemble. Instead, use is made of an apparently independent statistical assumption, one generalizing the Hypothesis of Molecular Chaos. We will explore the detailed structure of such hypotheses in the next few sections. Here we need only clarify their general nature. These hypotheses demand that at each moment of the evolution of the ensemble, the future development of the system can be obtained by assuming that probability over phase points is uniform *with respect to the values of the macroscopic parameters obtained up to that point.* That is, if the ensemble has developed in such a way that at a given time the parameters have a specified value, we can predict its future evolution as if we were starting with a random system having those parameter values. The past history of the system – how it got to the point it is at – is taken to be irrelevant for our future statistical predictions.

But it is not at all clear that such a "rerandomization" posit is even consistent with the underlying dynamical equations of evolution that fully determine the actual evolution of the ensemble from its initial condition. We shall see two approaches to resolving this problem, and we shall see each approach introduce its own need for foundational grounding.

One approach attempts to derive the legitimacy of the rerandomizing posit, or of its result, the kinetic equation, strictly from some posit about the initial randomness of the ensemble and the strict laws of dynamic evolution that govern each member of the ensemble. The other approach tries to show, from the laws of dynamic evolution alone, and without use of a posit about the nature of the initial ensemble, that in some appropriate sense the rerandomization posit is legitimate. This approach sometimes works by trying to prove that a kind of rerandomization (or, rather, its consequences) must hold "in the limit as time goes to infinity." In other cases, it proceeds by trying to show that even for finite times the rerandomization will be legitimate, at least with respect to consequences drawn from it for *some* macroscopic variables. How these procedures work in detail we shall explore in Chapter 7.

In the former kind of rationalization of rerandomizing, we are faced with the problem of justifying our new posit of sufficient randomness for the initial ensemble and of explaining why it holds when it does. In the latter kind of justification of rerandomization, there will be questions of several kinds. One series of questions arises out of the fact that there are severe limits on the conditions under which the results hold, limits that exclude most realistic systems from their scope. This is because of the KAM Theorem discussed in Chapter 5. Even when the results do hold, there remain issues that can only be resolved by once again facing the problem of the nature of the original ensemble, its justification, and the explanation of why it is the way it is. This is so because the results obtained from the dynamical evolution equations alone will not allow us to infer the kind of specific behavior with regard to characterizations of the system in specific parametric terms that we would like to justify. That is, these characterizations won't allow us to show that the system approaches equilibrium at a specific rate or in a specific manner when the system is characterized to a given degree of accuracy in certain chosen macroscopic terms. All of these issues will again be explored in detail in Chapter 7.

III. Approaches to the derivation of kinetic behavior

1. The kinetic theory approach

The first scheme for dealing with non-equilibrium that we will outline is one particularly applicable to the first cases treated by the statistical mechanical methods, systems composed of large numbers of molecules that are generally in free motion but that interact by means of an inter-molecular potential. Here, the aim is to go beyond the dilute gas considered in the derivation of the Boltzmann equation and to look for a general procedure by which the Boltzmann equation and its appropriate generalizations for the other cases (moderately dense gases, plasmas, and so on) can be derived. The aim is not just the practical one of deriving kinetic equations for the more difficult cases, but also of sorting out and making explicit just exactly what rerandomizing type statistical assumptions are needed in each case to derive the desired kinetic equation result. This is the first stage of what needs to be done if one is to ultimately justify or explain or otherwise rationalize the introduction of the rerandomizing posits.

A broadly applicable schematism is called the BBGKY approach, after Bogolyubov, Born, Green, Kirkwood, and Yvon, all of whom offered systematizations in this vein. Here, the formal scheme is usually accompanied by an informal rationale for it. This doesn't constitute a real justification from first principles for each move in the program, but rather a series of suggestions as to how to interpret the moves physically and a series of hints as to what an ultimately satisfying rationalization would require. Because Bogolyubov's account of these matters is usually thought the most enlightening, I will present the scheme and its explication from his point of view.

We look at the total set of "partial distribution functions" for the system. The function $f_n(p_1 \ldots p_n, q_1 \ldots q_n)$ is the probability density function that gives all the information about the probabilities of a particle being found in a particular region of position and momentum space given the distribution of all the remaining $n - 1$ particles. It is the complete description of the ensemble distribution, and giving this function is equivalent to stipulating the distribution function over the allowed region of Γ-space. The reduced functions $f_1, f_2, \ldots, f_{n-1}$ are obtained by

integrating over remaining variables, in a manner exactly analogous to
the way in which one can obtain the total probability of a 1 coming up
on one pair of dice by adding the probability of 1 on that die given a 1
on the other die, plus the probability of 1 on that die given a 2 on the
other die, and so on.

If we make some basic assumptions about the symmetry of the system,
plus the assumption that the intermolecular force is determined strictly
on a pairwise basis (that is, the force of one molecule on another is
dependent only on their separation, not on the location of any other
molecule), then we can obtain from the Liouville equation, the dynamical equation giving the evolution of f_n as determined by the underlying
microdynamics, a hierarchy of equations that give the evolution in time
of each of the partial distribution functions f_s. The equation giving the
rate of change of f_s will have on its right-hand side only components
depending on f_s and on f_{s+1}. If we could solve these equations individually, we would then have a route to deriving a kinetic equation.

Why? Most thermodynamic quantities we wish to calculate in the usual
way from the ensemble will depend only upon f_i's of low order. For
example, knowing f_1 will allow us to calculate our expected density for
the gas in local regions, and tracking the evolution of f_1 will allow us
to track the evolution of the density distribution of the gas as it goes
to equilibrium. For calculating transfer coefficients, f_2 will generally be
enough.

Furthermore, we expect the kinetic equation to be an equation for the
evolution of some such low-order partial distribution function. In
Boltzmann's original formulation, his evolutionary equation governed f,
the μ-space distribution function for the distribution of a single molecule
in a gas. In the ensemble transformation of the Boltzmann equation, it is
f_1, the probability derived from the ensemble of finding a molecule in a
specified region of position and momentum, that we expect to be the
proper surrogate for Boltzmann's f. So if we could generate a closed
equation for f_1 in the hierarchy described, we would hope that given the
right structural conditions for the gas (such as sufficient diluteness,
neglectability of the influence of the walls of the container on the gas,
and so on), this equation would turn out to be the Boltzmann equation,
now for f_1 instead of for f.

But the equations for the lower order f_i's are not closed. The equation
for f_1 requires knowing f_2, and generally the equation for f_i contains f_{i+1},
leaving us ultimately with only one closed equation, that for f_n, the
solution of which would require tracing the exact trajectory from every
possible initial condition. But that is exactly what we want to avoid in
statistical mechanics.

The trick is to close the lowest order equations by replacing, for example, f_2 in the equation for f_1 by some functional of f_1 alone. As we shall see in Sections 2 and 3, each of the approaches to describing non-equilibrium follows this general scheme of (1) finding a partial description to the ensemble that is adequate to allow us to calculate the macroscopic quantities we are interested in, and then (2) trying to formulate a closed dynamics for this partial description so that its evolution can be prescribed independently of the remaining, discarded, information in the ensemble. Bogolyubov's assumption is that we can distinguish three "time scales" in the evolution of the gas. The shortest is the time of the magnitude of the average time between collisions for the molecules. Here it is assumed that the f_i's can be arbitrary and can vary very rapidly. Next, there is the kinetic time scale. Here the assumption is that correlations among two particles vanish rapidly with the separation of the particles. So that if we are dealing with time scales large with respect to the collision time (but still small with regard to the time scale of the macroscopically observable change of the thermodynamic features of the gas), we can assume that f_2, for example, is reducible to just $f_1 \times f_1$ – that is, that the probability of finding molecule 1 in a region given that molecule 2 is in a region is just the product of the probabilities that each molecule is in the respective region when the molecules are not very near to collision. Similarly, it is assumed that on this kinetic time scale the higher order f_i's reduce to appropriate products of f_1.

Finally, there is the still longer hydrodynamic time scale, the order of magnitude of which is that over which macroscopic changes in the thermodynamic characterization of the gas (variation in its local density, for example) can be observed. The assumption of these three time scales here is a formalization of the idea that even if a gas starts in a wild and singular condition, it will, in a magnitude of time long with respect to the collision time but short on the hydrodynamic time scale, go to a condition that is at least of the locally equilibrium kind.

Next, Bogolyubov proposes that we can study the solution of the now closed f_1 equation by expanding the solutions in terms of powers of the concentration of the gas. Here he relies upon very successful techniques from equilibrium theory that derive the ideal gas constitutive equation in low-density limits and then derive modifications to it by treating the solution as expandable in a series the terms of which represent corrections that become successively important as the density of the gas is increased. Making the appropriate expansionary assumption and going to the dilute gas limit, one can convert the closed equation for f_1 into the ensemble version of the Boltzmann equation.

Alas, the derivation of the equation at this stage is hard to make rigorous, because subsequent investigation showed that the higher order terms in the density expansion behave very badly. This bad behavior has also made it very difficult to try and use the BBGKY technique to derive the hoped-for generalizations of the Boltzmann kinetic equation for gases denser than those for which the Boltzmann equation works. When one allows, for example, collisions in which three molecules become close in position and momentum space, many intractable correlation effects come into play. This is because the presence of one molecule effects the likelihood of the presence in various conditions of other molecules. The hope for a non-equilibrium expansion technique to parallel the very successful density expansion techniques applied in equilibrium theory turned out to be overly optimistic.

From our point of view, the crucial step in the derivation is of course the replacement of f_2 by $f_1 \times f_1$. This is the variant of the Hypothesis of Molecular Chaos in this approach to the derivation of the kinetic equation. It is the source of the closed kinetic equation and the origin of its time-asymmetric nature and the time-asymmetry of its solutions. It is this step that cries out for rigorous understanding. Exactly when is it legitimate to make such an assumption? And why is it legitimate when it is?

2. The master equation approach

The master equation schema for dealing with non-equilibrium is applicable in situations where the system can be considered to be composed of a large number of component systems that are coupled to one another only weakly. The energy function for the system then takes the form $H_0 + \varepsilon H_1$, with ε a small number representing the weak coupling. Typical examples might be gases that are nearly ideal, simple harmonic oscillators coupled by a weak energetic interaction or, a common example, radiation confined in a cavity in which the different modes (frequencies) of radiation are coupled to one another weakly by a "dust mote" that weakly absorbs and radiates energy in a range of frequencies.

Here the basic question is how the available energy is distributed over the array of almost, but not quite, independent modes. Thus, for example, we might start with radiation distributed over a range of frequencies in varying amounts in a cavity, and ask how, over time, the frequency distribution changes as radiation is absorbed and emitted by the dust mote.

The fundamental assumption is that the probability of energy leaving a region of modes will be proportional to the spread of that region, and that there will be a constant factor that gives the probability of energy from that one mode making the transition in a unit of time to any other region in question. A similar assumption is made about the probability of energy entering a region of modes from other modes as well. The net result is an equation that looks like this:

$$\frac{dP_t(\xi)}{dt} = \int [\omega(\xi,\xi')P_t(\xi') - \omega(\xi',\xi)P_t(\xi)]d\xi'$$

This says, essentially, that the distribution of energy changes over time in the following way: In an infinitesimal time, an amount of energy enters a mode depending upon what modes are occupied to what degree at that time and the constant fraction of these that will transit to the designated mode. And an amount leaves the mode depending upon how occupied it is and the constant probability of a system in that designated mode transiting to each of the other available modes.

Once again this is a kinetic equation that tells us that the change of the system at a time is determined solely by the condition of the system at that time. The past history by which the system arrived at its current state is irrelevant in determining its future evolution. As usual, all of this is to be understood, of course, as a kinetic equation for the ensemble, not for an individual system. The equation has, as usual, the appropriate time-asymmetry built in and, again as usual and as is desired, the appropriate "attractor" equilibrium solution as its only stationary solution.

But why should we believe the fundamental assumptions upon which the equation rests? Here again, a "picture" can be constructed that provides an understanding of the needed assumptions. But we must be careful in understanding what the picture provides. It makes it clearer to us what the nature of our basic assumption is. It in no way demonstrates to us the consistency of this assumption with the underlying dynamical equations that must govern the actual evolution over time of any chosen initial ensemble. Nor does it tell us what initial ensembles will in fact give rise to these results.

The picture presented is one in which the interaction of the systems is turned on for short periods of time and off for short periods of time of about the same length. In the off periods, it is assumed that the ensemble has time to go to its equilibrium condition. The transitions that actually occur when an individual system has its interaction turned on depend upon the particular condition of all the microscopic variables for each component at that time. Thus, for example, how much energy will flow from a simple harmonic oscillator of one frequency to one of another frequency by means of the weak coupling will depend upon the specific phase of oscillation each oscillator is in when the interaction begins. But the assumption that in the "off" periods the system goes to equilibrium allows us to "randomize" at each new interaction period over all possible initial conditions for the components in each mode. Here we use the usual equilibrium probability distribution for each mode at the time, given the energy in that mode at that time. It is this that allows us to be able to calculate the constants $\omega(\xi,\xi')$ that give the fixed probabilities of transition from one mode to another. Then we think of the off periods as going to zero, because the equation we derive by making the assumption of reequilibriation during each off period results in a transition equation (the one given) that does not contain the length of the off period explicitly.

The close analogy of this procedure to what we saw in the BBGKY approach to the derivation of the Boltzmann equation and its generalizations is clear. In both cases, irreversible approach to equilibrium shows itself by considering a function that reveals only part of the total ensemble probability distribution. In the BBGKY approach, this was f_1, rather than f_n, and in the present case it is the distribution function that restricts its attention to the distribution of energy over modes and ignores the remaining dynamical variables of the component systems. In both cases, a continuing rerandomization posit is used to derive our kinetic equation. In the BBGKY approach, it is the assumption that the f_i's can be construed as products of f_1. In the master equation approach, it is the assumption of a constant probability of transition from mode to mode.

3. The approach using coarse-graining and a Markov process assumption

A very general derivation of a time-asymmetric approach to equilibrium for an ensemble feature relies upon the idea of coarse-graining introduced by Gibbs and explicated so fully by the Ehrenfests. The generality of the method is matched, unfortunately, by the difficulties encountered in bringing it down to earth so as to make it applicable to the derivation of a specific kinetic equation in particular cases. But from the foundational point of view it provides the clearest understanding of what we are doing when we make a rerandomization posit for some partial description of the ensemble derived from the full ensemble description.

Let us look at a situation that is describable in a generalized microcanonical ensemble vein. A system is subject to certain initial constraints. An ensemble is constructed consisting of points in Γ-space that represent all possible micro-states of the system compatible with the initial constraints and a probability distribution over these points. The constraints are changed but the system otherwise remains isolated from the external world. For example, the partition confining the gas to the left-hand side of an insulated box is removed. How does the ensemble of points representing all possible initial micro-states evolve?

We know, by Liouville's Theorem, that its total phase-space volume remains constant. So it certainly can't evolve to the ensemble that would represent equilibrium for the changed constraints. But is there some sense in which it can approximate that equilibrium ensemble ever more closely as time goes on?

The suggestion is to divide the newly expanded phase-space into "boxes," subsets that are small relative to the total size of the now available phase space but that are still large enough to contain a substantial region of phase points. The equilibrium ensemble for the newly modified constraints will be such that if we measure the boxes in the standard measure, the fraction of ensemble points in a coarse-grained box will be proportional to the size of the box. This is because in the microcanonical case using the standard measure, the equilibrium ensemble for the newly fixed constraints is uniform over the newly available phase space. Although the old ensemble can't evolve to become the new ensemble, it can so evolve that the portion of it in each coarse-grained box becomes more and more proportional to the size of the box. In this sense, it can "approach" the new equilibrium ensemble.

As a means of representing the approach to equilibrium we observe for individual systems, the approach is rationalized by the familiar Gibbsian argument to the effect that there are limitations on our ability to macroscopically determine the micro-state of a system. The best we can do is to locate the state as being in some region of phase space, say the regions where the density in some macroscopic region of space is within some small variation from a specific value. An ensemble of systems that had approached the equilibrium ensemble for the new constraint values in our coarse-grained sense would then be, as far as observational determination is concerned, indistinguishable from the genuine equilibrium ensemble relative to the new constraints. This argument is one of some delicacy, and we shall have to return to it in detail in Chapter 7,III,3 when we discuss the rationalization of non-equilibrium approaches. For the moment we will let it ride.

But what guarantee do we have that this coarse-grained approach to equilibrium will even occur? After all, the evolution of the initial ensemble is fully determined by the micro-dynamics governing the trajectory in phase space followed by any system started at any point. What tells us that the distribution represented by proportion of the ensemble occupying a coarse-grained box will, for a given coarse-graining, approach the size of the box, leading to a coarse-grained kind of uniformity of distribution?

Once again, the result can be obtained in a very general way by making a posit about the continual rerandomization of the ensemble. Focus attention on a single coarse-grained box, box i. Imagine an ensemble of phase points distributed over i with the usual uniform microcanonical probability distribution. In a given time a certain proportion of those points will move, governed by the underlying micro-dynamics to another given box, box j. Call w_{ij} the "i to j transition probability." Now

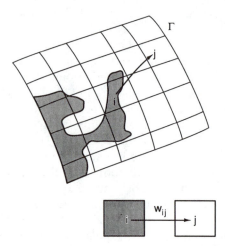

Figure 6-1. The Markovian postulate. The phase space has been coarse-grained into boxes. An ensemble has evolved in the manner illustrated. What is the probability that an ensemble point now located in box i will at the next observation moment be found in box j? The Markovian posit assumes that the conditional probability is a constant w_{ij} that can be found by asking what proportion of phase points uniformly distributed over box i will be in box j at the next observation moment. The probability that one of the given ensemble points will go from i to j in the next interval is, then, just this constant conditional probability of going from i to j multiplied by the fraction of box i filled by the ensemble points at the initial moment.

look at our original evolving ensemble. At a given time, a certain proportion of its points will have occupied box i, following the trajectories from the initial points determined by the underlying micro-dynamics. Assume that in the next time interval the same proportion of these points initially in box i will get to box j as was true in the uniform distribution over box i case – that is, a fraction w_{ij} of the points. With this assumption, the so-called "Markovian" postulate, we can once again prove that the function that describes portions of the ensemble occupying a given coarse-grained box will obey a kinetic equation that has a unique stationary solution, the solution corresponding to a coarse-grained uniformity of box occupation. (See Figure 6-1.)

Using the Markovian posit we can derive an equation analogous to the master equation of the previous section, an equation that equates the change in occupation of a coarse-grained box at a time to a sum composed of two factors: (1) the fixed transition probabilities from box to box acting on the proportion of systems in other boxes, giving the numbers that transit to the designated box, and (2) minus the fixed transition probabilities from the designated box to other boxes acting on the proportion of the designated box that is occupied at a time, giving the number of systems that leave the designated box. Another useful equation derivable is the Chapman–Kolmogorov equation. This tells us that we can compute the proportion of one box that arrives at another box at a later time by picking an intermediate time, taking the probability of transit from the original box to each of the coarse-grained boxes in that intermediate time, multiplying these by the probability of transit from each of these boxes to the designated final box in the remaining time, and adding the results together. So that if we have the fixed transitional probabilities for a unit time, we can calculate transitions for all multiples of that unit time (and we can then go to continuous time limits in familiar ways). Finally, using the Markovian posit we can define a coarse-grained entropy in the manner of Gibbs and the Ehrenfests and prove its monotonic non-decreasing property.

As usual, the assumption consists in the claim that "past history is irrelevant" in figuring out where the ensemble will go next. The region of box i occupied at a time by points in the original ensemble will be a very irregular sub-set of box i. The assumption is that only the amount of box i occupied by these points will matter in determining the likelihood that a point will be in box j at the later time because it got there from box i. How the points got into box i, and what the exact specification of the sub-region of box i occupied by the points is, are taken to be irrelevant. It is assumed that the same portion of points in the irregular region will get into box j at the later time as would the proportion of points in box i distributed uniformly over it – that is, the proportion w_{ij}.

As we shall see in Chapter 7,III,3, this broad approach to construing the approach to equilibrium of an ensemble will lead to many interesting contributions to the attempt at genuinely rationalizing and explaining the results of non-equilibrium statistical mechanics. That further justification is needed is clear for, as usual, it is a deep question remaining open as to whether any rerandomizing assumption is even consistent with the underlying deterministic evolution of the phase points.

IV. General features of the rationalization program for the non-equilibrium theory

The methods for deriving from the ensemble and its evolution an ensemble feature and a kinetic equation that describes the time-asymmetric evolution of that feature to its equilibrium condition are not the only such approaches. But the three techniques we have surveyed are representative enough to do justice to the larger variety of techniques. In the next chapter we will begin a systematic survey of the various attempts to show why the posits made by the various derivations are legitimate, when they are, and how the resulting kinetic equations and their solutions fit into the framework of the underlying dynamical evolution of the systems of the ensemble. Here, I want only to outline the spectrum of problems that must be faced in any such attempt at justification and explanation.

First, we note that each derivation works by focusing our attention on a function that represents only a portion of the information contained in the full ensemble distribution: f_1 in the kinetic theory approach, the distribution of energy among what would be stationary modes were the weak interaction eliminated in the master equation approach, and the proportion of coarse-graining boxes occupied at a time in the coarse-graining approach. The limitation here is supposed to be justified by a demonstration that all of the macroscopic phenomena that we observe and that display time-asymmetric behavior and approach to equilibrium depend only on this sub-component of the full ensemble distribution. A precise demonstration of this is called for, however, and is sometimes elusive – especially in the most general coarse-graining approach to the theory.

Next, we must note that it is a wide variety of results that we would like to obtain, and we must ask which of these results are actually obtainable from the approach in question. We would like to show that in some sense the non-equilibrium ensemble approaches an equilibrium ensemble. But the sense here is fraught with subteties. What exactly is the kind of convergence intended? Is convergence in some infinite time limit enough? We know that thermodynamic approaches to equilibrium are characterized by particular finite times, the so-called relaxation times. Can these be derived from the approach in question? The approach to equilibrium described by the thermodynamic laws is a monotonic approach to equilibrium. Can monotonicity be derived, or can the approach only give us convergence in "the long run" without guaranteeing the absence of "reversals" in finite time periods? The kinetic equation specifies a specific route for approach to equilibrium. Can our account not only derive the approach and the monotonic approach to equilibrium, but also allow us to derive the full, specific route we thermodynamically observe?

Each of the derivations depends essentially upon making a fundamental rerandomization assumption: that the f_i's are products of f_1's in the kinetic theory approach, that the transition probabilities from mode to mode are constant in the master equation approach, and that the box to box transition probabilities are constant and derivable from the uniform distribution over a box in the coarse-graining approach. To what extent can we show that such rerandomizing posits are in fact consistent with the evolution of the ensemble imposed upon it by the deterministic underlying micro-dynamical laws? To what extent is such a consistency derivation obtainable by utilizing the structure of the dynamical laws themselves? To what extent must we also invoke some special initial ensemble distribution to justify the introduction of the posit? If we do need to introduce such a special initial ensemble state, how can we explain why such a posit of initial states for the ensemble is so successful in giving us reliable probabilistic predictions? Or is the aim not so much to justify the rerandomization posit as to use the underlying dynamics and perhaps an initial constraint upon ensembles, to be able to bypass the rerandomization posit and derive directly the results obtained through its introduction?

As we shall see in Chapter 7, the program just described also involves discovering the *limits* of the legitimacy of the rerandomization posits. The equations derived by their use are frequently wrong in their predictions. Where a monotonic approach to equilibrium is expected, for example, surprising periodicity among non-equilibrium states is sometimes found. Just what determines the limits that govern those situations where the posits are applicable and the situations where their use leads to error?

Finally, there is that most famous foundational puzzle in statistical mechanics: How can we possibly reconcile the results obtained by any of these methods with their seeming demonstration of a fundamental time-asymmetry in the behavior of systems, with the basic time-reversal invariance of the underlying dynamical laws? Do the procedures for deriving kinetic equations and the approach to equilibrium really generate fundamentally time-asymmetric results? If they don't, how can these results really do the job we wanted of serving to characterize the time-asymmetric thermodynamic features of the world? Where, in that case, does time-asymmetry come in? If they really do result in time-asymmetric results, what was the trick that allowed them to do so? And what justifies *that* component of our fundamental theory? It is to all of these questions that we now turn.

V. Further readings

Some general remarks on non-equilibrium theory and its varieties are in O. Penrose (1979), Section 3.1. Jancel (1963) covers most of the standard formalisms. Zubarev (1974) treats the issues in an abstract and enlightening manner.

The basic elements of the kinetic theory approach through the BBGKY hierarchy are outlined in O. Penrose (1979), Section 3.2. Bogolyubov's original work is in Bogolyubov (1962). Jancel (1969) covers several approaches in this vein in Part II, Chapter V, Section V. Chapter 7 of Kreuzer (1981) is also a clear source.

The master equation approach goes back to Pauli (1928), p. 30. O. Penrose (1979) summarizes more recent work in Section 3.3. Jancel (1963) covers this approach in Part II, Chapter V, Section VI. Kreuzer (1981) outlines the master equation approach in Chapter 10.

A very clear early version of the approach using coarse-graining and the Markov postulate is van Kampen (1962). O. Penrose (1970) is a thorough exploration outlining the needed posits and studying the problem of their consistency with the underlying dynamics. O. Penrose (1979) outlines this approach briefly in Section 3.4. Jancel (1969) covers this abstract and general approach in Part II, Chapter V, Sections I–IV.

7

Rationalizing non-equilibrium
theory

In this chapter we take up the problems outlined at the end of the last chapter: How are we to justify the fundamental assumptions needed for non-equilibrium statistical mechanics? How are we to show their consistency with the underlying dynamics? Although most of the important answers to these questions are covered in this chapter, one approach, the approach that invokes cosmological features of the universe "in the large," is reserved for Chapter 8.

Because the organizational structure of this long chapter is a bit complex, it might be useful to outline here the route we shall follow through the issues. Part I deals with two important preliminaries from experimental science and from the use of computers to model the evolution of dynamical systems. These are results that later will be referred to a number of times in different contexts. Part II outlines the programs that have been offered to "back up" the three approaches to the derivation of a kinetic equation we examined in Chapter 6. Part III then takes up the various basic approaches that have been suggested to provide a physical explanatory context for the success of statistical mechanics in the non-equilibrium situation. First, a number of "unorthodox" approaches are discussed. Then the "mainstream" approach, which relies heavily on the derivations outlined in Part II, is taken up. Finally, some "radical" versions of the mainstream approach are critically examined. Lastly, in Part IV, the remaining fundamental problems are outlined. Here, the object is to survey the results obtained in Parts II and III once more, asking whether these results are truly sufficient to resolve the puzzles with which our investigation began.

I. Two preliminaries

1. The spin-echo experiments

A series of experiments dealing with the array of spinning, and therefore magnetic, nuclei in a crystal lattice provides an observational result that, if not of great importance in the ongoing discovery process of physics, is of importance as a test case for various foundational accounts of the origin of irreversibility.

The nuclei are first aligned with their axes of rotation parallel, by means of a strong, externally imposed magnetic field. An appropriate radio-frequency pulse then flips all of the axes ninety degrees so that they now point in a direction perpendicular to the imposed magnetic field but still parallel to one another. At this point the axes begin to rotate in the plane perpendicular to the imposed magnetic field due to the phenomenon known as precession. Variations in the perfection of the crystal lead to local inhomogeneities in the internal magnetic field experienced by the different nuclei, however, so they precess at differing rates. Soon, macroscopic observation indicates that the axes now are distributed "at random" in the plane – that is to say, more or less uniformly relative to angular measure. They are still perpendicular to the imposed strong magnetic field, but they point in every direction in the plane perpendicular to that field.

At a time Δt seconds after the nuclei had their axes pointing in the same direction in the plane perpendicular to the imposed magnetic field, a new radio pulse flips each spinning nucleus over in the plane so that the nuclei that had moved furthest in a given direction from the original common direction of pointing of all the nuclei are now furthest "behind" in the "precession race." The nuclei continue to precess at their varying rates. At a time $2\Delta t$ after the initial time, the nuclei now once more have their axes aligned in a common, initial direction. (See Figure 7-1.)

This reappearance of order among the axes can be detected macroscopically as a pulse using standard nuclear resonance techniques. The experiment can be repeated obtaining a series of randomizing and derandomizing motions of the directions of the axes of rotation of the nuclei. Gradually the pulses that indicate reordering of the spin axes diminish in intensity. This is due to the transfer of energy from the system of nuclear spins to the thermal vibrations of the crystal lattice, to spin-spin interactions among the nuclei, and to diffusion, which allows nuclei to migrate around in the regions of local magnetic inhomogeneity.

When these experiments were first performed, many commentators emphasized the fact that the nuclear spins behaved as isolated individuals, so that the situation was not really that of the "many bodies in interaction" type so familiar in statistical mechanics. Indeed, in the original experiments the spin-spin interaction of the nuclei was one of the factors destroying the reproducibility of the macroscopically observable coherence. More recently, however, spin-echo experiments have been performed where spin-spin interaction is taken into account. In effect, control is maintained at the macroscopic level so that the randomizing effects of the spin-spin interaction can also be dynamically reversed.

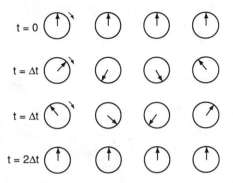

Figure 7-1. The spin-echo experiment. The top row represents a collection of spinning nuclei whose spins are all lined up in the same direction in a plane that is perpendicular to a magnetic field imposed on the crystal whose atoms have the nuclei in question. In the second row, a period of time Δt has elapsed. The directions of the spins have precessed at different rates around the imposed magnetic field, so that at this point the spins, although still in the same plane, now point in all directions "at random." In the third row, the spins have been flipped by a radio-frequency pulse. The spins furthest ahead in the precession race are now furthest behind. The result is shown in the last row. The time is now twice that elapsed from the first row to the second row. The spins have now all caught up with one another, leading to them all once again pointing in the same direction. From the third row to the last row there is the appearance of an equilibrium condition (randomized spins) spontaneously evolving into a non-equilibrium condition (spins all aligned).

Maxwell had referred to an imagined Demon who could lower entropy by sorting the fast and slow molecules of a gas by opening and closing an aperture at appropriate times. Loschmidt had objected to Boltzmann's *H*-Theorem by pointing out that systems that were the time-reverses of systems that had evolved from low to high entropy states would, by the time-reversal invariance of the dynamical laws, move from a high to a low entropy state even if kept isolated from the external environment. Putting these historical facts together, the experimentalists who performed the modified version of the spin-echo experiment just described speak of their process for reversing the dynamical evolution at the microscopic scale as the provision of a "Loschmidt Demon."

As they point out, their process allows us to restrict the region in which the points representing an ensemble are confined to a small region of the sort not usually determinable by the familiar kinds of macroscopic control. In at least this special case it is as if we could prepare a gas in such a way that an ensemble of gases so prepared would initially be uniformly spread throughout a box. But the overwhelming majority of the gases in the ensemble would then spontaneously flow to the left-hand side of the box. Obviously, these striking experimental results will provide an important test case for any foundational account of the origin of irreversibility.

2. Computer modeling of dynamical systems

Much insight can be obtained by using the power of high-speed computers to model the dynamic evolution of complex systems. An early discovery made by this method dealt a blow to a simple but too naive hypothesis about the origin of randomization in complex systems. At one time it was thought that even a small amount of dynamic interaction among otherwise energetically isolated component systems, even for a system whose components numbered in the tens, would be sufficient to guarantee randomization and approach to equilibrium. An early computer experiment by E. Fermi and others modeled the interaction of a number of simple harmonic oscillators coupled to one another by a small anharmonic term. It was expected that any energy distribution over the oscillators would, in short time order, begin a monotonic approach to a stable, equilibrium energy distribution. Instead, the energy distribution over time showed a non-equilibrium and periodic behavior, energy shifting back and forth among the oscillators in a regular pattern.

More recent experiments have been able to deal with systems of a very large number of components. Dynamical equations are modeled by numerical approximations to them, and the state of the system, beginning in a particular initial condition, is traced at small discrete time intervals. Many randomly chosen initial conditions are tried, giving a pattern of how the evolution of the system behaves qualitatively given the constitution of the system and its initial state. Care must be taken that an appearance of randomization is not introduced as an artifact through the necessity in the computation of throwing away all the decimal places in the numerical descriptions of the system at a given time that are beyond the ability of the computer's power to handle. But checks on this effect can be performed, and one can be confident that any appearance of randomness in the end result reflects the dynamics of the system and not the computational limits.

The technique used goes back to Poincaré. One considers in the phase space of the trajectories a two-dimensional sub-space. As a trajectory from an initial state crosses the sub-space, one notes the point of intersection, and then plots on the sub-space all of the subsequent intersection points. A system following a simple closed trajectory will intersect one of the sub-spaces at a finite number of points each time, and so will appear on the "portrait" as a finite set of points. A system that is multiply periodic – that shows periodicity in a number of variables but whose periodicities are not synchronized so that the trajectory never actually closes up – will intersect the sub-space in a series of points all of which lie along a one-dimensional curve. If enough intersections are plotted, the portrait begins to look like a continuous one-dimensional line in the sub-space. A system that is chaotic in its trajectory will intersect the sub-space in points that appear to be randomly distributed over some region. This region may be the whole available sub-space, or may be some sub-set of it. If one continues long enough, the points of intersection of a single trajectory that is chaotic appear to fill a two-dimensional region in the sub-space.

Such a diagram can provide only part of the insight we need. We still want an analytical understanding of why the regions of initial states have trajectories originating in them of the kind that they do. Further, even the appearance of a chaotic trajectory will not be enough to let us know what "kind" or "degree" of chaos is actually present.

Without the use of computers, but simply using his brilliant analytical powers, Poincaré was able to show that the phase portraits for even simple systems such as coupled oscillators with small anharmonic coupling or the restricted three-body problem in which three bodies, all moving in the same plane and which are such that two move about their common center of mass and the third moves under the influence of gravitational forces generated by the first two, are highly complex. Typical is the appearance of "islands" of stability where trajectories are multiply periodic, surrounded by regions of chaotic trajectories. If one looks on a more detailed scale, however, the situation sometimes becomes even more astonishingly complex. When one magnifies the regions of apparent stability, one discovers that on a more magnified scale it consists of a pattern of separate stable regions in a "sea" of instability, the pattern on this scale duplicating the pattern first seen on the original scale, and so on ad infinitum.

The unstable trajectories are, intuitively, the result of resonances, mutual interactions of the component systems that generate ever-increasing disturbing forces that tend to disrupt what would otherwise be stable trajectories. The surprising results of the early experiments with their unexpected non-randomizing behavior led to the view that the simple existence of resonances was not enough. Instead, regions of phase space where a multiplicity of resonance influences overlapped led to the chaotic trajectories. For the simplest systems, regions of instability can be "trapped" between regions of stability, in the sense that a trajectory started at a point in one of the regions will wander chaotically over that region but will never get into the others. For systems of many degrees of freedom, however, one expects the wild trajectories to wander throughout the chaotic regions.

For a given system, increasing the amount of the perturbing interaction will generally decrease the size of the stable regions, allowing the unstable portion of the phase portion to dominate. Again, for a given relative magnitude of the perturbation, increasing the complexity of the system by going to more and more degrees of freedom (more and more component sub-systems) will also increase the relative size of the chaotic region to the region of stability. This is not surprising because having many degrees of freedom opens the opportunity for many resonances and many regions of resonance overlap.

The appearance of these dominating regions of unstable trajectories is very suggestive for certain approaches to the grounding of the approach to equilibrium. For, as we shall see in Chapter 7,II,3, it is just such a condition of radical instability for the trajectory representing the evolution of a single dynamical system that will be sufficient to establish certain kinds of "spreading" behavior for an ensemble of such systems.

II. Rationalizing three approaches to the kinetic equation

1. The rigorous derivation of the Boltzmann equation

In the BBGKY approach to non-equilibrium, the ensemble surrogate for Boltzmann's f-function – that is, f_1 – was shown to obey the Boltzmann equation by an argument with two dubious posits. First, one had to assume that f_2 could legitimately be written as $f_1 \times f_1$. Then, one had to do an expansion in terms of density and "throw away" higher order terms. The second step was of doubtful rigor due to the bad behavior of higher order terms in the density expansion. The first step, the more crucial one, was a typical modern variant of the Posit of Molecular Chaos, and it was far from clear that such a posit could even be made consistent with the dynamical development of the ensemble as governed by the underlying micro-dynamical laws.

A rigorous derivation of the Boltzmann equation for f_1 has been provided by O. Lanford. It utilizes only the underlying dynamics plus an assumption about the initial ensemble, an assumption therefore at least consistent with the dynamical laws.

First, we should note that the results are proved not for realistic systems of a large but finite number of components, but in the so-called Boltzmann–Grad limit. This is a system of particles whose number, n, has gone to infinity but such that the product nd^2, where d is the diameter of a gas molecule, has stayed constant. We are dealing, then, with a number of molecules going to infinity and with a density going to zero, because nd^2 = constant implies $nd^3 \rightarrow 0$. Further, we are working in an idealized limit where the size of the molecules "goes to zero" on the scale of size determined by the size of the box confining the gas. Next, the original proof was given for a "hard-sphere" gas – that is, no inter-action until the molecules meet, and then perfect non-interpenatrability, although the results were later extended to more realistic potentials.

There is the very important randomness condition placed upon the initial ensemble. Actually, different versions of the theorem utilize subtly differing such initial conditions. The basic idea is to impose a probability measure on the phase space. One starts with a particular value of the one particle distribution function of Boltzmann, $f_{t=0}$, and chooses a probability measure so that high probability is assigned to the set of phase points, which are such that they are "near $f_{t=0}$" – that is, have the f_1 computed from them close to $f_{t=0}$. One then tries to show that if one looks at the set of phase points at some later time, $t = t'$, it will then be highly probable that the phase points into which the initial phase points have evolved, determined by the exact dynamical evolution, are now "near $f_{t=t'}$" where $f_{t=t'}$ is the f into which $f_{t=0}$ has evolved if the evolution of the f's is governed by the Boltzmann equation for the one-particle distribu-tion function. Lanford points out that it is "high probability of Boltzmann-like behavior" that he derives, and not, as in some understandings of the kinetic equation, behavior of the ensemble "on the average." We shall puzzle over this in Chapter 7,III,7.

In one version of the theory the initial probability distribution is, intui-tively, related to the idea that at the initial time the higher f's should factor into products of f_1 except to the extent that there is molecular overlap that would introduce inevitable correlation. Here, the constraint on an initial ensemble is, interestingly, time-symmetric. An ensemble that is like an ensemble that satisfies the constraint, except that each system in the new ensemble is the "time reverse" of a system in the original ensemble, would also satisfy the constraint. If the constraint itself propogated for-ward in time by the exact dynamics, this would lead to a paradox. For one could take the state of the ensemble at some time after $t = 0$, but within the limits for which the theorem holds, and consider the time-reverse of that ensemble. By the reversibility of the dynamics, that ought to evolve back to the initial ensemble, but by the theorem it ought instead to continue to show the irreversible behavior predicted by the Boltzmann equation. The paradox is resolved by the fact that this version of the initial constraint on ensembles does not hold at any time after $t = 0$ if it holds at $t = 0$. Notice that the constraint is a constraint on the initial ensemble only, and is, hence, not subject itself to the threat of inconsistency with the dynamic equations.

A probability measure that is closely related to the standard micro-canonical measure will do the trick in satisfying the condition on the initial ensemble, but the theorem is more general in showing that any probability measure that appropriately concentrates points of the initial ensemble sufficiently tightly around those that generate the right $f_{t=0}$ will assign high probability to the points at $t = t'$ being concentrated around $f_{t=t'}$.

Another approach to the theorem introduces an alternative constraint on the initial ensemble that is in fact propagated beyond $t = 0$. This constraint is "weaker" than the one described, but more complicated to formulate. The paradox noted here is avoided by the fact that this constraint is not time-symmetric. That is, if an ensemble satisfies the constraint, the ensemble of time-reversed systems generally will not satisfy the constraint.

It should be noted that the theorem guarantees Boltzmannian behavior of the ensemble from its special initial state only for a very short time period, approximately one-fifth of the time needed, on average, for molecules to transit between collisions! As Lanford says, "the theorem to be stated says only that the Boltzmann equation holds for times *no larger than one-fifth of a mean free time* and hence does not suffice to justify its physically interesting applications." He suggests that the way this time restriction functions in the theorem indicates that the theorem can be extended, or at least that the *result* should remain true for arbitrarily large times. He also notes, however, that the theorem relies crucially on the fact that the limit of zero density has been taken for a gas, and he is skeptical that this restriction can be overcome. This is for the same reasons that the Bogolyubov density expansion technique in the BBGKY approach is of dubious rigor – that is, because of the bad behavior of the higher order terms in the expansion.

We shall return in several places to the issue of the foundational role that these results can justifiably play. It will be useful, though, to make a few observations at this point. First, consider the version of the theorem that makes use of an initial constraint of ensembles that is such that if the ensemble satisfies the constraint, the time-reverse of that ensemble will as well. Because the initial ensemble shows "highly probable" Boltzmannian behavior into the future, mustn't the reversed ensemble show "highly probable" anti-Boltzmannian behavior – that is, f behaving in a time mirror image to the f behavior shown by solutions of the Boltzmann equation, into the past? Is it not reasonable to assume, in fact, that the ensembles that are such that a high probability is assigned to initial states where the f appropriate to a state is near $f_{t=0}$ are such that in both future and past it is highly probable (within the limits of the bounds imposed by the theorem) that the phase points will be such that the f corresponding to them will be near the f that would be predicted by the Boltzmann (resp. anti-Boltzmann) equation? But of course it is irrational to assume that final states were likely preceded by states closer to equilibrium.

Of course, we can avoid this apparent paradox by using the version of the theorem in which the initial constraint is not time-symmetric, so that an ensemble that satisfies it is not matched by a time-reversed ensemble that also satisfies it. But then there will be a different probability measure that is the time mirror image of the one chosen. This new constraint will then make anti-Boltzmannian behavior highly probable. The question will be, why assume one of these probability measures is the appropriate one to use in describing the world? Why is it appropriate to assume that initial states showing Boltzmannian behavior are highly probable and those showing anti-Boltzmannian not, rather than the reverse? Or, better, because the appropriateness of the assumption is shown by its success in the world, the question is rather why does that measure that is correct work as well as it does whereas its opposite fails entirely.

The problem here is the familiar one of justifying a choice of a probability distribution over a set of systems satisfying some macroscopic constraint. In equilibrium theory, we showed that a kind of "transcendental" deduction of the probability measure was possible, in the sense that there was a unique, absolutely continuous measure that remained invariant in time. Here, that is denied us because we are dealing with the non-equilibrium situation. We can of course "transcendentally" justify one of the non-time-symmetric measures as opposed to its mirror image by pointing out that the one gives Boltzmannian behavior, which we want, whereas the other predicts the incorrect time-asymmetry. But that doesn't fix uniquely one such probability measure as is the case (modulo sets of measure zero) in the equilibrium situation. More importantly, justifying the choice of a measure in this way goes no distance at all to offering an explanation as to why such measures work. In the case of the time-symmetric initial constraint on ensembles, a similar selection rule can be justified. The rule would be to use this constraint only to characterize the initial states of systems, not their final states. But, once again, that fails to provide an explanation as to why this procedure works.

The Lanford results certainly go at least some way in answering some profound questions in non-equilibrium statistical mechanics. For the first time, a rigorous derivation is obtained that uses the low density of the gas, in the limit sense of zero density, and the small size of the molecules, and an initial constraint upon an ensemble to derive, within severe time limits, the conclusion that evolution of the ensemble will be such that it will be highly probable that systems will behave in a Boltzmannian way. The repeated rerandomization condition in the form of ongoing molecular chaos, with its doubtful consistency with the exact dynamical evolution, is no longer needed. But the introduction of these special initial conditions on the ensemble, in the form either of a probability distribution that is time-symmetric but that is to be used only for initial states or in the form of a probability distribution that is time-asymmetric, raises the fundamental explanatory puzzle: Why are the micro-states of systems distributed in such a way as to give us parallel entropy increase into the future?

We should also note here that the "justification" of the Boltzmann equation given by these results may run into a curious conflict with alternative justificatory schemes, in particular with one that we shall discuss in the coarse-grained approach that is called mixing. We shall return to this point in Chapter 7,III,7.

2. The generalized master equation

From the Liouville equation to the generalized master equation. The aim of the generalized master equation approach is to transform the exact equation governing the dynamical evolution of an ensemble – the equation determined by the dynamical equations governing the evolution of each system in the ensemble from its initial micro-state – in such a way that one can see transparently what conditions must be met for a master equation to hold.

The procedure starts with the derivation of the Liouville equation. Here, one simply writes a formal equation, which on the left denotes the time derivative of the ensemble distribution function, ρ, and on the right introduces an operator, L, defined from the function, H, the Hamiltonian, that governs the evolutions of the individual systems, such that L acting on ρ generates the time evolution of ρ.

The next step is to decompose ρ into two parts. It is this procedure that generates the reduced description – that is, the function giving only part of the information contained in the full distribution function. This function is supposed to be shown to obey an equation of the master equation type, or at least to be intimately associated with the reduced description that does so obey the master equation. The trick, originally due to S. Nakajima and R. Zwanzig, is to introduce the notion of a projection. This is an operator that takes the full distribution function (thought of as a kind of vector) and decomposes it into two components, $P\rho$ and $Q\rho$, where $Q = 1 - P$. Suppose we wish to derive the master equation that gives the dynamics of the probability distribution over one variable. Then the projection works by averaging the distribution function over all the remaining "hidden variables" on the phase-space surfaces with fixed values for the variable we are interested in. Thus, if we had a number of oscillators coupled weakly, and wanted a master equation giving the evolution of energy distribution over the oscillators, we would take a surface of constant energy for each oscillator and average the value of ρ over all the possible values of position and momentum for the oscillators consistent with that energy distribution. The values of $P\rho$ are sufficient to allow us to determine the probability distribution over the designated variable, and the evolution of $P\rho$ is sufficient if we know it allows us to determine the evolution of the probability distribution over the variable with which we are concerned.

One can then decompose the Liouville equation into two coupled dynamical equations, one for P and one for Q. One can formally solve these two equations. Each solution contains two terms, one depending upon the initial ensemble distribution and one depending on an integral over all the values of ρ from the initial time to the time t at which the value of ρ is to be determined.

One usually works by assuming that one starts at time zero with the uncoupled systems each in equilibrium, and that the coupling perturbation is introduced at time zero. For example, radiation in a cavity has been allowed to reach equilibrium. Then the dust mote that allows energy to move from mode to mode is dropped. Under this assumption, $\rho_0 = P\rho_0$ and $Q\rho_0 = 0$. With this determination, one can then substitute from the equation for $Q\rho$ into that for $P\rho$, obtaining a closed equation for $P\rho$. This is called the generalized master equation. On the left it has the derivative of $P\rho$ and on the right, two terms. Like the Liouville equation, and unlike the master equation, it is time-symmetric. An ensemble evolution satisfying it will be matched by the time-reversed evolution of the time-reversed ensembles also satisfying the equation.

The first term on the right-hand side of the generalized master equation is like the master equation in that it involves only the state of the ensemble at time t. But the second term contains an integral over all states of the ensemble from time 0 to time t, so that the full information as to how the ensemble will evolve at time t requires not only reference to the state of the ensemble at that time, but to all the previous states in the history of its evolution. In this way the generalized master equation is not "Markovian," as is the master equation or as are the other kinetic equations we wish to rationalize.

From the generalized master equation to the master equation.
It can be shown that the first, unproblematic, term on the right-hand side of the generalized master equation is in fact identically zero, leaving the problematic "memory" term on the right-hand side. The aim is to show that in appropriate circumstances this term can be converted into one that refers only to the state of the ensemble at a given time, and not to its past history, thereby converting the equation into one that has the Markovian form of the master equation.

For some ideal models, one can show with some rigor that in various limits in which the coupling of the systems goes to zero, for times long after the beginning of the coupling, the solutions of the generalized master equation approach those of the master equation.

When the coupling is allowed to be small but finite, a rich body of techniques has been developed to allow for the expansion of the solution of the generalized master equation in terms of "correlation" diagrams for the ensemble. The procedure becomes extraordinarily complex for systems with realistic interactions, but various posits and moves to the thermodynamic limit combine to show that under certain appropriate circumstances, one can once again give long-time approximations to the generalized master equation by utilizing the solutions of an equation in which the memory term has been replaced by a term that refers only to the state of the ensemble at the single time from which its evolution is being continued.

It is clear that the posits needed to allow us to demonstrate the master equation solutions to be good approximations to the solutions of the generalized master equation for long times after the beginning of the perturbation are problematic in nature. This can be seen in the fact that the generalized master equation, even with the basic assumption made about the initial ensemble needed to derive it has been made, is a time-symmetric equation, whereas the master equation is not. Careful examination shows that at various points, assumptions are made about the behavior of the solutions of the generalized master equation for times after the initial time. Had the same assumptions been made for times earlier than the initial time, we would be able to derive the conclusion that the anti-thermodynamic analogue of the master equation suitably approximates the generalized master equation for past times, a conclusion we certainly don't wish to derive.

We also know that there will be many systems for which the master equation will not be a suitable surrogate for the generalized master equation. Those systems that fail to show approach to equilibrium – such as the coupled oscillators of the Fermi computer experiment, or systems that satisfy the constraints necessary for general stability theorems like the KAM Theorem to apply – will not have their generalized master equation approximable by a master equation. Although work has been done that goes some way to elucidating the conditions under which the master equation is usable and when it is inappropriate, the technical difficulties are severe.

Sub-dynamics. An interesting variation on the generalized master equation approach has been offered by I. Prigogine and others. The trick here is to look for a decomposition of the full ensemble distribution function into two components. One component will obey both the original Liouville dynamical equation and a Markovian kinetic equation that is not time-symmetric.

What conditions ought to be imposed when we look for a component of the full distribution function that is to obey a kinetic equation? (1) When the coupling is reduced to zero, the decomposition ought to reduce to the decomposition of the ensemble distribution into its parts determined by the operators P and Q of the previous sections, for in that case it is $P\rho$ that obeys a kinetic equation. (2) The operator that selects the kinetic component when applied to the kinetic component should leave it unchanged, because that component already obeys the conditions we want. (3) The decomposition of the ensemble into components should be independent of the time and of the stage of evolution reached by the ensemble. That is, as time goes on, the ensemble changes and so does its kinetic component, but the operation that selects the kinetic component out of the full distribution function ought itself to not depend upon time. (4) The component selected out by the operator should obey both the original Liouville equation and the master equation, at least for times long after the initial coupling.

A very interesting result claimed is that conditions (1) and (3) imply conditions (2) and (4). That is, if there is a time-invariant way of decomposing the distribution function that reduces to the P projection in the limit of zero coupling, then the component picked out by that decomposition must, perforce, obey a kinetic equation. The idea is then that the kinetic component is a "sub-dynamic" component that evolves both according to the general time-symmetric kinetic equation and a time-asymmetric master equation.

The formal problem then becomes twofold. First, one must find general conditions on the nature of the system (of its Hamiltonian or Liouville operators) that are sufficient to allow a sub-dynamics in the earlier sense to exist. Then one must show of interesting models that they do in fact obey the formal condition derived.

If a sub-dynamic component of the distribution function can be shown to exist, there are still a number of interesting conceptual problems to be explored. One is the question of whether we can show that the sub-dynamic distribution function is the one that contains the appropriate partial information that we need to compute the macroscopic quantities whose time-asymmetric evolution we wish to derive from our statistical mechanical model. Another is the fact that if such a sub-dynamics exists, there will be another component of the full distribution function, also projectible from it by an operator that is time-invariant and that reduces to P in the zero coupling limit, and that is such that this new sub-dynamic obeys both the Liouville equation and an anti-thermodynamic kinetic equation whose time-asymmetry is the reverse of the ordinary kinetic equation. We will still want to know why the kinetic description works as well as it does, whereas the anti-kinetic sub-description is useless and wrong-headed for predictive purposes.

3. Beyond ergodicity

In the last chapter we examined a general schematism for describing the approach to equilibrium that worked by dividing the phase space up into a large number of "boxes," each small relative to the total size of the phase space, but each containing a significant region of phase points. In this approach, the "reduced description" that is to provide the resources for calculating macroscopic quantities, and that is to be shown to obey a time-asymmetric kinetic equation despite the time-symmetry of the equation governing the evolution of the ensemble description as a whole, was the description that told us at any time what portion of a box was occupied by phase points that had evolved from the phase points representing the initial ensemble.

We saw that a posit sufficient to generate a time-asymmetric evolution toward an "equilibrium" distribution for such a coarse-grained description was a Markovian posit regarding the probability that a point in a designated box at one time would be found in a designated box at the next observation time. The Markovian posit was that such transition probabilities were themselves constant in time, and such that the probabilities could be calculated by assuming the first designated box filled with phase points and calculating what proportion of them would, under the dynamic evolution, move into the second designated box over the time interval. The argument presupposed that the same proportion of points would move from the irregularly shaped piece of the first box actually occupied into the second box, as would be the case if the first box were filled. In this way the past history of the evolution of the ensemble could be ignored in the manner characteristic of kinetic equations.

Here, we shall investigate what features of the dynamics could possibly be expected to lead to an evolution correctly described by making such a Markovian posit. And we shall ask to what degree the conditions necessary for this redescription to hold can be shown to follow from the constitution of a system and the underlying dynamical laws. We shall see, not surprisingly, that fundamental issues of the nature of initial ensembles will also, at some stage, arise and require examination. And we shall once again reflect upon the difficulties that arise when one tries to bring the general coarse-grained description "down to earth" – difficulties that arise when one asks what the numbers giving proportions of boxes occupied have to do with what we wish to calculate – macroscopic thermodynamic features of systems.

Dynamical instability. What, intuitively, must the trajectories of systems look like in order that ensemble evolution be ergodic? Ergodicity is equivalent to the fact that each trajectory (except for a set of measure zero) eventually goes through every sub-set of non-zero measure in the available phase space. This amounts to a strong restriction on the partitioning of the phase space into pieces that remain closed under evolution. Any such proper piece of the phase space must be of measure zero or measure one. So ergodicity is the insistence that each trajectory (except for a set of measure zero) "wander" all over the available phase space.

Is that enough, intuitively, to obtain a coarse-grained description of an evolution that matches our idea of approach to equilibrium from a non-equilibrium condition? No, it is not. Ergodic behavior is compatible with an initial ensemble that remains "coherent" as the individual points in it wander throughout the phase space. This would result in a probabilistic expectation that a system prepared in a non-equilibrium state might show a kind of regular periodicity. It is true that in certain senses discussed in Chapter 5 it will be highly probable that at a random time the systems will be found near to equilibrium. But this is compatible with a high probability that even into the distant future, if we select times appropriately, we can get high probabilities for the systems to be found far from equilibrium. This is true because the excursions from equilibrium from different initial states might be highly correlated with one another, as is shown by the "coherence" of the initial region as it wanders through the available phase space.

Intuitively, what we want for the model of approach to equilibrium is that any small region of phase space initially occupied by points will "spread out" over the entire phase space. Not, of course, in the sense of actually filling the newly available phase space, for this is impossible, but rather in the coarse-grained sense of approaching a "fibrillated" distribution that is such that the proportion of any coarse-grained box in the phase space that is occupied by points that were originally in our narrowly delimited region will be equal to the proportion of any other box so occupied. And that proportion will be more or less equal to the proportion of the entire phase space originally taken up by the initially filled region.

To get this result, what we will need, again intuitively, is that small regions of the phase space contain points arbitrarily near each other but from which trajectories will be generated that separate from one another in a rapid and randomlike way. Any small region will then, in short

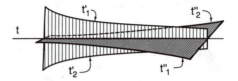

Figure 7-2. C systems. The phase trajectory t represents a trajectory whose instability is sufficient to generate the wanted ergodic features of ensemble evolution. It lies in one sub-manifold of the phase-space whose trajectories all, like t_1' and t_2', diverge exponentially fast from t as one goes toward the past direction of time. It also lies in another sub-manifold of the phase space all of whose trajectories, like t_1'' and t_2'', diverge from t at an exponential rate in the future time direction.

order, be found to be highly irregular in shape and dispersed throughout the entire newly available phase-space. So what we expect to suffice to demonstrate approach to equilibrium in our coarse-grained sense is a radical dynamical instability for the evolution generated by the underlying micro-dynamics. And it is the formalization of this notion that plays, for this new kind of evolutionary "chaos" of the ensemble, the role played earlier by metric indecomposability in demonstrating ergodicity.

A particularly important formal condition of dynamical instability is for a system to be a "C system" (sometimes called a U system, Y system, or Anosov system). Essentially, a system is a C system if each of its trajectories has the following nature: At any point on the trajectory, there exist two sub-spaces of the whole phase space. In one of these sub-spaces, trajectories arbitrarily close to the given trajectory diverge away from it exponentially fast in the future time direction. In the other sub-space, the arbitrarily-near trajectories all diverge away from the given trajectory exponentially fast in the past-time direction. This is a strong kind of fast divergence of nearby trajectories of the kind we would expect to allow us to derive spreading results for initially small and regularly shaped ensembles. (See Figure 7-2.)

There are, then, three problems. One is to give formal definitions of the notion of ensembles spreading out over the available phase space in the coarse-grained sense. Next, there is the problem of showing that a system that meets a formal condition of dynamical instability such as being a C system will also meet one or more of the formal notions of ensemble spreading. Finally, there is the difficult task of proving for reasonable ideal models of a system in the world that they meet the condition of dynamical instability.

While we are directing our attention exclusively to systems describable by classical mechanics, we might note here that there is little hope of extending the results we obtain into the realm of quantum statistical mechanics in any simple-minded way. This can be seen by reflecting once again on the fact that with classical systems we can have an ensemble whose members are finite systems and that is not almost periodic in its behavior, even if the individual systems are, by Poincaré's result, almost periodic. But in quantum mechanics, so long as we stick to finite systems, the ensemble will necessarily be almost periodic if the individual systems are, even if it contains an infinite number of member systems. This will certainly block any hope of proving a formal monotonic spreading-out result of the kind we are looking for. Of course, one can then move to the thermodynamic limit of a system with an infinite number of micro-components and so block almost periodicity for the ensemble, but the extension of the results we will describe to infinite systems, even in the classical case, is a non-trivial mathematical exercise.

Varieties of chaotic evolution. As it turns out, there is a hierarchy of notions that correspond to our intuitive notions of an ensemble spreading out in the coarse-grained sense or of the trajectories moving through the available phase space in a random or chaotic way. In some cases, the concepts can be ranked in order of their strength, one category being stronger than another if meeting its condition implies meeting the other condition but if there are systems that meet the weaker condition but not the stronger. In some cases, it isn't yet known if two concepts differ in their strength.

The weakest concept beyond that of ergodicity is that of weak mixing. Fix attention upon a single coarse-grained box in a coarse-graining of the phase space. Observe the proportion of that box filled at any time by points that were initially confined to the region of the initial ensemble. Starting at any time, look at the infinite time average of that proportion from that specific time on. If, for any coarse-graining and any box of it as time goes to infinity, this time average, also defined as a limit as time goes to infinity, becomes equal to the proportion of the allowed phase space occupied by the set that characterizes the initial ensemble, the system is weak mixing. Of course, all these references to "proportion of the coarse-grained box" and "proportion of the available phase space occupied by the points of the initial ensemble region" presuppose that one has selected a standard measure for the size of regions of the phase space. One can show that weak mixing is stronger than ergodicity.

A provably stronger condition still is that of mixing. Mixing is to weak mixing much as the strong law of large numbers is to the weak law. Once again, we give a coarse-graining and focus attention on a specific box of it. Once again, we look at the portion of that box occupied at a given time by points originally in the initial ensemble region. If, in the

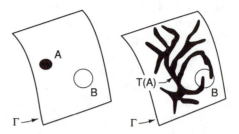

Figure 7-3. A mixing ensemble. A and B are two non-zero sized regions of points in the phase space Γ. B is kept constant. We track the evolution of systems whose initial micro-states are in the A region. The result is a series of T(A) regions as time goes on. For a system to be mixing, the A region evolves into a T(A) region "in the long run" (that is, in the limit as time goes to infinity) that is evenly distributed over the phase space in the coarse-grained sense. For this to be the case, it must be that in the infinite time limit, the proportion of any B region occupied by points that evolved from the A region is equal to the proportion of the original phase space originally occupied by points in the A region.

limit as time goes to infinity, this proportion approaches the relative proportion of the available phase space occupied by the original region, then the system is mixing. What this says is that if we focus on a fixed measurable region of the phase space and look at the regions into which the initial ensemble region is transformed by the dynamical evolution, in the limit as time goes to infinity the overlap of the transformed initial region with the designated fixed region will have its size given simply by the product of the size of the fixed region with the invariant size of the initial region and its transformations. Formally, for any measurable sub-sets A, B,

$$\lim_{t \to \infty} \mu(T(A) \cap B) = \mu(A) \times \mu(B)$$

where $T(A)$ is the set into which A is transformed by dynamic evolution at time T and is the standard measure. (See Figure 7-3.)

It can be shown that mixing is strictly stronger than weak mixing, but it is also known that a weak mixing ensemble can always be transformed into a mixing ensemble by simply changing the velocities with which phase points move along their trajectories.

An obvious generalization of mixing is to consider mixing to the nth order for each finite n. For example, second-order mixing would be defined by stating that for any three measurable sets A, B, C,

$$\lim_{t \to \infty} \mu(T(A) \cap T(B) \cap C) = \mu(A) \times \mu(B) \times \mu(C)$$

At this time it is still not known whether there are any systems that are mixing, but not mixing to every finite order.

A very powerful and interesting condition of randomization for ensemble evolution is for a system to be a K system. The formal definition is a bit abstract. Very crudely, a system is a K system if there is a coarse-graining that in a sense "generates" by its transformation over time the algebra of all measurable sets in the phase space. What one does is take the coarse-graining at a time. Look at the sets generated by what happens to its boxes at a later time. Take all the sets one can generate by including intersections of the original sets with the new sets. Keep the process up into the infinite time limit. In the limit, one gets by this process a collection of sets that contain all the measurable sets of the phase space if the system is a K system.

A consequence of being a K system, one that is actually equivalent to its formal definition, is more intuitive. Consider a coarse-graining of the phase space P (for "partition of the phase space"). Consider observations made from past time infinity to the present at fixed finite time intervals. For any specific system, locate its trajectory by identifying in which box of the partition P its representative phase point is located at a given time. The infinite sequence of such location numbers constitutes a "coarse-grained history" of the evolution of the particular system. Can one tell with probabilistic certainty – that is, with probability one – from this coarse-grained history in which box the trajectory will be observed to be located at the next observation? If so, we can say that the dynamic evolution is deterministic in the probabilistic sense. Only in the probabilistic sense, note, for the fact that given a coarse-grained history the probability is one that the trajectory will next appear in a specific box of the partition is compatible with a set of measure zero of such trajectories having that coarse-grained history but appearing with its members next in some other box of the partition.

A process (that is the combination of a partition and a particular dynamic evolution) is said to have the 0–1 property if the only sets that are probabilistically deterministic in the sense described above are those of measure zero or measure one. That is, the only "events" (next observation of the trajectory being or not being in a box) that can be predicted with probability one from the coarse-grained past history are those that correspond to observations that have probability zero or one on their own – that is, independently of any information about the past. It can be shown that the K systems are those systems that are such that for every partition and every time interval for measurement, the dynamic evolution is such that the resulting process has the 0–1 property. In this sense, K systems are radically indeterministic in the probabilistic sense. (See Figure 7-4.)

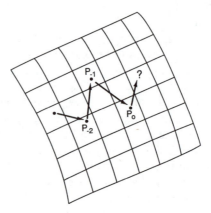

Figure 7-4. *K* systems. The phase space of the system is coarse-grained. At discrete time intervals, the trajectory of a system is looked at. At time 0, the system is at point P_0, at time −1 at point P_{-1}, and so on. Suppose we know, for each of an infinite number of past times, which box of the coarse-graining the point representing the state of the system was in. Can this tell us with probabilistic certainty – that is, with probability 1 or 0 – that the system will (will not) be found in some specific box - say box i – at the next observation? If the system is a *K* system, this will be possible only if box i is one with measure 1 or measure 0 to begin with.

It can be shown that every *K* system is mixing in every finite order, hence mixing and weak mixing and ergodic as well, but that there are systems mixing to every finite order that are not *K* systems.

At one time it was believed that being a *K* system was as randomizing as an ensemble evolution could be, but this is now known to be false. A strictly stronger randomizing character for a dynamical evolution is for it to be a Bernoulli system. Systems that are Bernoulli systems are, intuitively, in a coarse-grained sense, as unpredictable as the results of a sequence of totally probabilistically independent trials. This is so despite the fact that in the micro-sense, each trajectory is fully determined by its initial point.

A basic definition of a Bernoulli system is this. If the dynamical evolution of the system is Bernoulli, we can coarse-grain the phase space in such a way that the following holds: Make coarse-grained measurements at fixed time intervals from the infinite past to the infinite future. For each trajectory there is then a doubly infinite sequence of numbers that indicates which box the trajectory was in at a specific time. There will then be a probability distribution over those sequences that determines in the infinite collection of sequences what proportion satisfies a specific condition – say, having some finite sequence appear somewhere in it. If the system is Bernoulli, this set of sequences and the probability measure over it will be "isomorphic" to the sequences and probability distribution over them that we would get by (1) tossing a die, each of whose tosses was totally independent of the others, and (2) that had as many faces as there are boxes in the partition, with (3) a specific, fixed, probability on each toss of a given face coming up, and (4) then constructing the sequences by recording the results of a sequence of tosses from the infinite past to the infinite future.

To see how strong a randomizing condition this is, we should look at the notion of a generating partition. This is a partition that is sufficiently fine that the doubly infinite sequence of numbers that tells which box of the partition the trajectory is in at a given time will uniquely fix which trajectory the numbers represent up to a set of trajectories of measure zero. In other words, the sequence of coarse-grained observations (doubly infinite to be sure) will distinguish one trajectory with that sequence from almost all others. If the system is Bernoulli, then the probability that a trajectory will be in a box of even such a generating partition at a given time is totally independent of what boxes that trajectory was in at the infinite number of past observation times. In other words, learning all there is to know about the past coarse-grained history of the trajectory, even relative to a partition fine enough to be generating, will not allow us to "improve our odds" in betting on which box the trajectory will be in at the next measurement time. One could hardly ask for more randomness from a system that is, after all, at the microscopic level, deterministic.

It can be proven, with difficulty, that the Bernoulli condition is actually stronger than the K system condition, that is that there are K systems that are not Bernoulli. It is easy to show that every system that is Bernoulli meets all the other chaos conditions as well.

We now have a variety of increasingly strong notions of an ensemble spreading out in the coarse-grained sense over the available phase space and of the trajectories followed by its points showing chaotic or random behavior. But do any interesting physical models, at least of the idealized sort, display any such behavior?

First, it is necessary to find a condition of the dynamical flow sufficient to guarantee that a spreading or randomizing condition will hold. A theorem of great importance is that any system that is a C system, and that satisfies an additional mathematical constraint that is usually fulfilled, will be a K system, and hence mixing in all finite orders, mixing, weak mixing, and ergodic.

Next, one shows that certain idealized geometric flows are C systems. For example, flows that move along the geodesics (curves of least curvature) of Riemannian surfaces of constant negative curvature are demonstrably *C* flows. Finally, one shows, by dint of very arduous effort indeed, that certain idealized physical systems are *C* systems as well. Two or more hard spheres in a rectangular box, a single particle in a box with fixed convex sides, and other systems as well can be shown to be *C* systems. With even more effort it is possible to go beyond this and show such systems to be Bernoulli systems as well.

It is important to note that although we are concerned with systems that display instability in their evolutions, the feature of having such instability is itself "structurally stable." What this means is that a system having the features necessary to generate the kind of trajectory instability we have been discussing is itself insensitive to infinitesimal changes in the structural features of the system. In other words, systems with unstable trajectories are not rare curiosities that cease to have the instability property as soon as the smallest change is made in some parameter characterizing the structure of the system. Rather, a range of system parameters will all lead to the system being a *C*-system and having the needed trajectory instability.

The application of the randomization results. The results of abstract dynamical theory we have just seen are dramatic and important. Just how they are to be applied to give us the answers to the foundational questions in statistical mechanics that we are seeking is, however, far from clear. These are issues to which we shall return in later parts of this chapter, in Sections III,4–7 and IV. But it will be helpful at this point to at least itemize some of the obvious questions that need to be asked before moving on to deal with a few other important technical and formal results that must also be put in place before getting to the extended philosophical discussion of these issues. The questions itemized here differ markedly in the depth to which one must go in answering them. Some, as we shall see, immediately suggest at least programmatic answers. Others lead us to the most intractable foundational questions of all when we try to answer them.

(1) Some of the results obtained tell us only about what will occur "in the limit as time goes to infinity." How useful can such results be when what we really want to obtain are results showing us why systems display the finite time behavior that they do? These infinite time-limit results also fail to guarantee the monotonicity of approach to equilibrium that we would like to obtain, at least in the ensemble sense. Can this difficulty be overcome?

(2) In cases where finite time results are obtainable, a different problem arises. When we have shown a system to be a Bernoulli system, then we know, for a finite time interval between measurements, that some generating partition will show probabilistic independence when

measurements are made relative to its coarse-grained categories. But which such partition? Will its "boxes" correspond in any way to the kinds of macroscopic measurements we can perform? The results obtainable from dynamics alone are quite abstract, only guaranteeing that some such partition exists, and leaving us in the dark about just what the nature of the basic partition is.

(3) More generally, there are a host of problems encountered in trying to tie the results obtained here to experimental practice and, hence, to what we can macroscopically measure and to what features of systems we believe, on the basis of experience, show thermodynamic characteristics. The power of these methods is a result of their great abstractness, but that very feature makes their applicability to deriving and justifying specific results difficult to assess.

(4) We should notice that these results are all infected with the "except for a set of measure zero" problem encountered in equilibrium statistical mechanics and its justification. The results on mixing, for example, tell us that the size of the overlap region of $T(A)$ and B goes to the product of the sizes of A and B. This holds so long as the measure of size (or of probability) is invariant under the dynamic evolution. Because a mixing system is, perforce, ergodic, we know that the only such invariant measure that is absolutely continuous with the standard micro-canonical measure is the micro-canonical measure. But "absolutely continuous" means "assigns measure (probability) zero to sets assigned measure zero by the standard measure." By what right do we demand that restriction on an appropriate measure of probability? Here, again, we have a kind of "situations with probability zero in the standard measure can be ignored" posit that requires serious critical examination.

(5) The results discussed here are independent of the size of the systems constituting the ensemble. A system of two hard spheres in a box is mixing, for example. But, plainly, the large number of micro-components is essential for the thermodynamic behavior of systems. Exactly how must the results discussed here be supplemented by the facts about the large number of degrees of freedom of the systems when the results are to be applied to justify the statistical mechanical methods?

(6) All of the results discussed are completely symmetrical in time, as results obtained purely from the constitution of the system and the time-symmetric underlying dynamical laws must be. Parallel to the results about limits when time goes to the infinite future in the proofs of weak- and strong-mixing, are results about the limits as time goes into the infinite past. Parallel to the results about lack of probabilistic determinism from past coarse-grained history for K systems are results about lack of probabilistic retro-determinism from the coarse-grained future "history" of the system. And parallel to the independence of the probability of a

system's being in a box from any coarse-grained facts about its past for Bernoulli systems is an independence result about the probability of a system's being in a box now from conditionalizing on the future box occupations of the trajectory. How are we to go beyond these time-symmetric results to obtain the time-asymmetric results we seek for statistical mechanics?

(7) The questions asked in (6) lead us immediately to the most import-ant question of all: What role is to be played in the justification of non-equilibrium statistical mechanics by the initial ensembles from which ensembles evolve? The results just discussed make no reference to the nature of this initial ensemble. Intuitively, then, they cannot by them-selves provide the full picture of the structures that underlie the success of statistical mechanics. It seems likely that we will have to resort to facts about the initial ensemble to distinguish the results that help to justify statistical mechanics from their useless time-symmetric analogue. Even forgetting about the problem of resolving the time-symmetry problem, it seems, intuitively, that if the initial ensemble is chosen oddly enough it can, even if its measure is non-zero in the standard measure, result in behavior for the ensemble over finite times of the sort we don't want. How must initial ensembles be restricted to avoid this problem? What physical reason justifies the imposition of such a restriction?

(8) Finally, we must at some point explore the limits of the results discussed. We know that stability results of the KAM type will provide theoretical limits beyond which we cannot expect systems to be randomizing in the way these results describe. And we know experimen-tally that many complex systems do fail to show the monotonic approach to equilibrium predicted in statistical mechanics. What are the limits to randomizing behavior? And on which side of those limits do realistic systems (with, say, realistic intermolecular potentials instead of the sin-gular potentials of hard spheres in a box) fall? If they fall on the side of the non-randomizing systems, how can we evade this apparent failure of our justification? Will resort to the large size of the system and a move to the "thermodynamic limit" help us here?

We shall return to many of these issues in Chapter 7,III,4–7 and Chapter 7,IV.

4. Representations obtained by non-unitary transformations

Additional formal insights have been obtained by a method due to Prigogine, B. Misra, and their collaborators. In this approach, a method is sought that systematically transforms the original probability distribu-tion characterizing an ensemble to a form that reveals on its surface the randomizing character of the dynamics and its character as a process that drives the probability distribution to one that approaches equilibrium in at least the coarse-grained sense.

In operator theory, unitary operators play a role that generalizes the notion of a change of coordinates by rotation for ordinary vectors. A set of quantities transformed in common by some unitary operator essentially possesses all of the properties of the original set but now expressed in what is called a different "representation." The approach to statistical mechanics we are dealing with here treats the probability distribution as a generalized vector to be operated on by operators.

Next, a generalization of the notion of a unitary operator is introduced, the so-called star-unitary operators. A probability distribution transformed by one of these operators is a new function that still possesses the formal characteristics necessary for a probability distribution. In particular, it is still normalized (total probability adds up to one) and it is still non-negative everywhere – that is, no "probabilities less than zero" are introduced. Furthermore, this new operator has an inverse – that is, it associates with each of the original probability distribution functions, ρ, a unique transformed function, ρ^+. We can then, given ρ^+, reconstruct the ρ from which it is obtained in a unique way.

The importance of the new operator, Λ^+, and the new probability distribution function, $\rho^+ = \Lambda^+ \rho$, is that ρ^+ will, if the original dynamics is at least as randomizing as to be a K system, obey a new dynamical equation of evolution $\rho^+(t) = W_t \rho^+(0)$. This equation has the form of a master equation. It has no "memory" term and is Markovian in its nature. Furthermore, one can define a new fine-grained entropy in terms of the new distribution $S = -\int \rho^+ \ln \rho^+ d\mu$, and one can show that this quantity shows monotonic increase into the future under the dynamic evolution.

It might come as a surprise that such a transformation is possible. The new probability distribution seems to contain all of the information present in the original, because we can uniquely reconstruct the original from it. Yet although the original distribution function certainly had its fine-grained entropy constant, by Liouville's Theorem, the new representation seems to show a fine-grained entropic asymmetry in time. One can understand better what is going on here if one remembers the idea of a coarse-graining that is generating. The doubly infinite series of the box occupation numbers for a generating partition will determine a unique trajectory up to a set of measure zero. So, in a probabilistic sense, a coarse-grained description in terms of the generating partition is equivalent to the original description in terms of a continuous probability distribution. Yet a system can show asymmetry in time in terms of a coarse-grained entropy defined by the new partition even though its original probability distribution had its fine-grained entropy constant. The new operator technique shows that when the system meets the condition of being a K

system, one can go a little further, finding a new transformed probability distribution from which the original one can be reconstructed, and that represents the randomization of the original by obeying a strictly Markovian kinetic equation. The new representation reveals the spreading of the initial ensemble in terms of its own fine-grained entropy.

But what about time-asymmetry? The problem has certainly not yet been solved by this approach. For corresponding to the operator Λ^+ is an operator Λ^-. This operator transforms ρ into a new representation ρ^-, and this new representation can be shown to obey its own kinetic equation. If one defines the fine-grained entropy corresponding to this new representation, one discovers that it obeys an "Anti-H Theorem" – that is, the new entropy is provably monotonically increasing into the past!

Clearly, the resolution of this puzzle will once again require a foray into the problem of initial ensembles, and we shall later critically examine how the originators of this technique use special initial ensembles to "break the time-symmetry" we outline in Chapter 7,III,6.

5. Macroscopic chaos

For the sake of completeness, it will be useful at this point to note some additional results on stable and unstable dynamical systems that have been obtained in recent years, although we shall not be using these results in further explorations of the foundations of statistical mechanics.

We have noted how on the microscopic level both stable and unstable (and, hence, randomizing) kinds of motion can exist. This is true at the macroscopic level as well. Planetary motion is stable. So is the motion of fluids at sufficiently low velocities. But fluid motion that is stable at low velocities can show a marked and very sharp transition to instability as velocities are increased. This is illustrated by the transition from smooth, streamlined, laminar flow of liquid through a pipe into turbulent flow. The transition occurs at a certain critical velocity determined by the nature of the fluid and the geometry of the pipe. After the transition, the trajectory of even a very small piece of the fluid "breaks up" so that the piece in a very short time has its parts widely and chaotically separated from one another. Why does this transition occur?

Early models sought to find in the solutions of the equations of fluid flow – the Navier–Stokes equations – quasi-periodicities or periodic solutions that generated periodicities on top of themselves in higher and higher orders as velocities increased. These were to account for the transition. But all of the models failed, either by not applying in all cases where the transition was observed or because they failed to characterize the transition as sufficiently sharp.

Beginning with the work of M. Feigenbaum and others, equations were found that were very suggestive. They were simple equations containing a parameter. At one value of the parameter, a stable periodic solution of the equations existed. As the parameter had its value varied at a certain point, the stable solution became unstable and a new stable solution of twice the period showed up. Varying the parameter still further, the new stable solution became unstable, and one of twice its period appeared. And so on. Interestingly, the amount of variation of the parameter necessary to induce the transition decreased at each stage by a constant factor ("scaling"). The set of transition points of parameter values approached a limit point. Once the parameter reached this limit point, no stable solutions of the equation existed. Here was the kind of sharp transition to instability that had been looked for.

More directly related to the problem of turbulence was the discovery of "strange attractors." If a dynamical system can dissipate energy to the outside – say, through friction or radiative loss – then all the trajectories from a region may converge on a limited, self-contained sub-set of the initially allowed phase points. The simplest case is when all the trajectories from a region converge to a point, as would describe, for example, a ball rolling in a bowl with a central lowest point. The ball is started at any point in the bowl and loses its total energy through friction with the surface of the bowl, each ball coming eventually to rest at the single lowest point.

More interesting is the case where the system, although dissipative, has energy being supplied to it through some outside forcing. Here, an attractor can exist in the sense of a steady-state trajectory to which all trajectories in a region eventually converge. Then no matter how the system is started, eventually a state is reached where the systems all follow along the same one-dimensional trajectory in phase space.

Such attracting trajectories can sometimes show within themselves a very wild instability. The instability amounts to this: Pick any interval along the trajectory, no matter how small. Then there exist points in that region that will diverge from one another along the trajectory very quickly as time goes on. So that although the systems all converge to motion along this single trajectory, the trajectory itself has a wildly unstable nature. If one makes a Poincaré map by recording the intersections made by this trajectory with a two-dimensional sub-space of the phase-space, the points of intersection form "wild" sets and two systems that pass through the intersection space at points close together in the intersection set will find their future intersections far apart.

Whether or not such "period doubling" instability transitions and transitions to strange attractors correctly model the transition to turbulence of interesting physical systems is currently a matter of intensive research.

III. Interpretations of irreversibility

With the resources provided by the last two sections, we can now proceed to examine critically various accounts offered to explain the existence and nature of time-asymmetry in the world. I will begin with three "nonorthodox" approaches to the problem, and then move on to the mainstream approach, which seeks the origin of time-asymmetry in physical facts about genuinely isolated systems governed by time-symmetric micro-dynamical laws. One major approach – that which invokes cosmology as a part of the explanation of time-asymmetry – will be reserved until the next chapter.

1. Time-asymmetric dynamical laws

From the earliest days of reflection upon the paradoxes of irreversibility, the suggestion has been forthcoming that the source of time-asymmetry is to be found in some underlying asymmetry of the fundamental dynamical laws. Sometimes the suggestion is made that the standard microdynamics is itself non-time-reversal invariant when properly understood. Sometimes the source is sought in some specific body of laws of interaction – for example, electrodynamics, laws that are alleged to be timeasymmetric. Sometimes it is suggested that beneath the compositional structure with which our standard theories deal lies some additional microlevel of structure where time-asymmetry is to be found.

Let us first make two general observations: (1) Demonstrating that a set of laws is time-asymmetric is itself not a completely straightforward matter, as interesting issues in the nature of characterization of states play a role. (2) Even if a set of time-asymmetric fundamental laws is found, that is not by itself enough to solve our foundational problem in statistical mechanics. For we will also require a demonstration that the kind of lawlike asymmetry found can actually account for the kinds of asymmetry with which we are familiar from phenomenological thermodynamics and with which statistical mechanics deals.

To see the point of (1) we need only reflect on a familiar example used to show that an apparent asymmetry of a physical law is not really an asymmetry at all. If we explore the motion of a current carrying wire in the presence of the magnetic field generated by a bar magnet, it appears that the familiar "right-hand rule" for determining the direction in which the wire will move shows us that the appropriate mirror image of the experiment is a situation that cannot, as a matter of physical law, occur. But an appropriate understanding of the nature of the bar magnet – that its magnetism is the result of circulating electrical currents – shows us that the "mirror image" of a magnet is a magnet with its north and

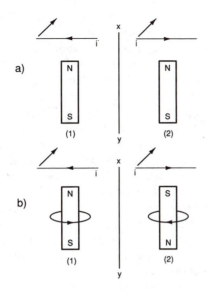

Figure 7-5. Symmetry of electromagnetic laws under reflection in space.
Fig. a1 shows a wire carrying a current i in the presence of a permanent magnet.
The force generated moves the wire into the plane of the diagram. The apparent
reflection of the situation in the mirror xy, as in Fig. a2, seems to show a situation
never encountered in nature. For in the situation pictured, the wire would suffer
forces moving it out of the plane. But Figs. b1 and b2 show the fallacy. For the
permanent magnet is generated by internal current loops, as illustrated in Fig. b1.
The reflected situation, as in Fig. b2, has the current loop reversed, and with it
the polarity of the permanent magnet. So the mirror image of the situation in Fig.
b1 is that of Fig. b2, once again the situation predicted by the electromagnetic
laws.

south poles reversed. With this change, the mirror image of the original
lawlike permissible situation is also lawlike permissible under the standard electromagnetic laws. So what seemed like "non-invariance under
reflection" for electromagnetism becomes an illusion. In any alleged
demonstration of the time-asymmetry of an underlying law, precaution
must be taken to assure one's self that one is not being similarly misled.
(See Figure 7-5.)

Time reversal symmetry means this: If a system prepared in state $S(0)$
evolves under the dynamical laws to a system in state $S(t)$ at time t, then
a system started in state $T(S(t))$, the "time reversed" final state of the
original system, will, governed by the same dynamical laws, evolve in
time t to a state, $T(S(0))$, the time reverse of the initial state of the
original system. But what is the time reverse of a given state, $S(t)$? Intuitively, "first-order" features in time such as velocities (directions of
motion) and rotational velocities (directions of spin) must be reversed,

along with perhaps other features of "odd-order" in time. But the existence of "hidden' motions (velocities or spins of sub-micro-components, for example) can mislead us into what appropriate $T(S(t))$ for a given $S(t)$ is the one we ought to choose.

To see the importance of (2) we need only reflect upon the one plausible candidate we have at present for a fundamental law that is not time-symmetric. This is the law governing weak interactions in physics. Experiment seems to show this interaction to be slightly non-time-symmetric. The experiment actually tests the non-symmetry of the interaction under the combined operations of mirror imaging and replacement of particles by their anti-particles. The inference to non-symmetry under time reversal comes about from a theorem, the CPT theorem, derivable from the axioms of quantum field theory. This theorem tells us that any system must be invariant under the combined operations of exchange of anti-particle for particle, mirror imaging, and time reversal. So if it isn't invariant under the combined operation of the first two, it cannot be invariant under the third either.

Whether the data and theory ought to convince us that weak interactions are indeed non-time-symmetric is an issue that we will not pursue. The point to be made here is just that no lawlike asymmetry of the alleged kind can possibly account for the familiar thermodynamical asymmetries with which we are concerned. Thoughtful computer experimentation and theory have been brought to bear on this issue, and there just seems to be no way that the small time-asymmetry of the weak interactions, if they do indeed exist, can account for the gross asymmetries of thermodynamic irreversibility.

An initially more promising attempt to found the thermodynamic irreversibility on irreversible fundamental laws relies upon a very general feature of wave generation and propagation as we experience it in the world. We see a stone fall into a pond generating outgoing waves around the point of impact. We don't see converging waves transfer energy to a stone at the bottom of the pond in such a way as to propel it out of the water. Similarly, we see accelerated charged particles generating outgoing waves. These waves carry energy and "damp" the acceleration of the particle. We don't see electromagnetic waves converging on charged particles from all directions in a symmetrical way so as to interact with the particle and generate an accelerated motion of it.

The nature and source of this asymmetry was the topic of an extremely interesting debate in 1909 between Einstein and W. Ritz. Ritz took the position that the existence of "outgoing" radiation and the absence of "incoming" radiation of the appropriately symmetrical kinds revealed an underlying time-asymmetry in the laws of electromagnetism. Einstein argued, on the other hand, that the asymmetry observed in the world was, once again, grounded on "initial conditions" and not a reflection of any fundamental time-asymmetry of the fundamental laws. From the Einstein point of view, the absence of the appropriate kind of symmetric converging radiation was not a fact ordained by the laws of electromagnetism, but rather the absence of a situation permitted by the laws but excluded by the actual distribution of initial conditions. The converging radiation was just like the anti-thermodynamic motions of the molecules in a box, permitted by the mechanical laws of molecular interaction but never (or with great rarity, as predicted by the statistical theory) observed.

A full understanding of electromagnetism, and of the difficult problem of the interaction of the fields with particles taken as pointlike, is one of some degree of subtlety. Even within the theory of quantum electrodynamics, questions remain. Indeed, there is even a current of thought that suggests that the now-orthodox approach of treating the electromagnetic field as a constituent of the world is a mistake, suggesting that the field theoretic approach be replaced by a subtle and ingenious "action-at-a-distance" account of electromagnetic interaction among charges. The general consensus is, however, that the fundamental laws of electromagnetism are symmetric in the time-reversal sense. The origin of the wave asymmetry noted here, then, is to be sought in the direction suggested by Einstein, as a constraint upon initial conditions.

From this point of view, the natural assumption for the behavior of charged particles and radiation trapped in a perfectly reflecting box would parallel the description of a gas trapped in a box offered us by Boltzmann. Over an infinitely extended time span, one would find the system almost all of the time in a condition close to equilibrium, with energy being exchanged "randomly" between radiation field and charged particles. Infrequently, one would find conditions far from equilibrium, say with energy highly concentrated in the particles. Such incidents would be preceded by situations in which unlikely converging radiation impacted on charged particles and succeeded by situations in which the energy once more dissipated into the radiation field by means of outgoing radiation from the now accelerated particles. The greater the divergence of the situation from equilibrium, the less often it would occur. The picture is perfectly time-symmetrical, as was Boltzmann's picture of the molecular gas over infinite time, and like the Boltzmann picture raises once more the question of the origin of the actual asymmetric situation we find in the world.

Under certain circumstances, one can give a nice formal rendering of the condition that must be the case in order that we experience, as we do, only outbound coherent radiation and not inbound coherent radiation. A time-symmetric condition of "conditional independence" has been formalized by O. Penrose and I. Percival. Essentially it states that correlation among radiation fields at spatially separated points is found only when such correlation can be traced back to a "common cause" in the radiation having its common source at a point charge. Such a condition parallels the kind of condition we might impose on the correlation of the motion of molecules in a gas, permitting only those that correctly correspond to increase of entropy and prohibiting their Loschmidt reversals. But what we would like is an understanding of why such a condition correctly describes the world of our experience.

One can imagine a program to try and account for all the irreversibility in the world in terms of radiation irreversibility. Such an account would take a posit of spatial non-correlation for radiation of the Penrose–Percival kind and try to offer an account of the origin of the thermodynamic irreversibility in terms of the electromagnetic nature of interaction among molecules. Alternatively, and this is the more common suggestion. one could try to offer an account of radiation irreversibility in terms of the asymmetry of the molecular behavior of the radiating and absorbing charges in the world. In Chapter 8 we shall return to these issues when we deal with cosmology and its role in attempts at understanding irreversibility. There we shall look briefly at the problem of the radiation asymmetry in the context where outbound radiation can "go off to infinity" instead of being absorbed by "nearby" charges. We shall see one ingenious account to explain radiation irreversibility in this cosmological context by resorting to just such a "thermodynamic" asymmetry among the charges in the world as mentioned earlier.

Although the possibility of finding the origin of asymmetry in the time-reversal non-invariance of the underlying laws is by no means a closed issue, I shall assume, along with the orthodox majority, that that route is not the one to pursue. So we will continue our exploration of the search for the origin of irreversibility making the common assumption that either the fundamental laws governing the micro-dynamics and the dynamics of radiation are time-reversal invariant, or that if they are not, their violation of this symmetry is too small to account for the gross asymmetry in time of the world we are trying to understand.

2. Interventionist approaches

From quite early days in the discussion of the foundations of the statistical theory, resort was made to the fact that the actual systems of interest with which we deal in the world are not, as they are in our micro-canonical idealizations, genuinely energetically isolated from the external world. It is surely impossible to shield a system entirely from the effects of energetic interchange by particle collision, stray electromagnetic fields, and so on. And even if these could be shielded for with perfection (as they can be well shielded for in practice), there is no possibility whatever of shielding the system from its gravitational interaction with the outside world.

We ought then, according to this stance, to treat systems not as genuinely energetically isolated, but as subject to perpetual intervention from the outside world. What could we hope to obtain by invoking such intervention? First, we might use it to explain away the "paradox" of the conservation through time of fine-grained entropy. If the system is truly isolated, the deterministic nature of the evolutionary dynamics assures us that no "information" about the initial molecular state is ever truly lost. In a gas confined to the left-hand side of a box, the restriction of the gas to a portion of the box is represented by an f_1 function that gives zero probability for finding a molecule on the right-hand side of the box. After the partition is removed, f_1 eventually takes on a form representing a uniform probability for finding the molecule anywhere in the box. But the initial restriction on molecular position is now contained in the complex f_n function. The information is "spread out" in the correlational facts about molecular probabilities. These are the correlational facts that tell us that if the ensemble were time-reversed it would revert to its initial form representative of the originally confined gas.

But if we allow for even small intervention into the motion of the molecules by effects from the outside, the sensitivity of the correlational function f_n, to even infinitesimal displacements in position and momentum of the molecules will "dissipate" the correlational information into the external environment and break the conservation of fine-grained entropy. Here, one usually relies upon the facts about the instability of trajectories used to derive mixing type results to argue that the initially macroscopically determinable information about the system does in fact become, over time, spread out among the correlations. The spreading occurs in a manner such that even infinitesimal disturbances will quickly make the ensemble representing, say, gas that was originally in the left-hand side of the box, and that representing gas that was in the right-hand side, indistinguishable from one another.

But it is not simply the non-conservation of fine-grained entropy for the system we would like to obtain. The irreversible nature of the evolutionary process also needs an account. Here, the idea usually is that if we started with an ensemble of systems initially coherent in the sense of representing a constraint on the members of the ensemble that has been removed, the continual causal intervention of the outside world would so modify the systems at a later time that no Loschmidt-imagined reversal of them would actually construct a set of systems that would all evolve back to their initial non-equilibrium state. Indeed, if the reversal were performed on a set of systems that had reached equilibrium at the later time, we would find, given the randomization of their molecular correlations accomplished by the intervention from the outside, that the overwhelming majority of the systems would remain in equilibrium. This would be in the manner predicted by equilibrium statistical mechanics with its time-invariant ensemble distribution. Essentially, the outside intervention would convert the fibrillated and merely coarse-grainedly spread-out ensemble into a genuine equilibrium ensemble, an ensemble spread out over the now available phase space in a genuinely fine-grained way.

Finally, we might like to go further and derive the details of the approach to equilibrium from the intervention. Here, the idea is to start with our initial ensemble, impose intervention from the external environment in the form of some randomized "forcing" superimposed on the otherwise deterministically fixed evolutionary motion of the ensemble, and derive from these two components an evolution of f_1, calculated from the now forced evolution of the ensemble, a behavior consonant with a kinetic equation such as the Boltzmann equation. The derivation is not to depend upon some special initial condition imposed on the initial ensemble, as is done in the Lanford derivation, for example, but is supposed to apply generally to any initial ensemble subject to the constraints imposed by the macroscopically construed initial non-equilibrium situation.

Typical of the problems of detail that the interventionist approach must resolve is to show that the time scales that will be derived are appropriate. The time taken by a system to reach internal equilibrium, for example, if the system is reasonably well insulated from its surroundings, is far shorter than the time taken for the system to come to equilibrium with the external environment with which it is so weakly connected energetically. J. Blatt provides reasonable, if not truly rigorous, derivations of the result that a plausibly construed randomizing intervention will "destroy the memory of the initial state" of the system – that is, drive its ensemble representation to the equilibrium ensemble – in a much shorter time frame than it would take for the system to come to equilibrium with the external perturbing environment.

Let us ask two questions. First, is it reasonable to credit intervention into the system by outside influences, due to imperfect energetic isolation of the test system, with all of the effects we normally subsume under the notion of entropic increase? Second, can the invocation of outside intervention resolve the fundamental problem of the non-equilibrium theory, the origin of asymmetry in time?

Insight into the first issue can be gained by reflecting on the results of the spin-echo experiments discussed in Section I,1 of this chapter.

Advocates of the interventionist account have sometimes used these results to bolster their case. After all, they argue, the results of the spin-echo experiments show us that without the outside intervention, in the case where the test system is genuinely energetically isolated over a period of time in a manner sufficient to block "randomization" of the micro-states in question by outside perturbation, the fine-grained entropy does *not* increase. Its theoretically predicted conservation is experimentally revealed to us by our ability to reconstruct the original macroscopic order out of apparent macroscopic disorder by the "Loschmidt reversal" of the spin flipping.

But from another point of view, and I think a persuasive one, the spin-echo results cast doubt on the reasonableness of placing all the responsibility for entropic increase on outside perturbation. After all, in the first stages of the spin-echo process there is an "entropy increase" of the familiar sort. That is, there is a dissipation of macroscopic order (the internal magnetization due to the alignment of the spins) into macroscopic disorder or uniformity (with the spins uniformly distributed in all directions in the plane into which they were originally flipped). Doesn't the spin-echo experiment really show us that at least in one case, there is the macroscopic behavior of the kind we normally take ourselves to be explaining in thermodynamics, despite the *provable absence* of outside intervention? This absence is revealed in the reconstructibility of the original order by the second flip. This kind of entropic increase is, plainly, merely "coarse-grained" increase. But doesn't the experiment convince us that even when fine-grained entropy is conserved, we should still expect an asymmetric increase of the kind of entropy familiar to us from thermodynamics? To be sure, the information about the original order of the system in question can't vanish from the system as a whole without something like outside intervention to allow it to dissipate into the outside environment. But what the spin-echo experiment shows us is that there is loss of information to be accounted for even when the full information has spread itself out into correlations among the micro-components of the system without truly disappearing altogether.

It is also important to remark that the interventionist explanation of entropic increase requires as one of its components just the reliance upon instabilities of trajectories and coarse-grained spreading of the ensemble that the advocate of internally generated coarse-grained entropic increase relies on. For it is only this that guarantees the spreading of the initial order into the correlations so extraordinarily sensitive to perturbation from the outside. Essentially, unless the evolution of the ensemble under its internal dynamic was from a coherent to a greatly fibrillated one, external perturbation could not succeed in converting the ensemble into a genuinely fine-grained equilibrium ensemble by slight perturbations of the trajectories of the member systems. So the advocate of coarse-graining as the source of entropic increase is likely to find the interventionist final step unnecessary. From his point of view, entropy has increased, in a form sufficient to explain the observed thermodynamic increase, even if the perturbation from the outside never occurs.

Can invocation of outside perturbation account for the asymmetry of entropic increase? At first, the answer might seem affirmative. After all, if we start with an ensemble representing some intrinsically ordered system, random external perturbation of the trajectories certainly will result in a future ensemble that represents a system that has lost some of its internal order. But here we must focus on an argument of a type that will often occur: Suppose we focus our attention on systems that share some kind of internal order at some point in their lives and that have been subjected to random perturbation from the outside in their earlier history. By an argument that exactly parallels the one used by the interventionist to show that ensemble entropy is increased into the future by random influence from the outside, we can argue that the entropy of the ensemble representing our systems ought to be higher in the past than it is at the time in question. This would lead us (or, rather, mislead us) to infer the false conclusion that it was highly probable that the systems were earlier in a more disordered state. But, of course, what we ought to infer is that at the earlier time the systems were more ordered still, although statistical mechanics won't tell us what that order was because many diverse ordered states approach the same equilibrium condition.

I can only imagine one way in which the interventionist can block this argument from parity of reasoning. It would be to argue that the intervention from the outside is itself time-directed. Because intervention is causation, and because causation is from past to future, the intervention can only modify the ensemble toward the future direction of time. But without some deeper understanding of what is being used here, some understanding of how causation is playing some role over and above lawlike correlation of states, this sounds more like an a priori restrictive instruction on when to use statistical mechanics and when not to. This is like O. Penrose's "principle of causality" concerning ensembles, which says that "the phase-space density at any time is completely determined by what happened to the system before that time, and is unaffected by what will happen to the system in the future." But is it an explanatory account of why the asymmetry holds in the world?

Once again, much more could be said to defend aspects of the interventionist approach, but I will assume for the most part (with a return to these questions in Chapter 8) that we want to account for the asymmetry of systems in time without reliance upon the necessity of continual interaction of them with the outside world.

3. Jaynes' subjective probability approach

We earlier explored Jaynes' proposal to found the rationalization of the standard probability distributions used in equilibrium statistical mechanics upon an interpretation of probability as subjective degree of belief. This approach utilized the Principle of Indifference, backed up by an appropriate invariance principle, to determine the probability distribution chosen. I argued earlier that the unpacking of this rationale, especially in the justification for choosing the standard measure over the phase-space as the one relative to which probability was to be distributed uniformly, implicitly rested upon the use of ergodic results that Jaynes had hoped to bypass. Jaynes has extended his approach to statistical mechanics to two considerations of importance in the non-equilibrium case, and it is to these proposals that we now turn. One proposal is to offer a general rule for the choice of initial non-equilibrium ensembles and to rationalize that choice. The other proposal is to found the derivation of the Second Law on certain elementary considerations of dynamics and of the method proposed for determining probability distributions and entropy. The desire is to vitiate the need for appeal to non-time-reversal invariant laws, external perturbation, coarse-graining, or to the subtle and powerful results of dynamics such as mixing.

In equilibrium statistical mechanics, Jaynes' prescription was to choose the probability distribution that maximized entropy (now thought of in the information-theoretic vein) relative to the known macroscopic constraints using, of course, the standard measure over the phase space to characterize the space of possibilities. The suggestion in the non-equilibrium case is to abide by this same general rule in determining the initial ensembles whose evolution is to characterize the evolution of a system prepared in a non-equilibrium condition.

Where the non-equilibrium state is one that was in equilibrium until a macroscopic constraint was changed, this rule seems at least as rationalizable as the prescription for the equilibrium case. The rationale can probably be motivated as well for those cases of non-equilibrium in which the system is at least close to equilibrium as a whole. The other common general case, where the system is describable as being locally in equilibrium even if far from equilibrium as a whole, is trickier. Here again, a "maximize information theoretic entropy relative to the standard measure over the phase-space" rule for determining the initial ensemble probability distribution can be constructed. But the rationale for it previously available, in which the prescription was backed up by the ergodic result that such a probability distribution was uniquely temporally invariant, now seems out of place. Perhaps a line of reasoning that argued that the locally equilibrium nature of the system gave us reason to assume that if we isolated a sufficiently small portion of it energetically it would remain in its local equilibrium condition could serve to justify the standard locally equilibrium ensemble as appropriate to the locally equilibrium macroscopic condition of the system.

In any case, Jaynes suggests his "maximize entropy" rule as a general rule for determining initial ensembles. His rationale for it is, again, that the choice of probabilities, being determined by a Principle of Indifference, should represent maximum uncertainty relative to our knowledge as exhausted by our knowledge of the macroscopic constraints with which we start.

Jaynes' "almost unbelievably short" derivation of the Second Law is more to the point here where we are seeking the ultimate rationale for the time-asymmetric success of statistical mechanical methods. The proof follows from the Jaynes method for dealing with non-equilibrium situations.

First, there is the injunction that these statistical methods are to be applied only when the situation is one of "reproducible results." That is, they are to be applied only when the macroscopic behavior, determined as it is by the microscopic state of the system with its varying possible initial state but dynamically determined evolution does in fact follow a lawlike (or at least statistically lawlike) course.

Next, one starts at a time with some known values of macroscopic parameters, o_i (or probability distributions over them). One determines the initial ensemble $\rho(t_o)$ by utilizing the principle of maximizing entropy relative to the standard measure – that is, by picking $\rho(t_o)$ so that $o_i(t_o) = \int \rho(t_o)O_i$ and $-\int \rho(t_o)\log\rho(t_o)$ is maximal, where O_i is the phase function appropriate to o_i. To calculate the time evolution of physical parameters o_i, one identifies the value of a parameter at a time as the phase average, relative to the evolved initial ensemble, of the appropriate phase function O_i. Next, one identifies the experimentally determined entropy at a time S_e as the entropy determined by the maximum entropy ensemble appropriate to the new evolved values of the o_i's.

So the initial experimental entropy, $S_e(t_o)$, is equal to the initial Gibbs entropy at t_o – that is,

$$S_G(t_o) = -\int \rho(t_o)\log\rho(t_o)$$

At a later time, t_1, the new, predicted, macroscopic parameter values will be $o_i(t_1)$ calculated as $o_i(t_1) = \int \rho(t_1)O_i$. Relative to these values will be a new maximum entropy ensemble, $\hat{\rho}(t_1)$, which is determined by the maximum entropy principle and the constraint that for each o_i we must have

$$o_i(t_1) = \int \hat{\rho}(t_1)O_i$$

We take the new experimental entropy, $S_e(t_1)$, to be $-\int \hat{\rho}(t_1)\log \hat{\rho}(t_1)$ by the identification of experimental entropy with Gibbs entropy of the maximal entropy ensemble relative to the new values of the macroscopic constraints. Then we argue as follows:

$S_e(t_o) = S_G(t_o)$ by the rule for experimental entropy.

$S_G(t_1) = S_G(t_o)$ by the consequence of Liouville's Theorem that fine-grained Gibbs entropy is invariant in time, where $S_G(t_1)$ is the Gibbs entropy at t_1 calculated using the ensemble $\rho(t_1)$ that dynamically evolved from that initially set up at t_o, $\rho(t_o)$.

Now define $\hat{S}_G(t_1)$ as the Gibbs entropy calculated by using $\hat{\rho}(t_1)$. It follows that $\hat{S}_G(t_1) \geqslant S_G(t_1)$ by the choice of $\hat{\rho}(t_1)$ as the distribution that maximizes Gibbs fine-grained entropy relative to the new constraint values and the fact that for both $\rho(t_1)$ and $\hat{\rho}(t_1)$ the constraint that o_i is obtained as the phase average over the ensemble of the appropriate phase function is met.

$S_e(t_1) = \hat{S}_G(t_1)$ by the rule for identifying experimental entropy with Gibbs entropy calculated by the maximum entropy distribution relative to the constraints.

So, $S_e(t_1) \geqslant S_e(t_o)$. QED

Do we have here the explanation of time-asymmetry we have been looking for? It would seem not. First, of course, the derivation completely ignores the search for the explanation of the *details* of the non-equilibrium process. Features of the evolution such as the relaxation time, the values of transport coefficients, and the detailed progression of the system as governed by the solutions of the kinetic equation are bypassed entirely. As several authors have pointed out, the "proof" even fails to demonstrate the monotonic increase of entropy we would like to derive. It shows only that entropies at later times are greater or equal to the initial entropy, not that entropy at one later time is greater than or equal to the entropy at some time intermediate between it and the initial time. It is crucial in the proof that one compares final entropy with entropy at the time in which the initial ensemble is constructed. The criticism here is interestingly reminiscent of one of the Ehrenfests' criticisms of Gibbs' proof of the increase of coarse-grained entropy. There the point was that although one could show that later coarse-grained entropies were at least as great as the coarse-grained entropy of the ensemble at the initial time in which the ensemble was set up, nothing in Gibbs' argument showed that the change in coarse-grained entropy had to be monotonic.

Even more disturbing is an argument due to Jaynes himself. He points out that the very same reasoning applied in the demonstration suggests the paradoxical result that the experimental, thermodynamic entropy at earlier times ought to be larger than that at the time the observations are made. All one need do is repeat the proof with t_1 taken as referring to earlier times than t_o and applying the principle of invariance of fine-grained entropy that follows from the determinism of the dynamic evolution in the past direction.

Jaynes' response to this is to argue that the various "rules" that generate the proof are rules that apply only to experimentally reproducible processes. These rules are the rule of predicting new macroscopic values by averaging over the ensemble generated by dynamic evolution from the initial ensemble, and the rule of taking experimental entropy to be fine-grained entropy calculated by a new ensemble that is maximum entropic for the new parameter value. But processes in one time direction are experimentally reproducible. The same non-equilibrium state always follows a statistically lawlike evolution to equilibrium. But this is not so in the other time direction, because many distinct routes from many distinct initial non-equilibrium states all lead to the same final equilibrium. This is certainly true. But it is just that fact that there is this parallelism in time of systems – that distinct systems show experimental reproducibility in one time direction and not the other and that it is the same time direction for all systems – that we want explained when we are trying to *account* for temporal asymmetry. From this perspective it is hard to see Jaynes' argument as anything but question-begging.

There is an important additional aspect of Jaynes' approach worth pursuing at this point. The demonstration of the Second Law works by assigning to the later state of the system the entropy appropriate to an ensemble constructed by ignoring the known past history of the system and instead acting as if all we knew about it were the values of the evolved macroscopic parameters. From the point of view of Jaynes' objective Bayesianism this is puzzling, because the general prescription it gives for ensemble construction is to maximize entropy subject to all of the constraints we know. What justifies this indifference to the known facts about the historical origin of the system in question?

This is an especially important question to ask given Jaynes' response to the results of the spin-echo experiment. Here, if we attempted to predict the evolution of the system by statistical mechanics after the system has apparently achieved equilibrium and then been flipped over, we would predict a stationary equilibrium state and not the evolution back to the original non-equilibrium condition actually observed. Jaynes points out that if we constructed our ensemble utilizing our knowledge of the historical origin of the system as (1) having evolved from a non-equilibrium condition, (2) kept its coherence in the sense of not having dissipated it to the outside environment, and (3) then been Loschmidt-reversed, we would get the correct results. In at least this case, then, correct results are obtained by *not* ignoring historical antecedents in constructing our ensemble. Of course, the fact that history can usually be safely ignored is just one more way of saying that usually a rerandom-ization trick of the kind used traditionally to derive the kinetic equations will be successful. But we want to know why such a procedure will be successful when it is. And we want some indication of the conditions under which it is and is not appropriate in constructing our ensembles at a time for making future statistical predictions to ignore the historical origin of the system and rely instead on its current macroscopic parameter values in choosing our constraints. And to this question the objectivist Bayesian (or, as Jaynes calls it, "subjective") approach provides no answer.

This is the appropriate place to make a few remarks about the introduc-tion of asymmetry into statistical mechanics suggested by Gibbs and described by the Ehrenfests as "incomprehensible" to them. Suggestions related to those of Gibbs have recurred later – for example, in some aspects of the approach to irreversibility of S. Watanabe. And they have been hinted at by E. Schrödinger.

Gibbs had suggested that the source of irreversibility was to be found in the fact that although it was frequently appropriate to infer the future on the basis of probabilistic posits over initial conditions, it would generally be entirely inappropriate to so infer into the past. Underlying this claim seems to be the assertion that although future events are not yet known to us, hence have as their "probability" only such probability as can be inferred by us from our statistical theorizing, past events, being known by the direct evidence of their occurrence, are not subject to having probabilities attributed to them in this way. This is certainly at least related to those doctrines that take the probabilities of the statistical theory to be subjective, the line being that knowing an event to have occurred gives it a subjective probability of one and precludes our as-signing it some inferential probability on theoretical grounds.

Although the Ehrenfests' characterization of "incomprehensible" for this line seems rather too strong, it certainly is an approach to irrevers-ibility fraught with difficulty. One problem is that past events are as frequently a matter of inference to us as are those of the future. To be sure, if we saw a gas evolve from the left-hand side of a box to fill the box we would be, from a subjectivist perspective, unreasonable in our inference of past state from present equilibrium state. That inference would infer a past equilibrium as well. This is because we already know that the past state was one far from equilibrium. But such an argument would still direct us to infer the *unknown* past state of a system found in equilibrium to be one of equilibrium and of a system found out of equilibrium to be, if its past was otherwise unknown to us, closer to equilibrium rather than farther away from it as we generally would, correctly, infer.

Such wrong-headed inferences to the past are blocked, of course, by our general knowledge that states of isolated systems proceed (in the statistical sense) from non-equilibrium to equilibrium in a temporal direction parallel to one another, the direction we call from past to future. But that claim, of course, is the posit of irreversibility in statistical mechanics whose justification and explanation was what we wanted in the first place. This seems quite on a parallel with the criticism of Jaynes that it is just the fact of parallel reproducibility of systems into the future that we wanted to understand.

In Chapter 10,III,3, we will examine briefly the claim that the very asymmetry of our knowledge of past and future, the fact that we can remember the past and have records of it but can only infer future events from direct knowledge of the past and inference founded upon generalization, rests upon the entropic asymmetry of the world. The claim, although highly plausible to many, is one that remains problematic and difficult to establish. But, in any case, it is hard to see how the entropic asymmetry itself can be thought to depend upon any relativization of thermodynamic notions or their statistical surrogates to our subjective states of knowledge combined with some given asymmetry of our knowledge of the world. Whatever subjective elements enter into the concepts of statistical mechanics through such notions of coarse-graining or the relative nature of entropy, it is hard to fill in any subjectivist theory in such a way as to convince us that the parallelism and asymmetry of entropic increase is only an asymmetry of our inferential applications of probability founded on asymmetry in our knowledge of past and future events.

4. The mainstream approach to irreversibility and its fundamental problems

If we reject the possibility of finding the origin of kinetic behavior and irreversibility in fundamentally asymmetric physical laws governing the micro-components behavior, in continual outside intervention into the evolution of the system in question from the outside, or in the structure of probabilistic inference understood from a subjectivist or logical theory of probability point of view, then we must seek the answer to the foundational questions somewhere else.

We wish to obtain from our account a variety of results of increasing richness. First, we wish to show how, in some sense, an approach to equilibrium arises, and we wish to understand why this process is asymmetric in time. We would like to go beyond this, however, and understand why it is that systems in the world approach equilibrium in a particular, finite time scale. That is, we would like to derive the length of the relaxation time from our account. Finally, we would like to go beyond that and understand the origin of the particular structure of the approach to equilibrium as described in the usual thermodynamic equations describing non-equilibrium or in their statistical mechanical counterparts, the kinetic equations.

Initially, we have as our resources the constitution of the system, its composition out of micro-components, the detailed form of the interaction among these components, the vast number of degrees of freedom in the system generated by its vast number of micro-constituents, and the existence of the system, in our anti-interactionist idealization, as genuinely energetically isolated from the world (at least in the case where we employ the micro-canonical ensemble).

We know that our account will be genuinely statistical, and that the evolution we describe will be not that of an individual system but that of an ensemble of systems. Or, less picturesquely, the evolution we describe will be that of a probability distribution over micro-states of systems compatible with the macro-constraints defining the systems of interest. We expect that for the description of the approach to equilibrium, and certainly for the characterization of the counterparts to the thermodynamic quantities that obey an evolutionary equation, we will be dealing with functions that are derivable from the full ensemble description but that individually reflect only part of the information contained in that full probability distribution. As we shall see, however, in some foundational approaches it will be something rather closer to the full probability distribution whose evolution is deemed demonstrably thermodynamic in kind.

We shall also need a way of understanding how to "read" the kinetic equation that requires something over and above the correlation of ensemble defined quantities with thermodynamic variables. Here I have in mind something like the Ehrenfests' reading of the solution curve of the Boltzmann equation as indicating not the most probable evolution of a system but, rather, the "concentration curve," the curve that plots the succession in time of overwhelming most probable intermediate states. Because, as we have noted, other attempts at rationalization, such as the Lanford derivation of the Boltzmann equation, read the kinetic equation as indicating an overwhelming probability of evolution, in contrast to the Ehrenfest approach, this interpretation of the equation of ensemble evolution, even in the statistical sense, is non-trivial.

Finally, there is one more crucial ingredient we expect to find in any orthodox attempt to ground statistical mechanics. We presuppose the underlying determinism of the dynamical laws governing the micro-constituents. We have eschewed relying upon interference in the evolution of the system from outside it. We also wish to avoid the invocation of some principle of rerandomization in our derivation. It is just such hypotheses of molecular chaos, imposed on the underlying dynamics from above, that we want to avoid. The very compatibility of such a posit with the underlying dynamics, much less its truth, is something that would have to be shown in any foundational rationale of the theory or in any explanatory account of its success.

But although we are not entitled to principles of rerandomization, we are entitled to posits about initial ensembles. The evolution of an ensemble is the joint product of its initial structure and the deterministic dynamics of evolution of the individual systems making up the initial ensemble. We fully expect that it is in the nature and constitution of this initial ensemble that we supply the crucial ingredients necessary to derive the kinetic behavior we want to extract out of the ensemble evolution. So the appropriate nature of such initial ensembles, the study of how their choice governs the results extractable from the picture, and most important of all, the crucial questions about why certain initial ensembles are the appropriate ones to choose and others are appropriate ones to avoid in a correct description of nature, will be critical components of any foundational account that attempts to rationalize non-equilibrium statistical mechanics.

5. Krylov's program

An extremely insightful and careful foundational study of non-equilibrium statistical mechanics is contained in the fragments of N.S. Krylov's *The Foundations of Statistical Physics*, published posthumously in 1950. Krylov believed that he could show that in a certain important sense, neither classical nor quantum mechanics provided an adequate foundation for statistical mechanics. His intended work was to begin with a detailed critique of orthodox attempts to found statistical mechanics on either of these underlying micro-dynamic theories, and then to provide the alternative positive account that would take their place. More correctly, he proposed to supplement them in a crucial way. Unfortunately, only the critical portions of the work were completed before Krylov's premature death, and we have only rather speculative reconstruction to go on in trying to formulate his positive program.

Part of Krylov's work constituted early contributions to the notion of ensemble spreading and fibrillation that later became formalized in the notion of a mixing ensemble. He emphasized the inadequacy of ergodicity by itself to ground the non-equilibrium portion of statistical mechanics, and he outlined in a thoughtful way the notion of mixing as it had developed from the Gibbs ink-in-water analogy through Hopf and others. The notion of mixing he has in mind is both richer and poorer than the formal notion outlined earlier. It is poorer in that it is not yet expressed in the full and subtle measure theoretical language developed by Kolmogorov and others. Again, although Krylov brilliantly emphasizes the role of dynamic instability in generating the mixing nature of an ensemble's evolution, he is not yet in the position achieved later of being able to formulate a precise condition, such as being a *C* system, sufficient to be able to prove mixing in the formal sense. Nor is he of course able to characterize those exact conditions on the constitution of a system sufficient to prove that it is indeed a *C* system.

It is a richer notion he intends, however, in the sense that Krylov is interested in discovering the foundational elements necessary to derive in statistical mechanics the actual thermodynamical behavior of systems, including the finite, real relaxation times exhibited by real systems. He would certainly not be satisfied with a demonstration that an ensemble was mixing in the mere "limit as time goes to infinity" sense.

Krylov's most important critical contribution is his emphasis on the importance of initial ensembles. Suppose we can indeed show a dynamic system to be mixing in some appropriate sense. Still we will be utterly unable to demonstrate that the correct statistical description of the evolution of a system will have the appropriate finite relaxation time, much less the appropriate exact evolution of our statistical correlates of macroscopic parameters, unless our statistical description includes an appropriate constraint on the initial ensemble we choose to represent the initial non-equilibrium condition of the system in question.

This is most easily seen by reflecting on Loschmidt's Paradox lifted up to the ensemble level. Suppose an initial ensemble evolves, in the appropriate finite time, to a fibrillated ensemble in the usual Gibbsian fashion. Then an initial ensemble constructed by taking the time-reverse states of the final ensemble will evolve in the same finite time from a fibrillated, spread out, configuration to the compact configuration of the time reverse of the original initial ensemble, and such an ensemble evolution will be of the anti-thermodynamic type. (See Figure 7-6.)

What must appropriate initial ensembles be like in order that their evolution properly describe thermodynamic behavior? They must have fairly simple shape, for allowing any degree of fibrillated complexity will plainly allow for the construction of anti-thermodynamic ensembles. And the initial ensembles will have to have a certain minimal size, for even a simply shaped initial ensemble if it is too *small* will, while spreading out into a fibrillated shape in the future time direction, do its spreading too slowly to represent the actual short relaxation time we experience in dealing with macro-systems in the world. Furthermore, within the boundaries of the simply shaped and sufficiently sized initial ensemble, probability must be uniformly spread. For a nonuniform distribution of

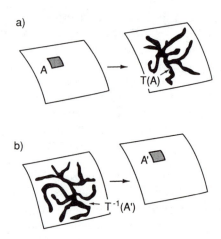

Figure 7-6. Reversibility at the ensemble level. Suppose, as in Fig. a, that A is a region of phase space that evolves over time into the fibrillated region T(A). One can then show, as in Fig. b, that there must be a fibrillated region of phase points, $T^{-1}(A')$, that evolves as time goes into the future into a compact, simple region like A'. Furthermore, both A' and the region from which it comes will be equal in size to A and to its fibrillated successor.

probability within the ensemble boundaries could once more allow the design of an ensemble either anti-thermodynamic or too slow in its thermodynamic behavior.

But then, Krylov maintains, we can see that neither classical nor quantum mechanics by themselves can provide an appropriate underlying dynamic framework for statistical mechanics. For neither of these theories provides any restriction on our ability to prepare systems in such a way that only the appropriate kind of initial ensemble is ever allowed as representative of the system's preparation. Indeed, the absence of limitations on possible initial ensembles would lead us to think that the use of sufficiently careful procedures in preparing systems could pin them down to their exact point micro-state. Each system would then have its evolution determined into the future in a completely lawlike way that specified a unique trajectory. This, Krylov believes, would be incompatible with the genuinely probabilistic nature of the evolution of states revealed to us in thermodynamics and its statistical mechanical underpinnings. So no foundational approach based solely on classical mechanics or quantum mechanics can hope to ground either the fundamental probabilistic nature of the world or the restriction to certain possible initial ensembles as representing the evolution of systems. And both of these are revealed to us by nature.

Krylov then takes up an attempt to avoid these problems by the invocation of what he calls "real ensembles" as opposed to "ideal ensembles." The ideas he is criticizing here are familiar ones, closely related to the insistence of H. Reichenbach, A. Grünbaum, and others that thermodynamics is the result of mere "de facto" regularities in the world, rather than a reflection of underlying laws of nature. The "theorists of real ensembles," as Krylov understands them, take it to be a "mere" fact about the world that initial conditions are so obtained that it is appropriate always in statistically characterizing a system to view it as subsumed under a uniform probability distribution relative to whatever macroscopic conditions are known to hold of the system.

Krylov has two major objections to this view. The first is that it would deprive the laws of statistical mechanics of their genuine lawlike status, reducing the asymmetry of the Second Law, the existence of uniform relaxation times, and the inevitable uniformity of kinetic behavior to mere happenstance, something he holds it impossible to believe. Second, he thinks the position fundamentally incompatible with what we know about nature.

What does the second objection come down to? The "real ensemble" theorist accepts the principle that initial states are so distributed in the world that whatever ensemble boundaries one picks, it is appropriate to view a system as subsumed under a uniform distribution within that boundary. But consider A and B, two intersecting but distinct regions on the energy surface. The probability density in A is the same as that in $A \cap B$. And that in $A \cap B$ is the same as that in B, by the uniformity of the distribution in A and in B. But then the distribution must be uniform over the energy surface – that is, the system must be subsumable under the equilibrium probability distribution, whatever its macroscopic constraints. But, of course, systems are usually not in equilibrium.

Consider, also, the following argument. Suppose the probability distribution at one time under which a system is subsumed is uniform relative to the region picked out by the macroscopic parameters at that time. Moments later the system will have new macroscopic parameters describing it. But the ensemble describing it at that time, if we mean the "real ensemble," *can't* be uniform relative to those parameters. For we know that given the history of the system, its present state must be confined to a sub-region of the totally allowed phase space relative to the new parameters. This is just another repetition of the argument from the conservation of Gibbs fine-grained entropy over time. So if the "real ensemble" was uniform at one time, it could not be a moment later. The principle that we ought to take it to be the case that systems just happen to be so distributed in the world so that we ought always to subsume them under uniform probability measures relative to whatever happens to be their present macroscopic state is not coherent.

What is Krylov's solution then? Unfortunately, here we are to a large degree at a loss to tell, because no full exposition of it was published. But the basic ideas are at least suggested in many places. Krylov sometimes speaks as though he is offering an alternative micro-dynamics to replace the usual classical or quantum dynamical basis, but more often it looks as though he wants to supplement the usual dynamics, which is, after all, still retained to describe the dynamic evolution of an initial ensemble once it is set up. The additional lawlike principle governs the appropriate structure for initial ensembles.

The basic idea is borrowed by analogy from the Heisenbergian interpretation of the Uncertainty Relations to quantum mechanics. In the early attempts to understand uncertainty, Heisenberg's idea – that it was the result of the physical interaction of measuring apparatus with measured system that "jolted" the measured system out of its exact micro-state in the very act of trying to determine what that state was – seemed a plausible reading of quantum uncertainty. This reading only later fell out of favor in the light of "measurement interference" on systems consisting of component parts causally separated from one another. That required a subtler reading of both uncertainty and the measurement process. Heisenberg's "interference" interpretation of uncertainty also became less plausible in the light of "no hidden variables" proofs. Krylov does not utilize quantum uncertainty to found the restriction on initial ensembles. Rather the idea is that the instability of trajectories and the ineliminable finite interaction of preparer with system being prepared guarantees that a collection of systems that we prepare, and whose thermal future is then watched, will be so "crudely" prepared that only the appropriate, standard, initial ensembles will statistically represent the systems so generated.

It is, then, the interaction with the system from the outside at the single moment of preparation, rather than the interventionists' ongoing interaction, that grounds the asymmetric evolution of systems. It is this ineluctably interfering perturbation of the system by the mechanism that sets it up in the first place that guarantees that the appropriate statistical description of the system will be a collection of initial states sufficiently large, sufficiently simple in shape, and with a uniform probability distribution over it. These are just the initial ensembles that when combined with the mixing instability of the dynamics, we can hope will lead to coarse-grained spreading out in the appropriate finite time. Such ensembles are then appropriate to the macroscopic parameters given at the *beginning* of the evolution of the system. They are not appropriate later on, or for arbitrary regions of phase space taken in general, as they would be in the "real ensemble" approach. The system is aptly so described only when it is creatively interfered with. Later on, the ensemble will not really be the uniform one relative to macroscopic parameters as they evolve, but of course we will be able to treat it as such to a degree, because it will rapidly coarse-grainedly achieve such uniformity because of the mixing nature of the system. It is in this mixing that the rerandomization posit of the kinetic equations finds its justification.

A general critique of Krylov's position is impossible, for we have only a sketch of a program to deal with rather than a fully worked out account. Exactly how does this initial interference lead to initial ensembles of just the kind we need? Here, a detailed treatment of preparation in all of its guises and of the effect of preparer on prepared system would have to be carried out. Could this difficult program really be carried out in such a way as to give Krylov the results he needs? It is hard to say. There are, however, a few brief remarks one can make, in the absence of the detailed program. These remarks will at least cast some critical doubt on whether Krylov has pointed out the correct route to travel to find the origins of kinetic behavior and temporal asymmetry.

One problem arises when we once again refer to the results of the spin-echo experiments. In many places, it looks as though Krylov is claiming that it would be *impossible* to prepare a collection of systems whose appropriate representative initial ensemble would be one leading to anti-thermodynamic behavior. But the spin-echo experiment shows us that we can, at least in certain exceptional cases, prepare initial ensembles that have just the fibrillated initial form evolving back to a simpler form posited in the Loschmidt Paradox thought experiment noted earlier. Of course, we have prepared these systems in this very special way, first doing an ordinary preparation of a highly isolated system, and then reflecting its time evolution at a later stage. But that this can be done at all indicates the need on Krylov's part to show us in detail just what it is about ordinary preparations that makes it impossible for them to result in the unwanted extraordinarily badly behaved initial ensembles.

Much more important are the considerations that arise when we reflect upon the important use of the notion of preparation that Krylov utilizes. It is the preparation of a system at the beginning of the time period over which its evolution is observed that so disturbs the system that the only possible statistical representation of it is in terms of a uniform probability distribution over a phase-space region that is sufficiently large and simple in shape. But what *counts* as a preparation?

Imagine a system constituted at one time by some process that leaves it for a short while isolated from its surrounding environment. At the end of that period, something is done that reintegrates the system back into the energetic whole of the surroundings. If at the earlier time the system was "prepared" in non-equilibrium, then the combination of the appropriate initial ensemble, restricted according to Krylov's prescription, and the mixing nature of the system, might be utilized to tell us at the later time that the ensemble representing the system at the time of "destruction" was one closer to the equilibrium ensemble. This leads us to the usual kinetic predictions.

But suppose we have a system that is far from equilibrium at the time of "destruction," and that is all that we know about it. Ought we not to assign to it the same large and simple ensemble region and then predict, wrongly, that it was to be found closer to equilibrium at an earlier time? If not, why not? I doubt that the intuitive response that Krylov would offer could be founded on something special about the particular nature of the creating ("preparing") and annihilating ("destroying") processes. Rather, I think, the intuition would have to be that the reintegration of the system into the surrounding environment could not be considered a "preparation" appropriate to representing the system by the Krylovian type of initial ensemble, because the reintegration doesn't "cause" the state of the system to come into existence at that time, the way the act of genuine "preparation" does. It is the intuitive idea that causation works from past to future that introduces time-asymmetry into the statistical account if that asymmetry is to be founded, as it is in Krylov's interpretation, on the basis of the special nature of initial ensembles.

But the problem of the asymmetry of thermodynamic behavior has then just been shifted to the mysterious nature of "causal asymmetry" in time. Just as the interventionist might be able to get asymmetry out of the statistical theory by invoking an ongoing, time-asymmetric, causal influence of the surroundings on the evolution of systems, Krylov can get time-asymmetry into his account only by relying on a temporally asymmetric notion of causation as distinguishing preparations of systems from their destructions.

This claim can be buttressed by making the following observation. Unless the distinction between creation and destruction rests upon an implicit assumption of temporally asymmetric causation, it would be hard to understand how Krylov would be able to generate the essential temporal *parallelism* of the entropic evolution of systems that he needs. For if the preparation and destruction acts are to be characterized in terms of some intrinsic difference of their nature that does not in itself imply a time-asymmetry, such as simply calling the preparation the act that is always on the same time side of the destruction as each other preparation is on the time side of its associated destruction it is hard to see *why* the time direction from act of preparation to act of destruction should always be the same time direction. And it is hard to see why that direction would always be the time direction we take as from past to future.

Krylov's interpretive account of the origin of kinetic behavior and time-asymmetry remains, then, a tantalizing partial proposal. Filling it out would require both a detailed explanation of how preparation results in the necessity for just the appropriate initial ensembles that when combined with the instability of trajectories generated by the dynamics, would allow us to extract the full finite time and detailed kinetic description we want. And it would require an explanation of how this account will get time-asymmetry out of the notion of preparation (and its distinction from destruction) without putting it in a question-begging way.

6. Prigogine's invocation of singular distributions for initial ensembles

I. Prigogine and his co-workers have constructed an elaborate and mathematically very sophisticated program aimed at the understanding of the legitimacy of the rerandomizing posits used to generate the kinetic equations. They also offer an explanation of the origin of the time-symmetry breaking that leads us from the reversible underlying micro-dynamics to irreversible thermodynamic descriptions.

First, they make a distinction between those cases that fall under the KAM Theorem and those that are to be treated by using the idealizations that allow for the proof of such things as mixing, K systemhood, or Bernoulli systemhood for the ensemble. Rather different treatments of the origins of kinetic behavior must be offered in the two cases.

For the systems for which it can be shown that there remain regions of the phase space within which trajectories remain stably confined – for systems in which the KAM Theorem holds – the approach is to focus on the conditions sufficient to show enough "chaotic" trajectory behavior in the remaining phase-space regions outside the KAM stable tori for the behavior in those regions to be describable in a chaotic manner. Generally, the program hopes to show that in the limit of a system of a very large number of degrees of freedom, in particular in the thermodynamic limit, this chaotic region will be overwhelmingly dominant as the stable tori shrink to insignificance in the phase space.

In these cases, we deal with systems that can be described by a Hamiltonian that corresponds to one obtained by neglecting all interaction among the micro-components but with a small perturbing interaction term then added. A mathematical condition is sought that will simultaneously perform two functions for us. First, it will characterize just those conditions where the orderly behavior of the trajectories in the regions of stability proved to exist by the KAM Theorem cannot be extended into the general phase-space region outside those tori. Here, the idea is that the multiple overlapping resonance regions of the dynamics results in a region that, if it cannot be shown to be mixing or of higher chaos in the measure theoretic senses, is still chaotic in the weaker sense that any existing additional constants of motion over and above the energy fail to determine "smooth" sub-manifolds in the phase space. It will still be true, then, that arbitrarily close to any initial condition leading to one trajectory will be found initial conditions that lead to wildly divergent trajectories. And we will not be able to partition the region outside the regions of stability into sub-regions of trajectories corresponding to systems of similar macroscopic behavior in any controllable manner. This can be the case even if the region of chaos fails to meet the strong measure-theoretic conditions, such as metric indecomposability and its strengthenings, needed to prove such strong chaos results as mixing, being a K system, or being a Bernoulli system.

The other task the mathematical constraint is to perform is to serve as the sufficient condition needed to demonstrate the existence of a decomposition of the ensemble in the manner described earlier as the program of sub-dynamics. That is, the condition is to be sufficient to show us that there is a unique decomposition of the ensemble into two parts. The decomposition must commute with the time evolution so that the evolution of the components can be described by decoupled dynamical equations. And it must be such that in the limit of the interaction going to zero, one component just becomes the dynamic description of the evolution of the ensemble of independent micro-systems. As described earlier, if such a sub-dynamical decomposition of the ensemble evolution is possible, it can be shown that one of the components of the decomposition will obey a master equation as its equation of dynamical evolution. That is, it will have the proper Markovian behavior, with further time evolution at a time dependent only upon its state at the time, to satisfy a kinetic equation type of dynamics. The macroscopic features of the system described by thermodynamics will all then be calculable from the behavior of this kinetic component of the ensemble alone.

The mathematical condition (actually the behavior of the so-called "collision operator" in the limit as its complex variable goes to zero) will then link together the "geometric" nature of the phase-space flow – sufficiently chaotic behavior of trajectories outside the regions of trajectory stability proved existent by the KAM Theorem – with the appropriate dynamical behavior of the ensemble as described by its differential equation of motion – decomposability in a time-invariant way into two components, one of which obeys a Markovian kinetic equation of evolution decoupled from the equation of motion of the other "correlational" component. Needless to say, proving that the condition needed is satisfied, in all but the simplest idealized cases, is a matter of great difficulty.

In the cases where the idealization leads to a provable measure-theoretic type of chaos for the entire phase space of the ensemble – that is, where such things as mixing, being a K system, or being a Bernoulli system are provable – the approach relies instead on the technique of showing that the ensemble and its evolution can be transformed by means of non-unitary transformations into a new representation that reveals the otherwise concealed kinetic properties of the evolution. The demonstration that such a transformation is possible has mixing as a necessary condition and being a K system as a sufficient condition. One can then show that there is non-unitary operator, Λ^+, that will transform the original ensemble into a newly described one. The new ensemble description will generate well-behaved probabilities and will reproduce the same phase averages for all phase functions produced by the original ensemble. Furthermore, the original ensemble can be uniquely reconstructed from the new representation. So, statistically, the two ensembles are equivalent.

Whereas the original ensemble representation will have its time evolution described by the time-reversible Liouville equation, the new representation will have its evolution described by a time-asymmetric Markovian kinetic-type equation. And one will be able to show that in the $t \mapsto \infty$ limit the new representation of an initially non-equilibrium ensemble will approach, monotonically, the new representation of the equilibrium ensemble. In fact, one can do more. A "time" operator can be devised that will characterize the "age" of an ensemble, essentially in terms of its deviation from the "fully aged" ensemble of equilibrium. Normally, ensembles will not have some definite age (they will not be "characteristic" ensembles of the time operator). But they will have an "average" age definable for them. Indeed, it can be shown that the "time" operator obeys a commutation operation with the Liouville operator, the operator generating time-rate of change of the initial ensemble representation. This is reminiscent of the commutation relations among the operators corresponding to "complementary" observables in quantum mechanics. And from this a kind of "uncertainty relation" can be generated. Although the physical interpretation of this relation is not transparently clear, it can be read as saying that the "age dispersal" of an ensemble is inversely related to its dispersal as being compounded out of periodic ensembles. In a sense, then, an ensemble of definite age is infinitely chaotic.

Despite the apparently different behavior of the two ensemble representations, we saw earlier why the possibility of representing a statistical ensemble in these two equivalent ways is perfectly understandable. The new representation doesn't put any time-asymmetry into the ensemble evolution that wasn't there already. It merely brings to the surface a time-asymmetry latent in the original ensemble description. From the usual viewpoint, of course, it is important to remember that this ensemble evolution in either of its representations is just the evolution of a probability distribution over possible systems compatible with the initial constraints on the system. From this orthodox perspective there remains the specific evolution of the individual systems determined by the dynamical trajectory from the point in phase space representing their original microscopic state. In at least some passages, Prigogine suggests that this standpoint is misguided. In conditions where even our most exacting, if not "perfectly exact," measurements cannot fix a phase-point region in such a way as to exclude there being in the region delimited by our measurements phase points whose future trajectories diverge wildly from one another – that is, to say in those regions of phase space outside the KAM regions of stability where they exist, or in the general phase space where measure theoretic chaos holds – we ought, he asserts, to deny the reality of the exact micro-state altogether. We should hold it to be a false idealization of the physical reality. This is an issue to which we shall return when we discuss the alleged reduction of thermodynamics to statistical mechanics in Chapter 9,III. For now we need only mention it, because the further development of the Prigogine view – the invocation of special initial ensembles as the account of time-asymmetry and irreversibility – doesn't really depend on this more radical view of the nature of individual systems.

Both approaches to the problem of the derivation of kinetic behavior from dynamical behavior we have just outlined leave the problem of time-asymmetry at least partly unsolved. Corresponding to the derived Markovian master equation, in the case where one is dealing with the chaotic region outside the regions of KAM stability, is a symmetrical anti-master equation. Whereas the former has monotonically equilibrium approaching solutions in future time, the latter has solutions that montonically diverge from equilibrium from past times. Corresponding to the Λ^+ transformation that takes the ensemble to a new representation in the K system cases, a representation that obeys a Markovian equation of monotonic approach to the new representation of the equilibrium ensemble in future time, is a Λ^- transformation that provides a new representation of the ensemble that diverges from the new representation of equilibrium from past times. How can we "select" the proper representation, and how, physically, can we account for the reason that one of these representations is descriptive of the nature of the world and the other not? Prigogine's approach to this problem is expounded primarily in the context where the idealization allows for the clearer and more powerful notion of chaotic evolution of the trajectories in the full measure-theoretic sense – that is, where the ensemble is taken to represent K systems.

Not surprisingly, the key to time-asymmetry is found in the notion that only specific kinds of ensembles are permitted as initial ensembles. Can we find a class of initial ensembles that will "pick out" one direction of time in preference to the other? These initial ensembles will, in some appropriate sense, approach equilibrium in one time direction but not the other. It is claimed that no initial ensemble of non-zero measure will serve to break the time symmetry for us. We are dealing with K systems. These are systems that are, perforce, mixing. Each non-zero measure ensemble will approach equilibrium (in the coarse-grained sense) in both the limit as $t \mapsto + \infty$ and as $t \mapsto - \infty$. And for each such ensemble that approaches equilibrium, so will the ensemble that is composed of all and only those systems that are time reverses of the systems in the original ensemble.

But this symmetry need not be true of *singular* initial ensembles, initial ensembles of measure zero. K systems have as sub-ensembles "dilating" and "contracting" "fibers." These are ensembles of measure zero. Because the evolution of the ensemble is measure preserving, they transform only into sub-ensembles of measure zero in either time direction. Yet there is an appropriate sense in which such a fiber can "approach the equilibrium ensemble" in one time direction, although it will not do so in the other time direction. And the time reverse of such an ensemble will not approach equilibrium in the future time direction. The sense of approach is, basically, this: Whereas the Λ^+ transformation of the initial zero measure ensemble is not the Λ^+ transformation of the equilibrium ensemble, the ensembles into which the original zero measure ensemble evolves, in one time direction but not the other, although still of measure zero, can be such that their Λ^+ transformation approaches that of the equilibrium ensemble. In this sense, these fibers "dilate" in one time direction but not the other. And their time-reverse ensembles show the reverse time-asymmetry with respect to this sense of approach to equilibrium. Figure 7-7, using the famous Baker's Transformation, a particularly simple instance of a K-system ensemble, makes it clearer what is going on here.

Furthermore, one can define an "entropy" for sub-ensembles. With this entropy measure, it becomes possible to show that initial singular ensembles that evolve toward equilibrium in the forward time direction have finite entropy, whereas the entropy, so defined, of initial singular ensembles that are represented by fibers that dilate in the past time direction and contract in the future time direction have infinite entropy. One can then argue that this shows that preparing a system in such a way that it evolves toward equilibrium (or, more exactly, preparing the relevant collection of systems) is possible. But preparing a collection of systems showing anti-thermodynamic behavior, because it would require an infinite amount of information, is physically impossible. Once again, Figure 7-7 will reveal at least the basic structure of this argument.

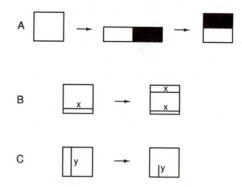

Figure 7-7. Singular initial states and symmetry breaking in time. Fig. A represents "Baker's Transformation." A square region is stretched to double its size in the horizontal direction and half its size in the vertical, preserving its area. Then the right-hand side is cut off and placed on top of the left-hand half. The process is then repeated ad infinitum. This transformation is one of the very simplest having radical randomizing properties such as being a K system and a Bernoulli system. Fig. B shows a horizontal line in the square. In the transformation process, this line continually gets stretched, then segmented and stretched again. It is a "dilating fiber" of the transformation. A vertical line, as in Fig. C, however, continually diminishes as a result of the transformation. It is a "contracting fiber." Even though both the horizontal and vertical lines are of measure zero, their behavior under the non-unitary transformations of Prigogine et al. is such that, in a sense, the transformation of x will approach the equilibrium ensemble in the future direction of time but not in the past direction of time. So the suggestion is made that such singular distributions can characterize the temporal symmetry breaking observed in nature.

Next, the formal devices of singular initial ensembles of the appropriate kinds and the new definition of entropy are connected to our physical experience. It is argued that a beam of particles in initial parallel motion, a beam that thermalizes as time goes on into the familiar collection of randomly moving particles, corresponds to a dilating singular fiber in the phase space. Corresponding to a contracting fiber would be the time reverse of this. We can prepare the former type of system (and collections of them) but not systems of the latter sort. So this distinction between singular ensembles of finite and infinite entropy (as entropy is here defined) not only serves formally to provide us with initial ensembles that asymmetrically distinguish the two directions of time, but represents in a reasonable way our physical possibilities and impossibilities of preparing ensembles of systems in initial states in the world. It is this distinction that selects out the possible thermodynamic from the impossible anti-thermodynamic evolution of systems that we experience.

This rich and highly developed study of the origin of kinetic behavior and of irreversibility is clearly a source of insight into the fundamental problems. But like the other accounts we have looked at, it has within it aspects worthy of some skepticism. As usual, the derivation of kinetic behavior, either in the form of a master equation or in the form of the Markovian equation for the Λ^+ transformed representation of the ensemble, rests on idealization and abstraction. As a consequence, we are left with the usual concerns about whether, in the case of realistic models, more can be inserted into the schematism that will do the job of demonstrating to us, first, that the kinetic behavior is attributable to those elements of the ensemble needed to derive the thermodynamic quantities' full irreversible and monotonic behavior that we wish to demonstrate, and, second, that the behavior so demonstrated will conform with the experimentally observed values of such parameters as the correct value of the relaxation time.

Next, there are several questions connected with the invocation of singular sub-ensembles as the appropriate characterizers of physically preparable initial ensembles. First, are they really necessary? Although it is certainly true that in a mixing system all non-zero measure sub-sets of ensemble will show identical "approach to equilibrium" in the infinite time limit, their behavior over any specified finite time interval can be very different indeed. Some will monotonically approach equilibrium in one or both time directions. Some will monotonically deviate from it, again in one or both time directions. Most will show non-monotonic behavior in both time directions. If we wish to "pick out" the future direction of time in terms of ensemble evolution, couldn't we do so without resort to the use of singular ensembles by focusing on those non-zero measure sub-ensembles that show a monotonic approach to equilibrium in the finite time periods characteristic of the experimentally observed approach to equilibrium of real systems? After all, as has been frequently pointed out here and elsewhere, it is this finite time behavior that really concerns us, and not the infinite time-limiting behavior.

And there are reasons for thinking the singular sub-ensembles inappropriate for describing most physical situations. Prigogine, of course, realized that our system preparation will not give us exactly measure zero ensembles. But he suggests that it is ensembles close to those of measure zero that ought to be taken as representing physically realistic prepared systems. But even this is doubtful. Some special cases (the almost parallelized beam that dissipates into random thermal motion) might be appropriately represented in such a way. But the more usual case – for example, a system in equilibrium relative to some imposed constraints whose constraints are then changed and that then evolves to a new equilibrium appropriate to the now modified macroscopic constraints – will not be so idealizable. The initial ensemble for this latter case, surely, is one of non-zero measure. It is one of those ensembles that however symmetrically it behaves in the positive and negative infinite time limits, evolves toward equilibrium for a finite time subsequent to preparation. This time will be the relaxation time experimentally observed for the system.

Nor is Prigogine's invocation of the "infinite entropy" of the singular states that behave anti-thermodynamically as ground for asserting the impossibility of their preparation without its questionable aspects. Prigogine's argument allegedly shows that ensembles that behave anti-thermodynamically are impossible to prepare, because it would require "infinite information" to set them up. But of course there are the spin-echo experiments to contend with. Sometimes, anti-thermodynamically behaving systems *can* be constructed. What exactly are the limits of the "impossibility" results that have allegedly been demonstrated?

Much more important is the fact that the underlying reversibility of mechanics, and the consequent time-symmetry of all the features of being mixing and being a K system, clearly allows us to develop a notion of entropy exactly time anti-symmetric to that invoked by Prigogine. Why is it that systems with infinite entropy of one kind are impossible of preparation, but systems of infinite entropy of the other kind are not? The most crucial question is the familiar one: What is it about the "preparation" of systems that engenders temporal parallelism of thermodynamic behavior? Why is it that the low and high entropy ends of system evolution are always, modulo fluctuation, in the same time order? Or, put another way, if having infinite Prigogine entropy is a restriction on the preparability of systems, what is it about preparability that makes the preparation end of the finite time evolution of a system always the same temporal end relative to the measurement or observation end? Why is the direction from preparation to observation the same for all systems? It is hard to see in any of the ingenious mathematics that lies behind the invocation of singular initial ensembles of the dilating kind as symmetry-breaking representations of initial states of systems any physical explanation of why it should not be the case that about half of systems ought to be represented by initial states that are dilating fibers in a given time direction, and the other half by fibers that contract in that time direction.

In summary, it isn't at all clear that the invocation of sets of measure zero as initial ensembles gains us any ground in understanding the physical questions that puzzled us. They seem both unnecessary and misleading as representing the situation for the usual evolution of systems from non-equilibrium to equilibrium. And although one can indeed invoke such singular states as mathematical devices to select one time direction out from the other in a formal symmetry breaking, it is hard to see how the invocation of such states advances our understanding of *why* parallelism of thermodynamic evolution is the rule of nature.

7. Conflicting rationalizations

One further issue deserves our attention before we move on to review the results that we have been surveying and to discuss their fundamental import. We have seen a number of approaches to solving the two fundamental problems of non-equilibrium statistical mechanics: the problem of the origin of "rerandomization" and the problem of the origin of temporal asymmetry. It is important to point out that some of these approaches to rationalizing the approach to equilibrium can conflict in ways that might not be immediately apparent. The conflicts I will focus on here lie at the very heart of the rationalization of rerandomization taken by the approaches that work by offering a rigorous derivation of the Boltzmann equation when contrasted with all of those approaches that rely upon the generalizations of ensemble chaos that go beyond ergodicity.

The Lanford approach to rationalizing the Boltzmann equation tells us that if we go to the Boltzmann–Grad limit, there will be an initial condition for the ensemble such that for at least a short time, almost all members of the ensemble will have their one-particle distribution functions evolve in the manner described by the Boltzmann equation. As Lanford makes perfectly clear, this is an approach that shows not that the Boltzmann equation will hold "on the average," but that it gives "an accurate description of the time development of 'almost all' initial points." In other words, this approach is an attempt to show that the Boltzmann equation describes, at least in the limit chosen and for a very short time, the evolution of "almost all" systems subject to the initial constraints.

As was pointed out earlier in the survey of the historical development of these issues, there is a radical incompatibility of such a claim of the monotonic Boltzmann equation describing the "overwhelmingly probable" evolution of systems with the claim of the Recurrence Theorem that almost all trajectories in an ensemble will return arbitrarily close to their initial point. The result is also in conflict with the results of mixing-type theorems. Crudely, the latter demonstrate that trajectories will not cohere with one another, but will diverge, so that every possible kind of evolutionary path (in a measure theoretic sense) will be followed by arbitrarily nearby initial points. But the Lanford result suggests, rather, a coherence of trajectories, so that almost all of them obey the condition of having their one-particle reductions remain close to that predicted by the solution curve of the Boltzmann equation.

Now, of course, the results of the original Lanford Theorem hold only for a very short time. And the recurrence and mixing results are results about what happens in the infinite time limit. But it may be a mistake to look for the mathematical reconciliation of the results in this direction.

For there seem to be good reasons for thinking that the results of the Lanford Theorem are still true even when extended (perhaps arbitrarily far) into the temporal future from the zero time, although the proof cannot demonstrate this truth. Rather, the formal reconciliation of the results may lie instead in the fact that the Lanford results hold only in the Boltzmann–Grad limit, which, we noted, entails a limit of zero density for the system. And there is the fact that there is little hope of extending these results to less idealized systems.

The important point here is this, as was made clear by J. Lebowitz. The "Boltzmann hierarchy" from which the Lanford results are obtained "does not have underlying it any flow in the phase space." Lebowitz continues ". . . there is no point transformation in the phase space that yields the time evolution of the correlation functions . . . This is yet another manifestation of the "loss of information" resulting from the use of the Boltzmann–Grad limit – a loss necessary to make dissipative macroscopic laws consistent with reversible microscopic dynamics." In other words, the going to the limit of $nd^3 \rightarrow 0$ while n goes to infinity was not a mere formal device for handling rarified gasses in a formal way. It is an essential component of the derivation of the rigorous Boltzmann results. It is in this limiting process that the mathematical consistency of these results with the Recurrence Theorem and the mixing-type theorems is hidden. For in this limit, the phase-space flow that generates recurrence and mixing ceases to represent the evolution of probability.

Mathematical consistency is one thing, but conceptual consistency is another. It seems clear that the rigorous derivation of the Boltzmann equation in the Boltzmann–Grad limit approach to seeking the origins of kinetic behavior is radically at odds with those attempts to find the origin of this behavior in such features as the mixing nature of the ensemble. The latter approach continues a tradition with its roots in the Ehrenfests' understanding of the solution curve of the Boltzmann equation as being the "concentration curve" of ensemble evolution. This is the curve passing through entropy values that are most probable at each time. This approach is also consistent with Gibbs' idea of seeking the approach to equilibrium in a coarse-grained notion of entropy. However, the former approach, Lanford's, returns to the idea of the kinetic equation as offering us, rather, a picture of how "almost every" system started in a microstate in an appropriate initial class will evolve monotonically toward equilibrium in the manner described by the Boltzmann equation.

Perhaps some further clarification can be obtained here by reflecting on the idealizations needed to obtain the various formal results under discussion. The Poincaré Recurrence Theorem requires, as the interventionists and Lebowitz have emphasized, at least the idealization that the system be genuinely isolated from its external environment. But that is impossible given the nature of gravitational interaction. The results that depend upon ergodicity, mixing, and their strengthenings apply only to those idealized systems (like hard spheres in a box) to which the KAM Theorem doesn't apply. Again, this is something not likely to be true of any realistic system.

Although the mixing-type results hold even for systems of only a small number of components – even two hard spheres in a box, for example – the Lanford result holds only for a system of an infinite number of degrees of freedom and vanishing density. From the mixing point of view, the only role played by the large number of degrees of freedom of systems is that of allowing us, in the Khinchin manner, to identify phase averages of ensembles with overwhelmingly most probable values for those quantities from which thermodynamic values are computed by the method of phase averages. But for the rigorous derivation of the Boltzmann equation, the limiting situation of infinite number of particles and zero density is absolutely crucial to the derivation.

But the crucial question for us is this: Which idealization is the appropriate one to choose when we are attempting to find a formal theoretical model to represent the observed thermodynamic behavior of systems in the world? It seems clear that even at the level of attempting to understand the rerandomizing aspects of the successful kinetic equations, much less in trying to go beyond this and understand the origins of time-asymmetry, we still are not at the point where a decisive answer to the question of which fundamental route to pursue can be given. And this is true even if we stick only to the more orthodox approaches to the origins of kinetic behavior, dismissing such less orthodox approaches as that from fundamentally time irreversible dynamical laws, from interventionism, or from subjective probability.

IV. The statistical explanation of non-equilibrium behavior

It will be worthwhile to conclude this chapter on approaches to rationalizing the non-equilibrium theory of statistical mechanics by once again surveying some of the technical results that have been obtained and that have been outlined here. This time, we will look at the results from the point of view of asking to what extent the various results can be utilized in providing an explanatory account of the non-equilibrium behavior of systems that fits the models of explanation, in particular statistical explanation, outlined in Chapter 4. Do the various formal and technical achievements of contemporary foundational statistical mechanics of non-equilibrium, obtained at such expenditure of time and brilliance, provide enough for us to reasonably assert that we now understand why non-equilibrium systems behave as they do?

It is a rich body of experience that cries out for explanation. Systems not in equilibrium can often still be described macroscopically in terms of a small number of parameters that obey a lawlike dynamics. These are the "local" thermodynamic quantities that appear in the hydrodynamical equations of evolution of non-equilibrium systems. It is our experience that with rare exceptions at least, the systems inevitably follow a path from their initial non-equilibrium configuration to the equilibrium state described by classical thermodynamics as appropriate to the macroscopic constraints on the system. This approach to equilibrium is characterized by a standard time-rate of equilibration, the relaxation time. And the approach to equilibrium is characterized by the lawlike behavior of the macroscopic parameters summed up in the hydrodynamic equations of evolution. Why do systems behave that way?

Convinced by the theoretical arguments directed against Boltzmann's first understanding of the kinetic equation of approach to equilibrium and by the observationally detectable existence of fluctuational phenomena, we presuppose that the explanatory account of the approach to equilibrium must, in some sense or other, be a statistical or probabilistic account. But what that comes down to remains open to debate. We also presuppose, although this, once again, is a presupposition of some ambiguity that can be unpacked in quite different ways, the subsumability of the behavior of the macroscopic system into a causal pattern of evolution. The causal evolution is governed by the state of the system at a time, as fixed by its macroscopic constitution and the dynamical states of the micro-constituents, and by the causal laws of dynamical evolution of complex systems derivable from the underlying dynamical, causal laws of the micro-constituents and their structuring into the macroscopic whole.

The most natural model of statistical explanation in which to try to fit statistical mechanical explanations of the evolution from non-equilibrium to equilibrium would be, then, the one in which individual systems are construed as having initial micro-states that deterministically evolve into future micro-states. Probability would appear as some distribution over these micro-states at one time, leading to a distribution over evolutionary trajectories. It would then be some statistical feature of these trajectories, considered as an ensemble or collection of individual trajectories, each evolving in its deterministic way, that would then be related to the observed hydrodynamic evolution of a macroscopic feature of the individual systems. Again, which statistically calculated quantity would be related to a macroscopic feature, and how this relation would be construed, would vary from one detailed interpretation of the model to another.

On the other hand, as we have seen, there will be other interpretations of the place of statistics in the explanatory pattern that differ from this standard one in fundamental ways. In particular, there will be attempts to show that the introduction of statistics into dynamics rests upon features of the world that when properly understood, require us to reinterpret the very notion of the micro-state of the system. According to this view, we must read the individual micro-state as some sort of "propensity" or "tychistic" statistical state and the evolutionary equations of statistical mechanics as somehow or another describing the evolution from probabilistic individual state to probabilistic individual state for any particular system. We must not think of the equation of evolution, then, as being descriptive of the collective behavior of individual systems, each of which evolves in such a way as to not require any statistical or probabilistic notions in the account of its individual behavior.

The view of the nature of statistical explanation in statistical mechanics one holds will of course be quite different depending on the view taken about the ultimate nature of non-equilibrium evolution. If one believes that this is to be accounted for by an essential continuous intervention into the system by the outside world, or if one believes in the existence of time-reversal non-invariant fundamental laws as essential to non-equilibrium behavior, then the methodological account of explanation offered will differ radically from those appropriate to the "orthodox" approach involving genuine isolation of the system as a legitimate idealization and positing time-reversal invariant fundamental laws at the level for micro-dynamics.

1. Probabilities as features of collections of systems

The most commonly held variants of the mainstream interpretation of statistical mechanics rely on assuming that individual systems have the exact classical micro-states characterized by the underlying dynamics. From this perspective, probability can characterize only a feature of some ensemble or collection of such systems. The macroscopic laws of evolution from non-equilibrium to equilibrium then, to be obtained by some association of the parameters in them with something constructed out of the probability distribution of statistical mechanics, must be thought of as characterizing individual systems only indirectly through some sort of "representative" behavior of the members of the ensemble.

Given a system in a non-equilibrium state, macroscopically characterized, the natural way to understand this model of statistical mechanical explanation is to characterize it in terms of a probability distribution over possible initial states of a system compatible with the initial macroscopic constraints. And it is natural to view the probabilities of evolution as generated by the underlying dynamics that fixes the evolution of each individual system from its initial micro-state. This seems inevitable given the idea of each individual system as having its particular evolution causally explainable by its micro-state at one time and its dynamical laws. Here, then, we would have a model of probabilistic explanation as characterizing the distribution of behavior of a collection of systems each deterministically causal in its evolution.

The aim of all such models will be to derive the characterization of systems in terms of the macroscopic parameters and the detailed laws governing their evolution from allowable posits. Then one will try to physically account for those posits. By "allowable posits" I mean to emphasize that whereas, from this perspective, a posited probability distribution over initial states of the system at a time is legitimate, if crying out for physical explanation, "rerandomizing" posits are not to be countenanced. This is because the dynamical evolution of the system is fixed by the dynamical evolution of its members, and that is a given from the underlying dynamics.

Perhaps the most realistic models of systems we can construct would be those where the conditions necessary to satisfy the KAM Theorem are met. We would then expect members of a collection of systems started in non-equilibrium to display quite different behaviors depending upon whether the initial condition for the system was in the region of stable trajectories near those that govern the system with the small interaction among components "turned off," or, instead, if the initial condition was in the chaotic region outside the tori of stability guaranteed to exist when the KAM Theorem holds.

At this point, the large numbers of degrees of freedom of the system would be invoked, the argument being that in such a case there would be a vast number of overlapping "resonance regions" in the phase space, and that, hence, the regions of stable trajectories would become "small" in the phase space. Although there is much plausibility here, the problem of making such claims rigorous is mathematically quite intractable. Even in the limit of an infinite number of degrees of freedom, for example, it is not at all clear that the stable regions can be expected to have "measure zero." Add to this the problem that even if the regions of stability can be shown to be "small," this means "small as measured by the usual measure." But we no longer have the "transcendental deduction" of the naturalness of the usual measure available to us in equilibrium theory, where that measure was the only time-invariant measure that gave zero probability to sets of measure zero in the standard measure. We want a physical reason for thinking that it would be likely that in a collection of systems we would find nearly all with their initial states in the chaotic region.

Suppose we assume that nearly all systems have their initial states in the chaotic region. We would still be far from having the full derivation that we would like of the non-equilibrium behavior we are trying to understand. We simply don't yet know enough about the kind of "chaos" in the chaotic region to be able to derive from it anything like the rigorous "randomization" results that can be derived in the less realistic, but mathematically more tractable, cases where the generalizations of ergodicity can be assumed to hold. There are highly interesting suggestive results, such as Prigogine's claim that the conditions necessary for such a chaotic region to exist are those necessary for the existence of a "sub-dynamics" that would extract a time-invariant component of the distribution function that showed the wanted kinetic behavior. But what we really would want would be some demonstration, for a realistic system, that in its chaotic region, with a "natural" distribution of initial conditions presupposed, an evolution of components of the statistical distribution over the states of systems evolved from that initial distribution would be properly kinetic. And we would want a complementary demonstration that these components of the overall probability distribution would be those "reduced descriptions," like the one-particle distribution function, sufficient to generate by association all of the macroscopic parameters that appear in the macroscopic equations of evolution. Here again, the vast number of degrees of freedom would be invoked in a Khinchin-type argument to assure us that the probability distributions for the reduced components of the ensemble probability function would be properly symmetrical and highly peaked. This would allow us to move with impunity, for example, from average to overwhelmingly most probable values of the macroscopic quantities.

Even if we had all of this, we would still have to contend with the "touchstone" problem of statistical mechanics, the origin of the temporally asymmetric thermodynamic behavior of systems. We want a *physical* explanation of why the "natural" distribution over initial conditions, needed even in the chaotic region to get the thermodynamic behavior of systems, is actually found in the world. If the distribution is such that when applied to final states of systems instead of initial states its application would lead us to expect anti-thermodynamic behavior of systems at the macroscopic level, we will want to know why such an application of it to real systems in the world is illegitimate. If it is itself a time-asymmetric condition, or, better, a distribution that when applied to states generates an asymmetry in prediction and retrodiction blocking any application of it to final states that would lead us to expect anti-thermodynamic behavior, we know that there will be another such distribution that would lead to exactly the opposite asymmetry. We would then want an explanation of why it is legitimate to expect one such distribution to hold and not the other, or more likely, why one of them holds and not some more "symmetric" distribution. The real issue here is the parallel time behavior of systems and not the fact that equilibrium is approached in the "forward" direction of time taken as a given.

If we move away from the realistic case of systems describable by the addition of a small interaction on top of otherwise independent components (the KAM Theorem case) to the less realistic idealization of singular interactions (like "hard spheres in a box") where the measure theoretic generalizations of ergodicity hold (mixing, being a K system, being a Bernoulli system), we obtain, as compensation for our greater idealization, rigorous mathematical results and a firmer grasp on just what "chaos" comes down to and just what can be derived from it. Closely related to these results are those that tell us that when the system is chaotic enough (being a K system), a new representation of the probability distribution governing the ensemble can be obtained, by a star-unitary transformation, a representation that shows manifestly kinetic behavior. The transformed ensemble obeys a Markovian equation of evolution of its face.

When results like mixing can be obtained, they can be obtained for idealized systems of even a small number of degrees of freedom. So the idealization of systems as having vast numbers of degrees of freedom is, in one sense, not a necessary condition for the explanatory scheme to apply. The vast number of micro-components of realistic systems does play a role, however, in once again allowing us to infer that the reduced descriptions of the probabilistic distribution have the Khinchin properties of symmetry and peakedness. So in the connection of the results obtained to the behavior of associated macroscopic quantities, the large numbers of degrees of freedom play a role.

The results of chaos that can be obtained are used to provide the rationalization of Gibbs' posited "spreading in the coarse-grained sense" of the ensemble that, the Ehrenfests noted, Gibbs had not really established. Indeed, they can go beyond this in getting the full results of a Markovian behavior of the coarse-grained ensemble description. This is needed to go beyond a demonstration of entropic increase to a demonstration that the route of evolution followed is that posited by the standard kinetic equation. Another virtue of this program is that at least some of the criticism directed against Gibbs' notion of coarse-graining as being "arbitrary," because the choice of coarse-grained boxes was not imposed by the dynamics or the nature of the system, is vitiated. Some results will hold of any coarse-graining (like mixing). Others (like being a Bernoulli system) will prove the existence of a coarse-graining having the appropriate behavior.

The connection between the behavior of the probability distribution as generated out of the dynamics and the observed behavior of macroscopic features of systems is made, from this perspective, through the route so clearly delineated by the Ehrenfests. The hydrodynamical equations are derived from the evolution of reduced probability distributions, the macroscopic parameters being functions of these. The kinetic equations governing evolution of the reduced probability distributions are interpreted as generating the "concentration curves" of the evolving ensemble. They don't describe "most probable" evolution, but rather the sequence of most probable states of the systems in the collection. Thus, the monotonic approach to equilibrium is reconciled with the Recurrence Theorem.

When, however, we ask how the results obtained by means of the method that generalizes the Ergodic Theorem are to be associated with the observable phenomena of the world we wish to explain, we see that the abstractness of this approach, the very root of its ability to provide us with rigorous mathematical results, hinders an easy understanding of its applicability to real phenomena of real systems. Some of the results tell us what happens "in the limit as time goes to infinity." But we wish to understand what happens "in the short run" – in particular, in time intervals of the order of the observed relaxation time of systems. And, unlike many of the cases where the thermodynamic limit of infinite numbers of degrees of freedom is used but where the loss of accuracy incurred in dealing with real systems having vast numbers of degrees of freedom but not infinite numbers of them can be assessed, the method provides us with no way of judging how long a time is "long enough" for the infinite time limit to be guaranteed to apply to some degree of accuracy. The "infinite" part of the time limit is essential to the result of the theorem.

In other cases where results are framed in terms of finite times, other related problems of tying the results to the observed phenomena complicate the understanding of the place of the results in our explanatory scheme. With the results on a system being a Bernoulli system, results that would, indeed, guarantee the kind of Markovian behavior among coarse-grained cells necessary to derive kinetic behavior for coarse-grained distribution functions and monotonic increase of Gibbsian coarse-grained entropy, the problem is that the method guarantees us that an appropriate coarse-graining will exist but doesn't help us to understand, for realistic systems, just what the constitution of such a coarse-graining would be. But we need to know this so as to allow us to infer the behavior we want for the macroscopic observables associated with the appropriate coarse-grained statistical quantities. It is one thing to show that because of the instability of the dynamics a probability distribution over an ensemble of systems has an appropriate representation that shows overt kinetic-like approach to equilibrium behavior, either in the coarse-grained sense or in the sense of Prigogine's star-unitarily transformed probability distribution. It is another thing to connect the provable dynamical behavior of the ensemble of this kind with real relaxation times and the real hydrodynamic behavior of local pressures, temperatures, and entropy densities described by the macrosopic hydrodynamic/thermodynamic equations of approach to equilibrium.

Because the results of mixing and so on, are obtained by reference only to the constitution of the system and to its underlying dynamics, it would seem at first glance that probability distributions over initial conditions, distributions that we have alleged to be an essential component of probabilistic-causal explanations of phenomena, are irrelevant to this approach to the explanation of non-equilibrium phenomena. But of course that is not true. It is at just this point that Krylov's insistence on the importance of the initial ensemble becomes pertinent.

Even if a system is mixing, the actual rate at which an initial ensemble will obtain a specified degree of coarse-grained spread-outness over the allowed phase space will depend upon that initial ensemble's specific nature. As Krylov emphasized, the initial ensemble would have to have a minimal size, a simple enough boundary, and a probability distribution spread out evenly enough within the boundary, to guarantee that it would spread in the manner appropriate to the observed relaxation time of realistic systems. Here, all the problems about the physical ground of the appropriateness of initial ensembles representing systems in the world reappear once more. What guarantees that systems could not be such – anomalies like spin-echo systems to the side – that the appropriate representative ensembles could not be "small," or have a boundary so fibrillated as to have the ensemble condense in a reasonable finite time, even if guaranteed to spread out once again as time goes to infinity? What guarantees that the appropriate probability distribution to pick within the boundaries will be uniform with respect to the usual phase-space measure, the inevitable natural choice? Indeed, without even the transcendental rationalization of the usual measure that we have available to us in equilibrium theory, what can explain to us physically the place played by the usual phase-space measure in our theory, a question we must answer even if we accept the uniform probability posit (relative to that measure) as given to us from some principle of indifference whose grounds rest in general axioms of inductive procedure? And we should note, once more, the degree to which the results we have been discussing rely upon an assumption that collections of systems of measure zero in the natural measure are of no physical concern, all of the results of this approach being couched in the measure-theoretic terms that systematically neglect such zero measure sets.

We have, of course, sketched some answers to these questions, in particular answers that rely upon the notion of preparation of a system such as are found in Krylov and Prigogine. But, as we have seen, the answers are not yet fully satisfying. Once again, the problem of the asymmetry of systems in time points out with transparent clarity the fundamental open problem here. Being derived from the time-symmetric underlying dynamics, all the results of the generalizations of ergodic theory are time-symmetric. Systems mixing in the limit as time goes to infinity are mixing in the limit if time goes to minus infinity. Systems probabilistically indeterministic in the future time sense (K systems) are probabilistically indeterministic in the negative time direction as well. Bernoullian coarse-grained behavior is also time-symmetric. So would be their consequences such as Markovian time behavior for the transformed ensemble. If a transformation to a new ensemble representation obeying a Markovian future time-directed equation is possible, so is a transformation to a new representation obeying an equation Markovian into the past. Why, then, are systems asymmetric in their temporal behavior? A representation of this time-asymmetry in our theory can only come about, in this perspective, from an appropriate choice of initial ensemble to represent the individual systems whose transformation from non-equilibrium to equilibrium macroscopic state we wish to understand.

Finally, within the general program of attributing the classical exact states to individual systems, and taking probability to be a feature of a collection of such systems, there is the approach that works through the rigorous derivation of the Boltzmann equation. In this approach, the connection between the results obtained and the observable macroscopic behavior of systems is more transparent than in the approach that generalizes beyond ergodicity. It is shown that under the conditions posited, the one-particle distribution function obeys the Boltzmann equation, and one can then use the usual association of quantities defined by means of this function and its appropriate short time behavior as governed by the equation to get a clear route to relaxation times and hydrodynamic behavior. Here, as we have noted, the understanding of probability with regard to evolution is a return to the idea of the theory as generating an "overwhelmingly probable" course of evolution, and not as generating a concentration curve through states overwhelmingly probable at specified times. This, of course, makes the approach conceptually at odds with both the Recurrence Theorem and the idea that something like mixing is at the heart of the approach to equilibrium.

As usual, idealization is a crucial component of the explanatory account. As we noted, the results hold only for very short times after the initial time, but there is some hope that they remain true, if not rigorously provable, for longer, perhaps realistic, times. In this approach, the large number of degrees of freedom is essential to derive the result. Indeed, the result holds only in the limit of an infinite number of particles, when density goes to zero and when size of particles relative to size of contain-ing box also goes to zero. It is important to remember that this idealiza-tion is unlike the usual one of going to the thermodynamic limit of an infinite number of degrees of freedom. In most cases, we do have a grasp of the degree of inaccuracy entailed by our idealization when we move to more realistic cases. But, as the originators of this approach make very clear, there is little hope that the results obtained in this idealization here can be extended to more realistic systems. Lebowitz has emphasized that the "Boltzmann hierarchy" from which the results are derived "does not have underlying it any flow in phase space," and Lanford emphasizes the difficulty in extending the result to the case where the molecules have small but finite (instead of zero) size.

Once again, a probability distribution over initial conditions is an es-sential posit of the approach. Here, it is needed to derive the main result, the Boltzmannian behavior obtained without resort to some continuous rerandomization posit. We noted that all the results about the high prob-ability of the system obeying the Boltzmann equation rest on picking the usual probability measure in phase space, and, as usual, that is some-thing that requires some more physical justification. More importantly, there is, once again, the physical puzzle about the rationalization of the initial probability distribution that is brought to sharp focus by the problem of the origin of time-asymmetry. If the initial probability distribution is one that is such that an ensemble satisfying it will have a time-reversed ensemble still satisfying it, then the same posit about initial probability distributions over initial conditions that leads to the prediction of thermo-dynamic behavior for systems in non-equilibrium in their initial states should lead us to predict anti-thermodynamic behavior for systems in non-equilibrium final states. And if we choose instead the constraint that is not satisfied by the time-reverse of an ensemble satisfying it, we need a physical rationalization as to why systems can be prepared in such a way as to be properly represented by such an ensemble, but not so as to be represented by the possible time-reversed but equally time-asymmetric condition on ensembles. None of this, of course, is meant to imply that the derivation of the result is in any way itself inconsistent with the time-reversibility of underlying dynamics. The consistency is obtained by the facts that the time-reversible constraints, if they hold at time zero, do not propagate forward in the time and that the time-asymmetric constraints simply won't hold of the time-reversed ensemble's initial conditions.

2. Probabilities as features of states of individual systems

As an alternative to the view of statistical explanation in statistical me-chanics just discussed, there is the view that takes the probabilities invoked to be characteristic of states of individual systems, and the statistical explanations as those appropriate to such systems. The general idea is that systems have no "finer" states than those associated to them by the probability distributions attached to macroscopic conditions by the statistical mechanical theory. The postulation of pointlike (in phase-space) macroscopic dynamical states, and the linelike trajectories deterministically issuing from them, made in the traditional account, is taken to be a false idealization. Along with this revisionist ontology for the states of systems comes, of course, a new understanding of what we are doing when we use statistical mechanics to provide us with an explanation of the dynamics of a system.

Here, the statistical-causal pattern of explanation is premised on the idea that a system at a given time has a definite "tychistic" state, a state characterized by the probabilities of transitions latent in it to other states of the system. Dynamical evolution is the transition of the system from state to state regulated by the irreducible probabilities of transition latent in the states themselves, and otherwise undetermined by some deeper level of "hidden" deterministic variables. It is not our ignorance of the underlying deeper deterministic state that forces an at best statistical explanatory scheme on us. It is the absence of those deterministic features, leading us to a statistical account as the deepest possible explanatory account of the dynamics as it is in itself.

Analogy is frequently made with the explanatory scheme of quantum mechanics. Here, systems are characterized by observable quantities. Determination of one of those quantities fixes for the system its quantum state, by the rule of associating "characteristic states" to measured values of observables. This quantum state then evolves in a perfectly deterministic way, according to the Schrödinger equation, in a manner fixed by the energy function for the system in question. It is in using this evolved quantum state to predict outcomes of new measurements on the system that the allegedly irreducible statistical nature of the system, prediction, and explanation comes in. For the quantum states determine only probabilities over the possible outcomes of further measurements of observable quantities. An important feature of the scheme is the so-called Projection Postulate, which specifies that if a complete measurement has been made on the system – that is, if the values of all possible compatible observables on it have been determined – the appropriate quantum system to attribute to the system for new predictive purposes is the characteristic state determined by the values of the measured observables. This rule, or its generalization, Lüder's Rule, that applies when the measurement is incomplete, posits that future statistical inferences about the behavior of the system are determined by the value obtained on the measurement performed, the way in which the system previously evolved to that state being irrelevant for future statistical predictive purposes.

The behavior of probability in quantum mechanics is of course fraught with conceptual concerns that are not ones for us to face here. In particular, the way in which probabilities "superimpose" on one another (to determine, for example, the probability of an outcome that could be brought about in more than one way as a function of the probability for the outcome by any one of the routes taken alone) presents us with a world of stunning perplexity. Followed out, problems of locality and reality of systems become endemic. Here, we wish to present only this fragment of the quantum mechanical scheme to see how closely statistical mechanics can be made to resemble it.

The analogy to the "observables" of quantum mechanics would be, presumably, the macroscopic features of a system. Here, the important question would be the analogy to the quantum mechanical notion of a complete set of observables. In statistical mechanics, this notion of "completeness" of a description would be given to us, presumably, by considerations of ergodicity. A system that is ergodic has a limited set of macroscopic conditions that can be imposed on it in a time-invariant way – that is, a limited set of global constants of motion that can be fixed by the macroscopic preparer or measurer of the system. Any additional global constants of motion would lead to metric decomposability and, hence, to a violation of ergodicity. This is important for us here, for otherwise one could imagine macroscopic conditions of ever greater complexity until the specification of all of them did constrain the system to a single phase-space point as its representative. That would be in contradiction to the idea we are exploring that the spread-out probability distributions of statistical mechanics are the finest representation of the actual state of the individual system imaginable.

Fixing a system in a macroscopic state would then determine transition probabilities to other macroscopic states. These would be determined by the ensemble dynamics, which, besides fixing the dominant course of dynamic evolution for the system (say as the concentration curve of ensemble evolution), would determine the much less likely transitions we could expect in the form of fluctuational phenomena. So the attribution of a probability distribution to the individual system of gas in the left-hand side of the box would tell us both to expect the gas to be in a much closer to equilibrium condition in the near future, but would warn us of the possibility of a rare fluctuational behavior to an even further from equilibrium macroscopic condition as a possibility and would tell us with what probability to expect such a rare occurrence.

There is even, in statistical mechanics, some thing like the Projection Postulate of quantum mechanics. This is the "rerandomization" posit used to derive the kinetic equations as discussed in Chapter 6 and whose rationalizations we have been discussing in this chapter. It was introduced in order that we could derive a statistical mechanical surrogate to the macroscopic dynamical equations of evolution. These macroscopic equations have the further evolution of a system depend in their nature only on the system's macroscopic condition at a given time, past history being irrelevant. In the statistical mechanical analogue, the rerandomization of the ensemble at each instant performed just the task given to the Projection Postulate in quantum mechanics, the task of readjusting the probability distribution to the new "observed quantity," and throwing away any information about how that quantity arose. Of course, the analogy is only that. In quantum mechanics, the projection occurs only when the system is "observed," a very problematic notion with which interpretations of that theory must come to grips. But in statistical mechanics, rerandomization (rationalized by proven mixing or by a demonstrated star-unitary transformation to a new ensemble representation obeying a Markovian equation or by some other means) is taken as legitimate even when the system is "unobserved" in any reasonable sense.

Despite these analogies, the disanalogies of probability distributions in statistical mechanics with those in quantum mechanics must not be overlooked, especially in our present context. In quantum mechanics, the mathematical apparatus posits no further underlying "point" phase-space state of the system beyond the probability distribution. Indeed, classical phase space is rejected altogether and is replaced by a phase space in which the mathematical representatives of the probability distributions (or, rather, the probability amplitudes from which the probability distributions are constructed by multiplying one of them by its complex conjugate – Hilbert space vectors) are the "atoms." Further, we are presented with alleged proofs to the effect that no supplementing of these probability distributions by a finer phase-space representation in which pointlike states existed and from which they could be derived as measures over collections of the points could be compatible with the statistics for connections of observable posited by quantum mechanics. These are the so-called "No Hidden Variables" proofs.

In statistical mechanics, we have, instead, built into the very mathematics from which the probability distributions are derived the underlying phase-space with its pointlike representatives of exact micro-states. Clearly, then here we can have no such "No Hidden Variables" results proved. But what of the claim that such mathematical artifacts of the theory misrepresent the physical reality – the claim, for example, that because we cannot, because of the instability of the systems, fix for ourselves the exact point state of a system that would guarantee some determinate future trajectory for it, the positing of such deterministic point states is only to be taken as a false idealization, misrepresenting physical reality?

Here, the results of the spin-echo experiment and related phenomena become crucial. Granted, such phenomena are rare in the world and that the preparation of the peculiar reversed states is a trick requiring an initial non-equilibrium ensemble to get the experiment going. But the fact that the initial non-equilibrium macroscopic state can be reobtained by a "time reflection" of the system would seem to indicate that the pointlike exact micro-dynamical state of the system existed at that time. Or at least, it shows that if a region of dispersion around a point state was the physical realization of the system, it was a dispersion much smaller than the spread-out probability distribution that statistical mechanics would attribute to a system having the macroscopic character possessed by the system at the time of reversal. All in all, the positing of genuine tychistic states for systems and of the probability distributions invoked by statistical mechanics as representing the individual states of such systems seems pretty implausible. We will return to this question once again in Chapter 9.

Improbable as a rendition of the ontology of the situation, and hence as a model of statistical explanation, as this account may be, its appeal does rest on something vitally important. The universality and apparent lawlike nature of the results of thermodynamics does seem to require of us, as Krylov emphasized, some understanding of just what the "in principle" limits on preparation of collections of systems must be. Some kind of "irreducibility" of the statistical account does seem to be implied here, even if the most radical way of viewing it we have just been discussing seems to go too far.

Finally, we need to emphasize once again the crucial role in the explanatory scheme played by the initial probability distributions. Prigogine introduced the idea of singular ensembles from the perspective of a state ontology of the sort we have been discussing. But, as noted earlier, there is no need to adopt such a radical notion of state and explanation in order to find for singular ensembles a role in the theory. As we also noted earlier, there are good reasons for thinking that such singular states are neither necessary for the theory nor sufficient for a complete theory. The fundamental problems, as always, are the reproduction in the theory of actual finite time behavior and the explanation of the time-asymmetry of the phenomena. For these purposes, appropriate non-singular initial ensembles will do. We need not resort to singular probability distributions to break finite time-symmetry, even if all non-singular ensembles behave on a par as time goes to plus and minus infinity. And even if we do invoke singular ensembles, we still need an account of why those need be invoked that give the right asymmetry in infinite time rather than those that give the wrong anti-thermodynamic behavior in the infinite time limit. Or, better, why one kind only ever

shows up in the world. Once again, reference to the "infinite information" required to prepare the anti-thermodynamic singular ensembles
begs the question in the definition of entropy chosen.

3. Initial conditions and symmetry-breaking

As we have seen, much still remains obscure about the general program
of rationalizing the kinetic equations of statistical mechanics. Exactly how
dynamics and idealizations are to function in a derivation of the appropriate statistical-mechanical ensemble evolution that is to represent the
approach to equilibrium is still a matter of grave controversy.

Most perplexing of all is the crucial role played by initial probability
distributions (initial ensembles). As we have seen, in some approaches
a posited initial probability distribution is essential to eliminate the objectionable rerandomization posit used in earlier derivations of the kinetic
equations. Even when rerandomization is obtained from dynamics alone,
initial probability distributions are still crucial in obtaining the short-term
and specific behavior of systems. Most crucial of all, initial ensembles are
essential in introducing time-asymmetry into the representation. Once
again, it is important to emphasize that it is not the fact that entropy
increases in the future time direction that is at issue here. That might be
obtained, ultimately, by some version of Boltzmann's posit that our very
concept of the future (as opposed to past) time direction is itself grounded
in the direction of entropic increase. It is, rather, the parallelism of time
direction of entropic change by the overwhelming majority of systems, a
parallelism presupposed by the Boltzmann argument, that can only be
obtained by a posit about appropriate initial ensembles that is not matched
by any corresponding posit about "final" ensembles. Or, more accurately,
we need a posit that systems all are appropriately represented by one of
these special ensembles at only one end of their course of evolution as
isolated systems, and that that end is in the same time direction for all
systems.

We have seen several approaches to rationalizing this posit of time-
asymmetric boundary conditions for ensemble evolution. One group of
approaches relies upon a notion of preparation. It is in how systems are
constructed or prepared (in particular, in the limitations imposed by
instability of trajectories on this preparation) that the source of the
appropriate initial ensembles (covering a large enough region of phase
space, being bounded in a sufficiently uncomplex or unfibrillated way,
having a uniform probability distribution over this largish, simple region)
is to be found. As we have seen, this proposal suffers from being, for the
most part, a suggestion rather than a worked-out theory, and, more importantly, from seeming to presuppose time-asymmetry by hiding it in

the time-asymmetric notion of causation. It is that notion of causation that leads us to say that boundary states at one and the same end of the evolution of each isolated system are brought into being by a preparation of the system, with the states at the other end generated by the systems evolution and not by some "destruction" that involves the interaction of the system with its surrounding environment.

Another systematic approach to the problem, we have noted, is to take seriously the radical distinction in the conceptual basis associated with the underlying dynamics between laws and initial conditions. This approach generally presupposes that initial conditions are totally unrestricted by any lawlike constraints, and argues that the time-asymmetric feature of initial conditions is just a "mere matter of fact" about the world, not susceptible of further explanation. This is the "de facto" account favored by Reichenbach, Grünbaum, and others. But, following Krylov, we have seen that this account, or, perhaps, this counsel of despair, is afflicted with many difficulties. Most important is its failure to ground the apparent "lawlike" status of the posited asymmetry. No one is claiming, of course, that the Second Law of Thermodynamics has true lawlike status as it was originally interpreted. Everyone is going to admit to the existence of fluctuational phenomena for individual systems. But the status of the statistical surrogate that replaces that law in our grounding of it in statistical mechanics seems too universal, too pervasive, and too inviolable to be simply dismissed as a grand cosmic accidental congery of innumerable purely contingent initial conditions of individual systems. Add to this Krylov's other pertinent observations about the incoherence of the view as usually formulated. If it is taken to be a fact about all macroscopically characterizable systems that the "real ensemble" has a uniform probability distribution appropriate to it, we soon obtain the conclusion that all systems ought to be inferred to be in equilibrium. For the only ensemble consistent with the posit is the equilibrium ensemble. And if we consider a system that started in non-equilibrium and that has remained isolated from the world, the posit of uniformity of probability over the phase space allowed by its later macroscopic condition is simply inconsistent with that posit applied to it at the initial time. This result is, of course, just one more application of the invariance of fine-grained entropy (phase volume) over time.

At this point, the suggestion is frequently made that it is an asymmetry of time itself that accounts for the asymmetry of systems. The trouble here is that this approach has one clear meaning and one vague one. The clear meaning is the posit that there are fundamental laws of nature that are time-asymmetric, and that this lawlike time-asymmetry underlies the asymmetries in time of thermodynamics and statistical mechanics. But, as we have seen, at present this seems implausible. The one area where time reversal non-invariance has been found in the laws of nature (per-haps) – in the theory of elementary particle interactions – seems to be grossly inadequate to account for the familiar pervasive asymmetries we are trying to account for. The vaguer interpretation of the approach ar-gues that something about the asymmetry of "time itself" expresses itself in the asymmetry in time of boundary initial conditions to which we have attributed the time-asymmetry of thermal phenomena. The problem here is that it is far from clear what such an asymmetry of time itself is con-strued to be. It is even more unclear how it would be supposed to govern the initial conditions of systems. The problem here is that the underlying dynamical theory seems to tell us that only the laws and initial conditions come into play when the evolution of the system is in question. No room is left, barring interventionist accounts of disturbance of the system from the outside, for any other factor. If there were some such "asymmetry in time itself," how would it find room to play a role in the dynamical evolution of systems?

But there is one other approach to examine in some detail. Here, the asymmetry in the boundary conditions for individual systems is traced to their place in the overall universe as a whole. This is the "cosmological" approach to time asymmetry in statistical mechanics, and to it we turn in Chapter 8.

V. Further readings

For a description of the spin-echo experiment, see Hahn (1953) and Brewer and Hahn (1984). Rhim, Pines, and Waugh (1971) describes the version of the experiment in which spin-spin interaction is allowed. See also Kreuzer (1981), pp. 330–336.

For the approach that derives the Boltzmann equation rigorously in the Boltzmann–Grad limit, see Lanford (1983) and Lebowitz (1983).

Early investigations into the generalized master equation and the tech-niques of sub-dynamics are Nakajima (1958) and Zwanzig (1960) and (1964). Later work can be found in Prigogine and Grecos (1972). Jancel (1969), Part II, Chapter VI, Section V, summarizes the approach. A lengthy and clear exposition is Balescu (1975), Chapters 14–17.

For the extensions of the measure theoretic approach beyond ergodicity, Arnold and Avez (1968) offers a terse and sophisticated introduction. Cornfeld, Fomin, and Sinai (1982) is an exhautive treatment of the math-ematical issues. Ornstein (1974) is compact and clear. Earman (1986), Chapter IX, outlines non-technically some of the aspects of this approach.

For the method of transforming the reversible ensemble into a statis-tically equivalent irreversible ensemble, Prigogine (1980), Chapter 8, offers a non-technical introduction. For more detail, see Misra, Prigogine, and Courbage (1979), Misra and Prigogine (1983), and Courbage (1983).

For macroscopic instability and chaotic motion, Gleick (1987) offers a non-technical introduction. Baker and Gollub (1990) is a good introduction to the technical aspects. Devaney (1986) is a superb treatment at the intermediate level of difficulty. A very illuminating introduction is Berry (1978). Lichtenberg and Lieberman (1983) is comprehensive and clear. Schroeder (1991) is an illuminating introduction to the wide variety of mathematical and physical results.

For a discussion of the time-symmetry of laws, see Davies (1974), especially Chapters 5 and 6. A brief discussion is Sklar (1974), Chapter V, Sections B and C.

For some "interventionist" approaches, see Blatt (1959) and Mayer (1961). Horwich (1987) offers an alternative interventionist approach combined with cosmology. Davies (1974) discusses interventionism in Section 3.5. See also, O. Penrose (1979), Section 3.6.

Jaynes' approach to irreversibility can be found in Jaynes (1983), especially in the piece entitled "Gibbs vs. Boltzmann Entropies." For Watanabe's Gibbsian remarks, see Watanabe (1969).

Krylov's critique of the traditional approaches as well as his "preparation" analysis of initial ensembles is in Krylov (1979). For two critiques, see Sklar (1987) and Batterman (1990).

Prigogine's introduction of singular initial conditions can be found in Prigogine and George (1983) and Courbage (1983). For critiques, see Sklar (1987) and Batterman (1991).

An alternative rationale for non-equilibrium theory not discussed in this book is found in Hollinger and Zenzen (1985).

8

Cosmology and irreversibility

I. The invocation of cosmological considerations

1. Boltzmann's cosmological way out

The late nineteenth century saw two attempts in science to invoke the conditions of the universe in the large to explain phenomena that had previously been thought to be purely "local" in their nature and hence explainable by reference to only nearby facts about the world.

One of these novel approaches to explanation was Mach's introduction of the overall structure of the universe as a component of his proposed solution to the problem of absolute acceleration in Newtonian dynamics. Newton had accounted for the difference between inertial and non-inertial motion, a physical difference revealed by the presence in the latter case of effects due to the so-called inertial forces, effects absent in the case of inertial motion, by positing that genuine accelerations were accelerations relative to "space itself." Later, related theories posit the inertial frames of space-time, in the case of neo-Newtonianism or the Minkowski space-time of special relativity, or the local time-like geodesic structure of space-time, in the case of general relativity, as the reference objects relative to which motion is absolutely accelerated motion. Mach found the postulation of "space itself" methodologically illegitimate, and sought for an explanation of the asymmetry between inertial and non-inertial motion in some theory that would involve only the relative motions of material objects to one another in its explanatory account.

Mach found his proposed explanation of the inertial effects in the relative accelerated motions of material objects, a relative motion that, he believed, could generate the forces appropriate to generate the effects. But inertial effects are not explicable, as Newton decisively argued, by the relative acceleration of a test object to nearby material objects. The forces are indifferent to the nature and disposition of those nearby objects, and they seem to appear everywhere and at every time in the universe. Mach's resolution of this problem was to suggest that the forces generated were highly independent of the distance between the relatively accelerated objects, but highly dependent on the masses of the objects. The relative acceleration of the test object to the average smoothed-out distant cosmic mass of the universe (the "fixed stars" in Mach's terminology) could then account for the ubiquity of the effects and their independence from local variation in the test object's surroundings. The theory could also account, in an elegant manner, for the observed agreement of the rotation rate of the earth as measured by dynamical effects with that determined visually by its relative motion with respect to the "fixed stars."

It will not be to our point here, of course, to follow out the long and fascinating career Mach's speculations have had in the space-time physics of the past century. I introduce it only to point out that more than one attempt existed in the late nineteenth century to escape from a foundational scientific difficulty by introducing cosmological facts as surprisingly relevant to local phenomena.

Much more relevant for our purposes is the fascinating, if briefly sketched, set of proposals made by Boltzmann and noted by us in Chapter 2. Boltzmann was faced with the following objection to his later probabilistic version of the theory: Boltzmann wished to account for the existence of the equilibrium state – the attractor state to which all systems converge and that is itself stable over time – by claiming that state is the "overwhelmingly most probable" combination state for a system, the state obtained by far more permutations of the micro-components than any other combination state. But then, how is he to explain to us why the world in which we live is so grossly out of equilibrium, so grossly improbable on his terms?

His suggestion was that the universe as a whole *is* in equilibrium. Given the universe's vastness in space and its vast duration in time, the theory of fluctuations would imply that we ought to expect large spatio-temporal regions of non-equilibrium to be found as isolated bits of the sea of overall equilibrium. But if the equilibrium regions so overwhelmingly dominate the non-equilibrium, why do we find ourselves in a portion of the universe far from equilibrium? Wouldn't it be much more probable that we would exist in an equilibrium region? No. Here Boltzmann invokes a "selection by the observer" argument. This kind of argument is now often called, unfortunately, an application of the "weak anthropic principle." We simply could not exist as observers in the equilibrium regions, for an observer must be a fairly complex, stable, macroscopic system. Such systems could only be sustained over time by energy flows generated by the existence of local non-equilibrium.

Sometimes it is said that Boltzmann's argument "explains why our region of the universe is in non-equilibrium," but we must be a little cautious here. It certainly is a persuasive argument to the effect that local non-equilibrium, being a necessary condition for our existence, will inevitably be found by us to be the case in our vicinity. But it is not, of course, some substitute for a causal explanation of why this region of non-equilibrium exists. The statistical-causal explanation for the regions of non-equilibrium existing, and, hence for the possibility of observers like us existing, is the distribution over initial conditions of the sort proposed in standard equilibrium statistical mechanics combined with the vastness of the system that leads to high probability for regions of great size to obtain entropies far below maximal values for extended periods of time.

It is sometimes objected to Boltzmann that because a much smaller fluctuation from equilibrium could also sustain evolution toward sentient beings, his argument is fallacious. But that criticism may be thinking of his argument as trying to do more that it intends. We can show that the conditional probability of our existing in a local universe such as the one we experience is much higher than the conditional probability of our existing in an equilibrium region of the universe, for that latter probability is zero. But it is certainly true that the probability possessed by regions that fluctuate less from equilibrium than our local region is much higher than the probability of a region like this. And the probability that sentient beings will exist in more modest fluctuational regions than ours is only somewhat smaller than the probability of sentient beings in a region like ours. So it is reasonable to suppose that the probability, given that a sentient being exists in a region, that the region has a fluctuational low entropy that is very low but not nearly as low (or extensive) as the low entropy of our region, is substantially higher than the probability, given sentience, that the region is like ours.

But the Boltzmannian had a reasonable reply. The universe is filled with fluctuational regions. Many have less divergence from equilibrium than ours, but enough to sustain sentience. Many of those probably do have sentient observers in them. Above and beyond that it is still the case that regions as extensive and with as low entropy as ours will be found, and they will very likely sustain sentience. Indeed, lots of different evolutions to sentience probably occur in them. We happen to be sentient creatures in a fluctuational region that is of a kind less common than many fluctuational regions able to sustain sentience but of higher entropy and smaller extent than our region has. But our region is of a kind that is more common than other even more extended and lower entropy fluctuational regions. Our actual existence has by this argument been placed in a statistical pattern that is compatible with the alleged actual probabilities of situations in the world. And this is one version, we saw, the liberal version, of what it is to explain a phenomenon statistically. The statistical pattern posited is compatible with the claim that equilibrium states are overwhelmingly most probable, even when probability is interpreted by some version or other of an "actual frequency" account or something like it.

Finally, there is Boltzmann's important response to the query as to why we find ourselves in a region in which entropy *increases* in the future time direction rather than increasing in the past time direction. This is his important claim that our very notion of which direction of time is the future direction is grounded in entropic phenomena, a claim we will discuss in Chapter 10.

The major contemporary objection to Boltzmann's account is its apparent failure to do justice to the observational facts. As far as our observational resources have allowed us to explore the universe, we find it to be, overall, far from equilibrium. And, as far as we can tell, the parallel direction of entropic increase of systems toward what we intuitively take to be the future time direction that we encounter in our local world seems to hold throughout the universe. Quite a different picture of the overall structure of the cosmos now has overwhelming acceptance from the scientific community than the picture Boltzmann offered of a vast sea of equilibrium inhabited by islands of non-equilibrium whose sparseness increased with their deviation from the equilibrium state. But, as we shall see in Section 8,II,2, some elements of a Boltzmann-like picture do re-appear in some speculative cosmologies still under consideration.

2. Big Bang cosmologies

Relying on a combination of important observational data and systematic theorizing, a view of the overall structure of the universe (or, as we shall see in Section 8,II,2, in some accounts just of our "component" of it) has become orthodox in contemporary cosmology. The picture of the universe in this view is undoubtedly familiar to most readers, and I will sketch only its most important structural elements here.

Observationally, we discover that the distant galaxies seem to have the light we receive from them red-shifted in an amount that indicates a velocity of recession of them from our galaxy. Such an effect is just what would be expected from a uniform expansion of the matter of the universe, and of space-time itself if one thinks of the galactic locations as mapping out moving points in the space-time. In the past direction of time, we see what would appear to be convergence of all the matter to a single point in a time period of some tens of billions of years. It certainly looks as if all the matter of the universe is expanding outward from an initial singular pointlike concentration in what we call the past time direction.

A very important consequence of this picture of the cosmos is its denial of the so-called "perfect cosmological principle." Although the "cosmological principle" assumes that each point in a spacelike slice of the universe is (in a smoothed-out sense) like any other, and that all spatial directions are alike, the perfect cosmological principle assumes that the universe is structurally alike at every moment of time as well. Both of these views assume that one can partition the space-time of the world into spaces at a cosmic time, a subtlety we will ignore, although such simple worlds as rotating universes would be incompatible with it. The perfect cosmological principle is the foundation on which steady-state models of the universe were built, and if we can trust the data of the Hubble red-shift and the standard interpretation of it, that cosmological view, at least as originally understood, becomes untenable.

The model of a dynamic universe fits nicely with our theoretical account of the universe's dynamics as well. General relativity, like Newtonian gravitational theory before it, pictures gravity as a totally attractive force. Because all the matter in the universe is being attracted to all the other matter by the gravitational attraction, a stationary model of the cosmos would require an overall countervailing "cosmic pressure" to keep the system in static equilibrium. Einstein invoked such a force in some early cosmological models within general relativity in the form of the cosmological constant term of the gravitational field equations. But he was delighted to be able to drop the term when the dynamic model became observationally preferred. Crudely, the various dynamical models can be understood in terms of throwing a ball up in the air on the earth's surface. The ball might reach a maximum height and fall back down. If thrown fast enough, it might "go off to infinity" with its velocity never slowing to zero. Or it might, exceptionally, be at just the transition between these two possibilities if its kinetic and potential energies just balanced at the moment of its release.

To get a tractable theoretical model of the cosmos, we must make certain simplifying assumptions. Even if they are wrong, they are where theoretical exploration ought to begin. Assume that one can partition space-time into spaces at a time, so that events can be labeled by a global "cosmic time." Assume that the spaces at each time are locally isotropic. Assume that they are homogeneous as well. Then the worlds are all described by the so-called Robertson–Walker metrics. These metrics are characterized by a constant, k, which can take one of the three values $+1$, -1, or 0; and by a scalar parameter that varies with time, $R(t)$. The case $k = +1$ corresponds to a universe that has "closed" (finite volume) spaces at a time. These are three-spheres of constant curvature. These universes are the analogues of the ball that returns to earth. After a period of expansion, they reach a maximum spatial volume and then begin to collapse. The case $k = -1$ corresponds to the ball that goes on forever, retaining a non-zero velocity even in the infinite time limit. The spaces at a time are of constant negative curvature and have infinite volume – that is, they are Lobachevskian three-spaces. The case $k = 0$ corresponds to the ball that when tossed goes to infinity but has a limiting velocity of zero, the dividing case between the first two. The spaces in such a world are of infinite volume and flat – that is, they are all Euclidean three-spaces. $R(t)$ is the parameter that characterizes the "size" of the universe at a time, going from zero at the initial singularity, reaching a maximum and returning to zero in the $k = +1$ case, and going off to infinity in the other two cases. Just how R will vary with time depends on how the universe is constructed – whether, for example, it is filled with pressureless "dust" (a model for ordinary matter), radiation, or some mixture of the two, or with some combination of the various fields posited by quantum field theory, and so on.

The existence of an initial singularity in these models – of an initial state in which the universe was concentrated in a point where all genuine physical quantities became divergent, including the scalar curvature of space-time itself – was long thought to be a defect of the models, because it was thought to be merely an artifact of the unreasonable choice of an idealized perfect symmetry. More recent work has shown, however, that any world obeying general relativity that has a convergence like that of the universe of the Robertson–Walker models backward in time, must converge to a singularity, whether that universe is symmetrical or not. At that point, the theorist's mind naturally turns to the fact that general relativity, being a non-quantal classical theory, is probably incorrect in the very small anyway, and that a proper treatment of the initial singularity must be in terms of a quantum state of space-time with its manifold of evolutionary possibilities governed only by probability distributions. We will have a little more to say about such matters as this chapter goes on, but the philosophical exploration of the issues here would require a full-length book of its own. We will keep our attention focused as narrowly as possible on the issue that concerns us – whether the invocation of cosmology can solve the local irreversibility problem of statistical mechanics.

The most common models of evolution of the universe from the initial singularity assume that the matter of the world is, in periods not too far removed in time from the initial time, describable in a smoothed-out way as homogeneous and isotropic in the spaces at those times and as existing in thermal equilibrium. Some of the consequences of a posit of this kind have been marvelously confirmed by contemporary observational data. The relative abundance of the elements that would be formed in the primordial matter "soup" – in particular hydrogen and helium – is that predicted by the standard theory. The theory predicts that at the point at which the primordial matter becomes cooled down enough so that matter becomes mostly neutral and radiation becomes decoupled from matter, the universe should be filled isotropically with radiation. As the universe expands, lowering the frequency of this radiation and increasing its wavelength, this radiation should now be apparent to us as an isotropic distribution of radiation with a black-body frequency distribution. Such cosmic background radiation has been found. The standard model would lead us to expect that the observable matter of the universe at present should, if we use a large enough scale, be found to be homogeneously and isotropically distributed in space. Although there are now some doubts about this due to the discovery of non-uniformity in the distribution of galactic matter at astonishingly large scales, the matter of the universe perhaps being distributed with vast "bubbles" of empty space devoid of matter, there is some reason to think that the expected uniformity in the large can still be accepted.

How would thermodynamics and statistical mechanics apply to the description of such a model universe? To that we turn in the next section.

3. Expansion and entropy

As a whole, the universe (or, again, that part of it presently open to our observational inspection), has its entropy increasing in the forward direction of time. Sometimes it is alleged that the universe, which may be an infinite system, ought not to have an entropy attributed to it at all. But this can be handled by dealing with finite portions of the universe of larger and larger size, each containing its smaller predecessor in the sequence, and taking the behavior of the whole as the limit of the behavior of the finite sub-systems. Or it can be handled by using one of the devices from contemporary rigorous statistical mechanics that allow quantities like entropy to be directly assigned to systems with infinite numbers of degrees of freedom. More problematic, as we shall see in Section 8,II,1, is taking account of entropy resident not in matter but in the gravitational field – that is, space-time itself. Why does the universe as a whole have its entropy increase in the forward time direction?

A number of investigators have attributed the entropy increase to the fact that the universe is expanding, and expanding at a non-zero rate. In ordinary thermal processes, expansion at a finite non-zero rate (say, of a molecular gas in a container bounded in part by a movable piston) is accompanied by an entropic increase. Couldn't the situation be the same with the universe as a whole? First, it is important to notice a crucial difference between cosmic expansion and ordinary expansion. As R. Tolman showed in a decisive way, a universe filled with pressure-free dust and a universe filled purely with radiation both expand at the non-zero rate with no entropic increase whatever.

If the universe has both dust and radiation, it is true that non-quasi-static expansion can result in entropic increases. The dust and radiation may respond to the expansion differentially in such a way that a joint system, in equilibrium at one time, may be driven to non-equilibrium by the expansion. Then, any newly obtained equilibrium will involve the kind of heat flows and other relaxation processes that normally result in entropic increase of a system. Could this then be the ground of the universe's entropic increase in the sense that any expanding universe must have its entropy increase?

The answer is pretty clearly, no. Consider a universe that expands and then contracts. The same general assumption that told us that entropy increased in the expansion – basically, that the usual Second Law behavior holds – tells us that entropy will increase in the recontraction stage of cosmic evolution as well, just as the entropy of a molecular gas increases under non-quasi-static compression. Because we are, as usual, assuming time-reversal invariant laws, this would mean that the time-reverse of the universe in question, a universe that had entropic increase throughout expansion and recontraction, would be a universe that had an expansion with entropic decrease followed by a recontraction also with entropic decrease. If such a universe is lawlike allowed, then expansion by itself cannot be the origin of entropic increase.

The conclusion of this argument can be evaded if we posit, as have H. Gold and others, that the entropy-increasing expansion of an expanding and then recontracting universe would be followed by a recontraction in which entropy decreased. But the very Second Law type of posit that makes us expect entropy increase in the expansion phase leads us to expect entropy increase in the contraction phase as well. In order to avoid this, one would need a "delicate contrivance" of the microscopic states of the matter and fields of the universe being just so correlated that at the moment of termination of expansion and initiation of recontraction the entropy began to decrease instead of increase. Now we know that such astounding correlations can be induced in some ways. They show up in the spin-echo experiment, for example. But there the causal mechanism that reflects the correlation induced by an initial low entropy state from which the system evolved is clear. Each micro-component of the system is acted upon directly by the reflecting radio pulse. The absence of such a mechanism in the case considered here is clear. How do the micro-constituents of the universe "know" that the turning point of expansion-contraction has occurred? How could such a reaching of the point of maximal expansion serve as a "Loschmidt Demon?"

Whereas the implausibility of the scenario just outlined makes it implausible to associate a strict necessary connection between expansion and entropic increase, the possibility of a model of this kind can't be decisively dismissed. As we shall see in Sections 8,II,1 and 2, an "improbable" condition must be imposed somewhere in order to get the observed cosmic entropic behavior we do find in the world. But there is a distinction between the posit we are rejecting here and the posits of the kind we shall explore in Sections 8,II,1 and 2. These latter depend on an initial constraint on the system characterizable in a simple, local, macroscopic way. Low entropy is "put in" to the system much as we are accustomed to putting it into temporarily isolated sub-systems of the universe, by constructing them with fixed macroscopic constraints. In the case we have been considering, it is the necessity for a delicately contrived n-particle correlation function (where n may in fact be infinitely large) at all space-time points simultaneously, that makes the posit of a universe whose entropy begins to decrease at the moment of cosmic expansion turn-around so implausible. But nothing in physical law makes the posit rejectable out of hand.

Let us assume, however, that even if the universe begins to contract, stars will still radiate heat to outer space, cream will still stir into coffee and not out, and, in general, entropic increase will continue in the time direction we now call the future time direction and not in the anti-parallel direction posited by the account just discussed. If expansion by itself is not the relevant factor, what is?

4. Radiation asymmetry and cosmology

A brief look at the irreversible behavior of radiation in the cosmological context is in order at this point. Once again the claim may be made that the fact that we see accelerated charged particles emitting coherent outbound radiation and having their accelerations damped, but never see coherent inbound radiation spontaneously inducing an acceleration on a receiving particle, indicates a fundamental lawlike irreversibility in the nature of radiation. We have seen how, in the case of radiation confined to a bounded environment, the solution accepted by most is that the existence of outbound and non-existence of inbound coherent radiation is to be explained, once again, in terms of an asymmetry in initial conditions. This makes the asymmetry a symptom of the general asymmetry treated by statistical mechanics, rather than a grounding explanation for that asymmetry.

In the case of radiation in the universe as a whole, the only modification comes about because of the possibility of radiation that is either emitted from a source and never absorbed, or absorbed but that "comes in from infinity," never having been emitted, or radiation that exists from negative to positive infinity never emitted or absorbed. Actually, this presupposes an open universe of infinite time duration in past and future. In the broader general relativistic context, things will be a lot more complicated, but we shall ignore the refinements needed.

In a spatio-temporally open universe we might experience coherent radiation emitted from an accelerated particle that goes off forever, never being absorbed. But we know not to expect such coherent radiation coming in from infinity. It is sometimes argued that the existence of such inbound coherent radiation can be dismissed on deeper grounds than the mere resort to an asymmetry of initial conditions. But it is hard to see why this should be so. As far as the laws of nature go, the two situations, coherent radiation going off forever from an accelerated particle and coherent radiation coming in from past-forever to accelerate a receiving particle, are on a par. Of course, we can make the former situation occur by a local action of causing a particle to accelerate in a situation where some of the radiation it emits may be unabsorbed for all time. But we cannot "set up" the latter situation, for that would require us to control "at past time infinity" the delicate coherence of radiation over an infinite spacelike domain. But this is just the familiar fact about initial conditions in our world that we experience in ordinary statistical mechanics. Barring spin-echo-like tricks, we can induce future delicate correlations in particle motions by a local, macroscopic manipulation of the system at an initial time. But we can't set up the delicate correlations at the initial time that would cause the system to run anti-thermodynamically backward from macroscopic equilibrium to macroscopic non-equilibrium. It isn't clear why the greater spatial scope of the conditions needed to set up inbound coherent radiation, either from time minus infinity or over a finite spatial region from a finite past time, differs from the familiar thermodynamic impossibility. But, of course, it is the reason for that asymmetry of possibilities that we are seeking.

If one posits that no radiation either escapes to infinity or comes in from infinity, and that all radiation is bounded by particle emission-reception at both ends of its life, then the reduction of the electromagnetic asymmetry to a statistical mechanical asymmetry is more direct. For now, constraints on the distribution of initial conditions over all the matter of the universe other than the test particle (the emitter of coherent radiation or the receiver of it) will do the job of selecting outbound coherent radiation as allowed and inbound as forbidden. A self-consistent solution in which only outbound coherent radiation appears follows from the universe being constituted as a "perfect absorber" of the outbound coherent radiation posited. Basically, what one needs to posit is no coherence in the behavior of the absorbing particles other than that expected *after* it receives the radiation.

An interesting variant is obtained if one drops radiation as a substantial field existing in its own right, as is done in "action-at-a-distance" electrodynamics introduced to handle some of the problems of field theory that result from the interaction of the particle with the field it emits in the standard theory. To make such an action-at-a-distance theory compatible with relativistic constraints one needs the action of particle 1 on particle 2 to be "retarded" by what, in the standard theory, is the propagation time of the wave from the one to the other. The Wheeler–Feynman version of this theory posits both retarded and *advanced* action of each particle on the other. Here, once again, the appearance in the world of observed retarded interaction and the absence of apparent advance interaction is explained away in terms of an asymmetry in the initial conditions of the "absorber," the bulk of the universe at large with which the test particle is interacting. Once again, one can obtain, for example, damping of the accelerated motion of a charged particle. This is now explained not by the interaction of the particle with its local self-generated radiation field, but, instead, as the complex of advanced and retarded interactions of the particle with the rest of the particles in the universe. But one can also obtain the absence of anti-damping that would spontaneously put unaccelerated charged particles into accelerated motion. Ingenious improvements in the arguments allow for self-consistent temporally asymmetric solutions from either the electromagnetic field view with time-reversal invariant laws or from the temporally symmetric retarded-plus-advanced direct interaction view even if, from the field point of view, the universe is "transparent" and some radiation is allowed to "escape to infinity."

Let us assume, then, that the source of temporal asymmetry on the cosmological scale is not to be found in any primitive and lawlike asymmetry in the behavior of radiation (or of action-at-a-distance) but that this asymmetry is to be accounted for along with all the others by a general asymmetry of the type familiar to us from statistical mechanics. Once again, we need to find some appropriate ground for this asymmetry in the cosmos as a whole.

II. Conditions at the initial singularity

1. Initial low entropy

A direction in which to explore when seeking the explanation of the temporally asymmetric entropy increase of the universe is suggested by the fact that we can, if current cosmology is correct, trace the course of the universe back to a localized initial state, the initial singularity and the "small" universe developing immediately from it. Can we find a way of characterizing the universe at those early times that would, using methods of reasoning familiar from the thermodynamics and statistical mechanics of ordinary temporarily isolated systems in the world, account for the one-way direction of entropic increase?

Ordinary systems show entropic increase when they are started in low-entropy initial conditions. The gas is confined on the left-hand side of the box and the partition is removed. Given the usual assumption, with which we are now so familiar, that the micro-state of the gas at that initial condition is not one of the very small (in the usual probabilistic measure) collection of micro-states that would lead the gas to evolve to an even more non-equilibrium condition, we expect immediate entropic increase of the gas. That is, we expect the gas to flow toward a condition of more uniform density. Making the same argument at each further time (invoking the rerandomization posit), we expect a monotonic increase in the system's entropy toward equilibrium. There are, as we have seen, many questions about the justification for such an expectation, and even about what, in probabilistic terms, the expectation comes down to. But we have been through this. If we could characterize the condition of the universe at (or "instantaneously after") the initial singularity as being a state of low entropy and a state without "improbable" delicate correlations among its micro-components, couldn't we then explain the cosmic entropic increase in the usual manner of non-equilibrium statistical mechanics?

At first glance, the usual description of the cosmos just after the Big Bang is unpromising from this perspective. For matter is usually described as being in a thermal equilibrium state in the early stages of the universe. Perhaps the expansion combined with a multiplicity of phases could generate some entropic increase in the manner described by Tolman. But an important proposal, made by R. Penrose and others, suggests that another source of low entropy is far more important. The early universe is spatially homogeneous and isotropic. But such a uniform distribution in space-time is actually a low-entropy state when gravity is taken into account. The purely attractive force of gravity is such that such a uniform distribution is (from the usual statistical mechanical perspective) grotesquely *improbable*. As the universe develops, matter clumps. How this clumping takes place (into superclusters of galaxies, into galaxies, into stars), in what order it clumps (big or smaller structures first), what the fluctuational structure of the early universe is that generates the clumping by a positive feedback effect in which slightly clumped matter gravitationally attracts more matter to itself and so sustains a self-generating growth are unsolved problems of descriptive and explanatory cosmology.

But, it is generally supposed, some such transition of the universe from a low-entropy smooth space-time state to a clumpy distribution of matter, and, hence, of local space-time curvature, takes place. It is this clumping that provides the low entropy necessary to gather matter, originally in thermal equilibrium, into the radically uneven distribution we experience when we look at the cosmos, with hot stars radiating into cold space generating an entropic increase in the process.

The evolution of a universe that expanded and then recontracted would follow a pattern of entropic increase in both phases of its evolution from this perspective. In the expansionary phase, matter moves from its initial thermalized state to one of thermal non-equilibrium, the transition being paid for out of the enormous budget of gravitational initial low entropy as the uniformly distributed matter clusters into stars. The now hot stars radiate into cold space, adding more entropy to the universe in the process. The processes of local clumping followed by thermal radiation continue even after the expansion has reached its maximum and the universe has begun to recontract. In its final stages, just prior to the "Big Crunch" when the universe collapses to its final singularity, entropy continues to increase. The initial and final singularities differ in their entropy by a vast amount, the entropic increase of the universe over the two-phased process. If one is unhappy in attributing an entropy to the two singularities themselves, one can deal, of course, with the sequence of nearby non-singular universes just after the Big Bang and just before the Big Crunch.

The picture is completely consistent with all the laws governing its evolution being time-reversal invariant. The time-reverse picture of the universe is of one starting in a Big Bang that immediately goes to a new universe of extraordinarily high entropy. As it evolves, once again expanding and contracting to a Big Crunch, entropy steadily decreases. Radiation flows out of cold space into hot stars, and, even more importantly, a space-time initially clumpy becomes uniform, approaching perfect uniformity as a limit as it nears its final singularity. In this picture, as in the standard view of temporal asymmetry in finite systems, it is the special nature of the initial condition, the low entropy of the Big Bang, that determines the actual entropic course of events, not some time-irreversibility written into the evolutionary laws.

2. Accounting for the initial low-entropy state

Suppose the pervasive entropic increase of the universe is then attributed to the special nature of the Big Bang – that is, its being characterized as having low entropy in the form of an initial smooth distribution of matter and space. This is combined with a distribution of the other microscopic conditions of the sort to be expected from a system in equilibrium and not containing any extraordinary correlations of the kind that would lead to anti-thermodynamic behavior. Then the question of why the universe has its entropy increasing is pushed back to the question of why the initial state has the features it has.

It is frequently alleged that there is a very special problem in understanding why the initial state has its low-entropy condition. After all, low entropy is identified in the statistical mechanical picture with "improbability." Why should such an improbable initial state exist? The idea here is to imagine the collection of "all possible initial states of the universe," all possible initial conditions for space-time and matter in a universe having the fixed parameters of our world (total mass, for example) and obeying its laws. A "natural probability metric" over such a collection would give the class of worlds like ours that have such "unusual" initial states in small measure. Why, then, should such an improbable initial state have occurred in our world?

This "improbable" condition of the initial state is often associated with some other peculiar features of it. The standard cosmologies picture a world in which portions of the universe remain outside of each other's domains of causal influence, causal propagation being assumed to be limited by the speed of light as maximally fast causal signal, even when we trace the regions backward in time toward the initial singularity. Although distant galaxies get closer together backward in time, the amount of time since the initial singularity that a causal signal had to get from one region to the other diminishes as well. In the world as we find it, assuming that the rate of expansion we now experience can be extrapolated backward in time in the standard manner all the way to the Big Bang, we find that the universe has a structure in which portions of the world remain forever causally unconnectible no matter how far back in time we go. How then can we explain the uniformity of the early space-time, a uniformity that argues a coherence among its states at all spatial positions at early enough times? If the regions were in causal connection, some causal process might be invoked to "relax" non-uniform portions into the general uniformity, but if the regions are immune to causal contact as far back as we go, how does one region "know" the conditions at all the others? Doesn't this make the posited initial uniformity even more mysterious?

Other "improbabilities" are noted as well. If we try to determine whether our universe is closed, open and with negative curvature, or open but at the bounding condition of having its spaces at a time flat, we discover that it is at least fairly near to flatness. Scaling conditions on the features as they trace them backward in time indicate that something reasonably close to flat now would have had to have been flat to an incredible degree of accuracy at early times. But flatness is grossly "improbable," just as it is improbable that a ball tossed up in the air with a "random" impulse will have exactly the kinetic energy needed to achieve escape velocity, no more or no less. What accounts for this surprisingly improbable condition holding in the world?

When we start seeking "explanations" of the sort of facts just outlined, we have obviously entered a problematic methodological area replete with pitfalls for the unwary. The exploration of the sort of questions confronting us here hardly only began with the discovery of the Big Bang and its somewhat surprising features, though.

Consider the most obvious question of all about the initial state of the universe: Why is there an initial state at all? Why, for example, is there something rather than nothing? Putative answers have been given, ranging from the invocation of a deity as "uncaused cause" of the universe to speculations that the universe might be a "quantum fluctuation" out of "nothing," to such amusing observations as that because there are so many ways a universe can be, but only one way in which nothing can exist, the existence of something is a priori more plausible than the existence of nothing (R. Nozick). There is also a long tradition of denying that any coherent question can be asked here, much less answered. If we take it that the only explanation we can give for the existence of something is a causal explanation accounting for its existence by reference to the existence and nature of other ordinary objects and their features earlier in the history of the world, and if we accept the claim that grounds for causal inference are based upon experience of causal regularity in the ordinary world, then we might argue that the search for the explanation of the existence of "everything" is simply an illegitimate extension of the very notion of what can be explained.

Kant, for example, in a persuasive series of arguments in the *Critique of Pure Reason* and the *Prolegomenon to Any Future Metaphysic* argued in just this way. Although his arguments occur in the framework of a theory in which causal reasoning is applicable to the world only because causal structure is imposed on the world by the mind of the subject to whom the world is given as a world of experience, the general structure of his account will appeal to many who reject his views of causation as being "transcendentally ideal." Kant argues that finding relative causes of phenomena in the world, and believing, correctly he thinks, that everything has a cause, we are driven inexorably to ask the ultimate cause of everything. But such a demand is unfulfillable. For causes are always "in the world" demanding for themselves further causal explanations. The very idea of the "First Cause" of everything, whether called "God" or "vacuum fluctuation" or whatever, is incoherent, he maintains. Nonetheless, according to Kant, it is a good thing that we do illegitimately act as though such an ultimate explanation of the existence of everything could be found, for it is the unending search for this non-existent that goads us to go further and further in our search for deeper causes in the world itself.

Paralleling arguments to the existence of God as uncaused first cause of the existence of the world – so-called Cosmological Arguments – there were also arguments for God's existence framed on an inference from the *order* of the world as well – so-called Teleological Arguments. Usually these were based on the existence of the order of lawlike regularity in the world. Impressed with the system and order of the world summarized by the laws of nature, it was asked how such an order could exist without an intelligent designer to account for it. If one found a watch upon the beach, the argument frequently went, would not that imply a watchmaker? Here it is the "improbability," a priori, of there being any lawlike order at all to the cosmos that is taken as requiring an explanation for the order that we do find, and in the traditional version as implying an orderer.

In the arguments closer to the nub of the topic we are concerned with, the existence of the lawlike regularities governing the world as we find it is frequently taken as a given. It is now the existence in the early universe of the "improbable" conditions such as the low entropy of a uniform space-time or the near flatness of the space-time that are taken to require an explanatory account. Here it is argued that even given the laws of nature, a continuum of possibilities is compatible with those laws for the early universe. The condition we find is astonishingly improbable from an a priori point of view. Hence, some explanation is required as to why this condition, and not one of the far more "probable" ones we would expect, actually obtains.

But of course there have been many who, in the context of critiques of the traditional Teleological Arguments for the existence of God, have denied the very need for anything like "explanation" of the phenomena in question, whether that explanation be sought in God-the-Designer or elsewhere. If the issue is the explanation of regularity, it is claimed, regularities can only be explained by their subsumption under broader or deeper regularities evidenced in the world. Demands for the explanation of "regularity in general" are, it is alleged, fundamentally misconceived. They once again try to extend the demand for explanation from phenomena in the world to a transcendent explanation of the basic phenomenon of the world itself. If the claim is that it is "improbable" initial states that cry out for explanation, the critic will frequently deny the very propriety of attributing "probabilities" to these initial or overall conditions of the world at all. Attributions of probability, the claim often goes, depend upon observed relative frequencies in the world (or some generalization of that notion) from which probabilities are inferred. To talk about the probability of a universe is, from this point of view, incoherent. For there is no "real collection" of systems from which this universe is "chosen" that can serve as the needed reference class relative to which a probability can be grounded. As Mach said, in a somewhat different context, "The Universe is given to us only once . . ."

Yet, skeptical of the very coherence of the demand for an account of the universe's order as some may be, others have offered "explanations" of why the improbable initial state of the universe is as we have found it. Some have resorted, although usually in a sufficiently evasive way as to avoid direct criticism, to positing a "benevolent" creator in the familiar religious vein. Frequently, such arguments, of the old-fashioned sort that such a design implies a designer, are accompanied by arguments to the effect that the improbable initial state is a necessary condition for the ultimate existence in the universe of sentient life, thereby implying that the Designer was designing with us in mind as Its(?) ultimate aim. Of course, why that ultimate aim might not have been the existence of rutabagas, which also have the improbable conditions as a necessary condition, is never made clear. Other arguments are somewhat more ingenious and worth noting.

One approach is to argue that relative to certain specified constraints, such as the standard laws of nature holding and the total energy of the universe being like that of this world (or things of that sort, there being difficulties in general relativity of really defining such global quantities), it is "highly probable," using the standard a priori probability measure over the class of possible universes obeying these constraints, that there be a great entropy *difference* between one end of the universe and the other. This argument takes its most natural form in the case of a universe that has expansion and then recontraction from and to singularities, but it might be made coherent even in the case of a universe with a singularity at only one end. The claim would then be that it is highly probable that entropy will be greatly different at the singularity end and at the limit as time goes to infinity in the non-singular end.

The explanation as to why entropy is low in the past direction and high in the future is then accounted for in the manner of Boltzmann's argument that what we count as the future end is the high-entropy end. This obviates the Loschmidtian objection that in the probability measure chosen there will be as many universes with high entropy at the initial singularity and low at the final as the other way around, for the former are just the time-reverses of the latter, and the collection of time-reverses of an ensemble of systems (even whole universes) is of the same natural measure as the original collection. This is now admitted, but it is argued that in all these cases, entropy increases in the future time direction simply because the future direction *is* the direction of time in one of these possible universes in which its entropy increases "by definition." If we think of the collection of possible worlds as a collection of imagined co-existing worlds, and we allow ourselves the possibility of "rigidly" (as the philosophers put it) taking the notion of what is future time in one world and talking about "the same direction of time" in another possible world, then it will be the case that in half the worlds, entropy increases in a given time direction and in the other half it increases in the other time direction. But in each world, observers will say that entropy increases in the future time direction.

This Boltzmann "way out" will of course have its critics. One objection to this solution of importance is to argue that even relative to the constraints imposed, it is unlikely that a universe will have the *magnitude* of entropic difference, or the *degree* of low entropy at one singularity end, that we find in this world. If we take it that the Big Bang really has the "perfect" space-time smoothness credited it in many familiar models, can that really be said to be "probable" in a standard measure over the class of possible universes compatible with the imposed constraints? It seems doubtful.

Some solutions look to taking the class of imagined universes just discussed and viewing them as all "equally real." Or the proposal might even go further and allow for a broader class of existing universes, including those with constraints in such form as total mass, and so on, that vary from world to world. Or, even more extravagantly, we might allow the laws of nature themselves to vary from world to world. One version of this approach involves quantum mechanics, and the curious attempt to resolve its so-called "measurement problem" that is called the "Many Worlds" interpretation of quantum theory. In this understanding of quantum mechanics, an understanding that is fraught with conceptual problems of great difficulty and intractability, one envisions the universe as a multiplicity of "branches" continually splitting into a greater and greater manifold of alternative universes. Could it not be that most of these universes have a more "probable" nature, in some natural probability measure, than ours, and that the universe we find ourselves in, with its puzzling low overall entropy, is merely an exceptional "branch" or "sample" of the whole collection of universes most of which are in more probable high-entropy conditions?

Here the analogy with Boltzmann's "way out" is clear. Although he posited that our local region of the universe is a rare low-entropy fluctuation in a vast sea of high-entropy overall equilibrium, now our entire universe is taken to be a rare exceptional case in a sea of co-existing universes, almost all of which are of the more expectable kind. Not surprisingly, such speculative proposals frequently invoke the other aspects of Boltzmann's solution as well. It is not to be wondered at that the universe we find ourselves in is of low entropy, it is argued, for only in such universes does the structure of the universe allow for the evolution of sentient beings. Nor should we wonder about the particular time direction entropic increase takes in our particular universe, for, once again, the future time direction is by definition the direction of time in which our universe's entropy increases. The usual problem, that even a much more probable universe would be enough to sustain sentience, and that the probability that the universe would be as exceptional as it is, even conditional to the existence of sentience, is low, often appears as an objection to this account. Needless to say, the whole model of a collection of real alternative universes also comes under fire as a wild speculation ungrounded on any serious empirically supported scientific understanding of the world.

One variant of the approach just described is particularly interesting here. We have remarked that the initial singularity's uniformity is puzzling given the causal unconnectibility of regions of the space-time for all past times in the usual cosmologies, and that its seemingly near perfect flatness at early times also constitutes a puzzle. An attempt to solve both of these puzzles simultaneously is found in the so-called "inflationary" cosmologies. Here, features of quantum field theory are invoked to give a characterization of a universe that, starting from a singularity, has a period of slow expansion (or even stasis) in which a vast store of latent energy is accumulated through a field-theoretic "phase transition" that holds off until the energy of transition is released in a fast and enormous "jolt" to the universe. This occurs in a manner analogous to the fast crystallization that occurs when a supercooled fluid finally has its transition to a solid that has "held off" become triggered, say by the introduction of a seed crystal. When the phase transition is finally completed, the space of the universe can expand at an exponential rate. Such a period of rapid expansion will allow for early periods of the universe in which causal contact could have occurred throughout the spatially separated regions of the space-time and would also account for the near flatness of the currently observed universe without assuming the almost perfect flatness of the universe near the singularity of the standard cosmologies. Variants of the inflationary scheme would take our universe to be only one "bubble" in a vast sea of inflationary bubbles, each of which constituted its own observationally closed universe. Once again, one could try to handle the low-entropy problem for our world by some combination of Boltzmannian arguments of the kind noted. Once again, the familiar objections would hold.

An alternative general approach to the problem of the apparently surprising "improbable" conditions at the beginning of the universe is to try to explain these away as features of a model in which not low but high entropic conditions are posited at the Big Bang. Perhaps the universe starts off not with a smooth space-time structure but with the erratic, clumpy structure associated with high space-time entropy. But perhaps this initial irregularity is smoothed out over the course of time by a process of thermalization due to the interaction of space-time with the particle fields. Perhaps, that is, the initial condition is "chaotic" and the appearance of uniformity, the absence of chaos, is due to the usual process by which irregularities are smoothed out in thermalization.

But, as R. Penrose has emphasized with great clarity, this would lead us to expect a large entropy of the matter fields. The entropy of matter in the universe can be estimated, for reasons having to do with the application of thermodynamical-statistical mechanical concepts to quantum field theory, by the ratio of the number of photons to the number of baryons in the universe, a ratio estimated at being something like a billion to one. Sometimes this is said to be a large value, so perhaps the appearance of low entropy for the initial space-time is misleading. Perhaps it started out in high entropy but fed that disorder into the matter fields. But, as Penrose makes clear, the entropy of matter is absurdly small given the entropy contained in space-time itself.

In whatever way we look at the cosmological situation, we must admit that we have entropy increase in the evolution of the world. When we consider how high the space-time entropy could be in the Big Crunch at the end of a recontraction cycle in a universe that does recontract, we find that we must posit a low entropy of space-time at the Big Bang compared with the vastly higher entropy at the Big Crunch, and that the existent present entropy of matter cannot be enough to allow us to get away with a posit of initially high entropic space-time. Basically the problem is that if we wish to account for the evolution of entropy toward ever higher values in the universe in the way in which such an evolution to higher entropy is standardly treated in statistical mechanics of finite systems, we must posit for the universe the same kind of initial low entropy in the form of a macroscopic constraint (like the gas all being on the left-hand side of the box). And we must posit a high probability micro-state underlying the macro-state to avoid anomalous anti-thermodynamic evolution from the initial non-equilibrium condition. In cosmology, the most plausible macro-constraint to invoke is initial "improbable" uniformity of the spatial structure at early times.

Finally, one could try to obviate the appearance of "low probability" for the initial condition by positing constraints in the form of laws of nature such that, relative to these constraints, the condition no longer appeared as an improbable condition selected from a large pool dominated by much more probable kinds of initial states. A new law of nature can restrict the initial conditions permitted by the now expanded totality of lawlike constraints and hence make the condition we actually do find a probable or even inevitable one. If this law is to permit the much higher spatial entropy expected at the Big Crunch of a universe that cycles back toward a singularity, though, it must have a time-asymmetry built into it, permitting the extreme spatial irregularity at the final singularity and banning it at the initial.

Just such a proposal has been made by R. Penrose. He suggests a new law of nature governing space-time singularities. The particular one he suggests is that Big Bang-type singularities be governed by the condition that their conformal space-time structure, described by the Weyl tensor in the differential geometry of space-time, be smooth – that is, that the Weyl tensor be identically zero. Because the Weyl tensor, in a certain sense, represents the structure of space-time over and above the structure imposed on it by the distribution of mass-energy in the world of a non-spatiotemporal nature, this comes down, intuitively, to a kind of assertion that initial space-times are determined, as a matter of law, to have their intrinsic structure smooth (and, hence, of extraordinary low entropy).

Naturally, this invocation of a new "law" has a terribly ad hoc flavor to it. If there were lots of white holes in the universe – that is, lots of "mini Big Bangs" within the space-time generated out of the big Big Bang of the cosmos – it might be possible to see if, indeed, all of them conformed to the Penrose law of initial smoothness. But it is far from clear that there are any such white holes, the theoretically allowed but problematic time-reversals of the familiar black holes predicted by general relativity. Indeed, Penrose suspects that they are impossible. So how methodologically reasonable is it to invoke such a "law?" Or does the posit simply come down to a posit about the initial structure of space-time at the Big Bang with a denial that any further explanation can be expected? This approach, we might note, takes the past-future distinction as given and simply invokes a time-reversal non-invariant law. In these respects it is rather a denial of the general program for finding "ways out" of the statistical mechanical paradoxes that Boltzmann suggested in his replies to his critics and that we have seen repeated in metamorphized form in most of the other attempts at accounting for initial low entropy.

A thorough exploration of the issues here is clearly beyond our scope. The profound questions on the very possibility of explanations of what are posited as cosmological or "basic" conditions of the universe that ground all normal explanations is one crying out for deeper exploration. This need is made crucial by the recent tendency of practicing science to try to deal with such ultimate conditions of the universe. But this is something we will have to forego here. Our question, rather, will be whether the entropic behavior of the universe just discussed will serve the purpose for which it is frequently invoked in foundations of statistical mechanics. That is, can it serve the purpose of accounting for the irreversible behavior of ordinary systems? Can the clearly observed gross entropic increase of the universe as a whole, or at least of the region of it available to our observational access, when it is introduced into statistical mechanics, resolve the remaining gaps in trying to go from probabilistic reasoning in non-equilibrium theory as a component of a time-symmetric theory to probabilistic reasoning as explaining the origin of the notorious thermodynamic asymmetric behavior of systems in the world?

III. Branch systems

1. The idea of branch systems

In whatever way we are to account for the observed global entropic increase of the universe we observe, we must accept that such an increase is a fact of nature. Can we invoke this fundamental fact as a basic posit, and use it to account for the most problematic fact of the world we want to account for in statistical mechanics, the Second Law behavior of "small," individual systems? Can we get the asymmetric thermal time behavior of systems that has so far eluded us out of the given asymmetric time behavior of the universe? That the introduction of this global asymmetry is promising is clear. Here, at last, is a time-asymmetric fact that might "break the symmetry" imposed by the time-reversal invariance of the laws. This symmetry could only be broken in the earlier accounts by "putting in" a distinctive probability distribution over micro-states of the system at the "initial" state and never at the "final" state of the system.

Those who hope to find the key to irreversibility in cosmology often invoke the notion of "branch systems." The real systems of the world with which thermodynamics deals are not the idealized systems existing in splendid energetic isolation from past time infinity to future time infinity. Whereas consideration of such permanently isolated systems may be useful for some purposes – studies of global stability of trajectories or investigations into the Recurrence Theorem, for example – perhaps the idealization misses exactly what it is about real systems that is needed to account for their time-asymmetry. Real systems are constructed by energetically isolating a piece of the energetically interacting universal whole from its surroundings, and then letting the system exist in energetic isolation for some finite period of time. During this time, the system's thermal behavior is watched. At the end of this period, the system's isolation is ended, and it becomes, once again, a mere energetically interacting fragment of the whole cosmic system. Call such a temporarily energetically isolated fragment of the universe a "branch system." Can we use the facts about cosmological entropy increase and the fact that systems studied in thermodynamics are branch systems to account for the irreversibility whose explanation has previously eluded us?

Now, one might try to prove one of three different things using cosmology and branch systems. First, one might try to show that it is overwhelmingly probable that entropy will increase in branch systems in the intuitive future direction of time. Next, one might try to show that the overwhelming probability is of entropic increase in branch systems in that direction of time in which the entropy of the universe as a whole is increasing. Finally, one might only try to show that it is overwhelmingly probable that the branch systems will have their entropic increases "parallel" to one another – in the same temporal direction – whether this be the intuitive future time direction or the direction of time in which the entropy of the global system increases or not. Those who seek a demonstration of the latter two results usually try to extract the former as well by an application of Boltzmann's "argument by definition" – that is, by his claim that our intuitive notion of the future direction of time is grounded on that time direction's being the time direction in which with overwhelming probability entropy of systems increases.

We start with the following facts: (1) The systems with which we deal in thermodynamics are branch systems. They live in energetic isolation from the bulk of the universe only for finite periods of time, being in energetic contact with the remaining world before and after their beginning and ending points as isolated systems. (2) The entropy of the universe, or at least of our region of it, is extraordinarily low (note that this is true both at the beginning of a branch system's existence and at its end). (3) The entropy of our local region of the universe is increasing in one time direction. The hope is that we can use these facts to derive one of the versions of the statistical surrogate of the Second Law noted earlier.

Usually, as we shall see in the next section, it is admitted that the derivation requires an additional probabilistic or statistical posit as well. The hope is, however, that this additional posit will not itself be one that puts the desired result of temporal asymmetry into the conclusion. That, after all, is supposed to derive from the time-asymmetry of the universe's entropic increase. Rather, the probabilistic posit should be something of a temporally neutral kind. Can the derivation be carried out? To that we now turn.

2. What cosmology and branch systems can't do

The notion of branch systems was introduced by H. Reichenbach in his seminal book, *The Direction of Time*. Sections 12–16 of that book also

contain the most thorough discussion in the literature of the attempt to
derive Second Law behavior from the cosmological entropic asymmetry.
Reichenbach discusses the origin, in Boltzmann's late work, of the idea
that thermodynamic behavior of individual systems is "inherited" from
the direction of entropic change of the main system from which the
individual systems are temporarily isolated. He agrees with Boltzmann's
fascinating suggestion that the cosmic entropy might be directed in
opposite directions of time in some spatial portions of the universe at a
given time, or at different times even in the same place, and that such a
difference of direction in time of increase of entropy of the main system
will result in counterdirected thermal behavior in those regions of the
universe whose cosmic entropy increase is counterdirected. He calls this
the "sectional" nature of irreversibility.

Reichenbach says, with justice, that Boltzmann just assumed that the
branch systems would (in the probabilistic sense) have their entropy
increase in time parallel to each other and parallel to that of the main
system. But, he claims, again with justification, that to infer this requires
a posit to just that effect (Assumption 5, p. 136). Now, for Reichenbach's
main purpose in his book, invoking such an assumption is just fine.
Reichenbach is mainly concerned with the claim, to be discussed in this
book in Chapter 10, that our intuitive notion of the past-future distinction
can be grounded in the distinction between the two directions of time,
that in which entropy decreases and that in which it increases. So if
Assumption 5 is true, it describes the world and it characterizes the
sense in which there is a direction of entropic increase in time that
will serve this ultimate Reichenbachian purpose. This is the direction in
which the branch systems have their entropies increase parallel to one
another. But of course it leaves unexplained why the systems behave,
probabilisitically, in parallel to one another and to the main system,
because it just puts that in as a posit.

Reichenbach suggests, though, that Assumption 5 can be derived from
other assumptions and to this extent "explained." Assumption 1 is just
that the universe now has low entropy and is situated on a "slope" of
the entropy curve – on a place where its entropy is higher in one direction of time and lower in the other. That is, it isn't currently in a local
maximum or minimum condition of entropy. Assumption 2 is just that
branch systems exist that are connected to the main system at both ends
of their existence, but energetically isolated from it for the duration of
their finite lives.

Assumptions 3 and 4 are more important. Assumption 4 is that, of the
branch systems, the vast majority have one low-entropy endpoint and
one high-entropy endpoint. This would seem to exclude the possibility
of there being lots of branch systems that are cut off from the main

system in equilibrium and rejoined to it in the same condition, which seems unjustified as an assumption. But one can just read Assumption 4 as delimiting those branch systems with which we will be concerned, the ones that suffer a serious entropy change during their brief lifespan. The aim is of course to get their changes parallel to one another and to the main system (and, perhaps, to get it increasing in the intuitive future time direction).

Assumption 3 is the crucial posit, the one that provides the core of the premises from which the otherwise question-begging Assumption 5 is to be derived. Assumption 3 is that "the lattice of branch systems is a lattice of mixture." What does that mean?

Consider a large collection of similar systems that evolve over identical periods of time. Keep track of some macroscopic feature of them, noting its value at small, equal time intervals as the systems evolve from their starting point. Keep track, for example, of their Boltzmann individual system entropy values. Make a square array of the values obtained:

$$
\begin{array}{cccc}
a_{11} & a_{12} & a_{13} \cdots\cdots\cdots & a_{1n} \\
a_{21} & a_{22} & a_{23} \cdots\cdots\cdots & a_{2n} \\
\cdot & \cdot & \cdot & \cdot \\
\cdot & \cdot & \cdot & \cdot \\
\cdot & \cdot & \cdot & \cdot \\
a_{m1} & a_{m2} & a\,a_{m3} & a_{mn}
\end{array}
$$

Assume that a large number of systems is involved (m is large) and that many measurements are made (n is large).

Postulate *row independence* for the array. What that demands is that the probability of a_{k1} having a specific value conditional on $a_{k(1-r)}$ having some specific value is not changed by conditioning a_{k1} on some $a_{k'p}$ where k' differs from k, or by any combination of a values taken from other rows than k. In other words, the probability of going from a value of a at one time in a row to some other value of a at some other time in that row is independent of the a values at any combination of times in any combination of other rows. The rows are probabilistically independent of one another.

Next, postulate *lattice invariance* for the array. Suppose we find the probability that a certain value of a, say a^*, is found conditional on another value, a^{**}, immediately preceding it in a row. It is assumed that because all the systems are identically constructed, this probability – the conditional probability that $a_{k(i+1)} = a^*$ given that $a_{ki} = a^{**}$, is the same in each row. Now look at two consecutive columns, j and $j + 1$. We can find the probability that in the reference class that now consists of all pairs in any row in which one member is in column j and another in column $j + 1$ (rather than our previous reference class, which was all pairs in a single row one column apart), a has the value a^* in column $j + 1$ given that it had value a^{**} in column j. Lattice invariance is the generalization of the demand that this probability be equal to the probability of a^{**} being immediately followed by a^* in a row. Intuitively, it demands that the column-to-column conditional probabilities be given by the similar conditional probabilities within rows.

These two conditions define what it is for the lattice to be a *lattice of mixture*. From them, one can prove, among other things, that no matter how the values of the a's are fixed and distributed among the left-hand most column of the array, the distribution of a values in the right-hand column will, in the limit of a very large array both horizontally and vertically, be given by the probability of a's having the values in question over any of the rows. If, then, we started with ergodic systems, almost all of which were at their initial states as represented by a values in the left-hand column in states of low entropy, we would find that the distribution of entropy values in the final states, as represented by the distribution of a values in the right-hand column, was the usual equilibrium distribution with the overwhelmingly greatest portion of the systems at high entropy values.

From this, Reichenbach argues, we can see that Assumption 5, which expresses the parallelism (in the probabilistic sense) of entropy change of the branch systems with each other and with the direction of change of entropy of the universe as a whole, can be replaced by the postulation that we are dealing with a set of branch systems with high- and low-entropy ends and that the array of their entropy values constitutes a lattice of mixture. Is he correct?

The basic argument goes like this: Even if we don't have any preconceived notion of which direction of time is the future direction, we can determine that end 1 of system A is in the same time direction from end 2, its other end, as end $1'$ of system B is from end $2'$ of system B. A collection of such systems, all of which have low entropy at their 1 ends, and almost all of which have high entropy at their 2 ends, forms a lattice of mixture. And the same holds if it is the 2 ends that have the low probability and the 1 ends the high probability. But a collection of systems, half of which have low and half of which have high entropy at their 1 ends and that have the same proportion of low- and high-entropy states at their 2 ends, is *not* a lattice of mixture. For such a lattice always has its right-hand column in the equilibrium "almost all high" distribution of entropy values. So, systems evolving with their entropy change in parallel obey the lattice-of-mixture assumption. A collection in which half went one way and half the other does not.

The impact of this argument rests on the apparent innocuousness of the lattice-of-mixture assumption. Doesn't this assumption just come down to the joint assertion that individual systems are probabilistically independent of one another, and that the transition probabilities in the ensemble ought to duplicate those in an individual system? What could be more natural? And because the probabilities in the individual system, as we have known since the Loschimdt-like objections to Boltzmann's statistical interpretation of the H-Theorem, are time-reversal invariant, how could we have begged the question in favor of parallelism by making the innocent assumption of mixing for the array?

To see that the assumption is not as innocent as it first appears, consider the following "paradox." Once again, consider a vast collection of branch systems each with a low- and a high-entropy end. Label the ends of the systems the 1 end and the 2 end, with each 1 end in the same time direction from its systems 2 end as in every other system. Assume that half these systems have the 1 end at low entropy. What does the missing posit tell us to infer about them? Answer: That they are almost all at high entropy at their 2 ends. Now let the other half of the systems have low entropy at their 2 ends. What does mixing tell us to predict? Answer: That they have high entropy at their 1 ends. But if we consider the *total* collection, now a collection of systems half of which have low and half of which have high entropy at their 1 ends, what does mixing lead us to infer? Answer: That almost all have high entropy at their 2 ends (and similarly with 1 and 2 interchanged).

There is no contradiction here. In an individual system, we infer from low entropy at a time to high entropy at an earlier *and* at a later time in the familiar way. From high entropy, though, we infer high entropy at the earlier and later times. All of this holds because of the familiar line that "high-entropy states are overwhelmingly dominant in the history." So all the mixing inferences are legitimate. But now consider the ensemble composed of the systems with their low entropy at the 1 end with their states lined up starting from the 1 end and the systems with low entropy at 2 with their states lined up starting with the 2 end! The left-hand column of the array will consist of all low-entropy states. Mixing will then lead us to infer that the right-hand column will consist of nearly all high-entropy states. But now the two groups of systems have their right-hand column states in opposite time directions from their left-hand column states. If we arbitrarily call one direction of time "later," then, the mixing posit applied to *this* array will lead us to infer that if one group of states "starts" with low-entropy states and "goes" to high-entropy states "later," then the other group must be expected to "go" from low-entropy states "later" to high-entropy states "earlier," or, what amounts to the same thing, to go from earlier high-entropy to later low-entropy states. In other words, with the array arranged this way, the positing that it is a lattice of mixture predicts *anti-parallelism* for the behavior of the two ensembles of systems, with entropy increasing for them in opposite directions of time!

It was in setting up the array with the left-hand states always in the same time direction from the right-hand states that a "bias" toward parallelism of systems was built in when the apparently innocuous posit that the array was a lattice of mixture was assumed. It is in arranging the states of the system in this way, always from earlier to later, or, at least, always in the *same* time order, and applying the probabilities that hold for a single system to the array so arranged, that parallelism is built into the system. Parallelism in, parallelism out. We are back to the familiar problem. Given a macroscopic characterization of systems, if we apply the usual probabilistic reasoning to the distribution of micro-states compatible with that macro-condition at the initial state of a system, we get predictions confirmed in experience. But if we use the same probabilistic reasoning to a "final," but still low-entropy, state of a system we get the wrong results. Yes, that is true. But what we wanted to know is why the asymmetry holds in the world. Reichenbach's posit of the array being a lattice of mixture, combined with the array being constructed the way it is, just imposes parallelism on the systems without explaining it.

But, to be fair to Reichenbach, he never really claimed he was explaining it. Yet the appearance that an explanation has been given has, I believe, misled many readers. Worse yet, the appearance is that one has explained the parallelism of systems by reference to the direction of entropic increase of the main system. (Witness the claims that time direction is "sectional" and that a change in the main system's entropic direction will be accompanied by a change in the direction of parallel entropy change by the overwhelming majority of branch systems.) Notice, though, how reference to the direction of entropy change in the main system plays no role whatsoever in the derivation of parallelism from the posit that the array is a lattice of mixture.

On page 138, Reichenbach says this:

It is perhaps possible to show that in the history of a system, such as the universe, slopes satisfying Assumptions 2–5 are much more frequent than slopes that do not satisfy these assumptions. Hence, we would have a high one-system probability that if Assumption 1 is true, then Assumptions 2–5 are true.

He goes on to argue that this probability "wouldn't help us very much" because it refers to very long time periods in the history of the universe, and that the best reason for thinking 2–5 true is empirical confirmation of them.

Now, no doubt, we do have reason to believe that Assumption 5 is true. Branch systems do parallel each other in their direction of entropic increase, and they parallel in this the direction of entropic increase of the universe as a whole. But, the question is, does the direction of entropic increase of the universe as a whole *explain* the direction of entropic increase of the parallel evolving branch systems?

Reichenbach never says what the argument might be that "perhaps shows" Assumptions 2–5 are highly probable. Remember that Assumption 5 posits both parallelism of branch systems and their parallelism to the main system. As we have seen, the parallelism of branch systems of Assumption 5 is obtained by Reichenbach only by use of Assumption 3 (that the array is a lattic of mixture). And, as we have seen, it is obtained by a hidden assumption about how arrays must be constructed that, with Assumption 3, amounts to positing parallelism. If we simply posit parallelism, then a kind of argument that the evolution of the main system will be in the direction of the parallel branch systems might simply be that if we assume that it is highly probable that any given system will have its entropic increase in the same direction as that of the bulk of other systems (that is what positing parallelism says), it follows that it is highly probable that the main system, being a system, will parallel the other systems, which is equivalent to their paralleling it, because parallel direction of entropic increase is a symmetric relation.

But that hardly constitutes what we thought we were going to be given. What we thought we were going to get was an explanation as to why systems evolved parallel to one another and parallel to the main system, an explanation that was grounded in the fact of the entropic increase of the main system. What we get is an argument for parallelism that comes down to positing it, and a reverse-direction inference to the parallelism of branch and main systems.

One can see what is going on here if one looks at a simplified version of Reichenbachian arguments due to P. Davies in Section 3.4 of *The Physics of Time Asymmetry*. First, note that the universe is a vast pool of low entropy. If we find a branch system in low entropy, it is much more likely to have been separated from the main system than to have spontaneously fluctuated to that low-entropy state from a higher-entropy state while isolated. Now, consider a system in a low-entropy initial state, just cut off from the main system. Assuming the time-neutral posit that the probability of a micro-state compatible with this low-entropy macro-condition is given in the usual way, we ought to infer with high probability that later states of the system will have higher entropy. We would also be impelled to infer the same thing about earlier states of the system if it had any, but having been just created, there are no earlier states of the system, breaking the time-symmetry of the inference. So, Davies says:

If all branch systems are formed in *random* low entropy states, then it has been seen already that the entropy of these new isolated systems will almost certainly increase, independent of our convention about the direction of time. That is, it may be asserted confidently that almost all branch systems will show *parallel* entropy change. It is the asymmetry regarding the formation of branch systems which brings about the parallel increase in all (nearly) branch system entropies. . . . If the branch systems are regarded as not existing in the *past* then the entropy of the overwhelming majority of these systems will *increase with time*. It is through branch systems that the customary intuitive notion that entropy increases with time is derived.

He continues:

With these considerations an entropy law can be formulated which is valid even in the case of a totally permanently isolated system. For if there is an ensemble of such systems which on a particular occasion are in randomly selected states of entropy $S << S_{MAX}$ then from the forgoing arguments it is clearly overwhelmingly probable that all these systems will be in molecular chaos at an entropy minimum, and so all will increase with both greater and lesser t, i.e. in both directions of time. If any two systems A and B are selected, and their entropies at the first mentioned occasion called S_A^1 and S_B^1 respectively, and on another occasion (and too widely separated from the first) called S_A^2 and S_B^2, then the product:

$$(S_A^2 - S_A^1)(S_B^2 - S_B^1) = 0$$

whichever way it is chosen to increase, for both A and B will almost certainly change in a parallel direction as seen. If $S_A^2 > S_A^1$ then $S_B^2 > S_B^1$, but if time is measured the other way and $S_A^2 < S_A^1$ so $S_B^2 < S_B^1$ and the product is positive in both cases. This is Schrödinger's formulation of the entropy law, and it reduces to the usual form if A, say, is allowed to become the system of inference, and B becomes the rest of the world. Then if the increasing direction of time is *defined* as the direction in which the entropy of the whole world increases, [the product rule above] becomes $\Delta S \geqslant 0$ where $S = S_A^2 - S_A^1$. This equation is just the law of Clausius.

This is confirmation of what is normally regarded as obvious, that the direction of entropy change of the branch system is parallel to that of the outside world (the main system).

Now it is certainly correct that the universe being in such an extraordinarily low-entropy state, we ought generally to infer, if we find a system isolated for a finite time, that its low entropy is a result of its interaction with the main system and not an improbable fluctuation from equilibrium. But, as far as I can see, that is all that cosmology by itself can really give us.

The crucial argument is the argument for parallelism of entropy change of systems. Once again, it is from that that the parallelism of branch systems with the main system is derived. The direction of entropy change of the main system doesn't *explain* the direction of entropy change of the branch systems. Rather it is just that any two systems being expected to evolve in parallel, any system ought to be expected to evolve in parallel to the main system. Then the Boltzmann trick is performed of saying that what counts in future is the direction of entropic increase of the main system (a little unlike Reichenbach, as we shall see in Chapter 10,III,3, who would take "future" to be defined, rather, by the parallel direction of entropic increase of branch systems), and we have the usual version of the Second Law.

So everything hinges on the "demonstration" of parallelism. Once again, it derives parallelism by a posit that seems innocuous but isn't. To be sure, if we take the 1 end of systems A and B to be at the same time, or even if we take for both A and B the 1 end to be in the same time direction from the 2 end for both systems, and we then invoke the usual hypothesis of random micro-state at the 1 end, taken to be of low entropy, we get the prediction that the 2 end will be of higher entropy, and so the two systems will have a low and high entropy end in the same time order. But if we take two systems of short equal lifetime, with the 1 and 2 ends of the systems in *opposite* time order from each other, and with the 1 end again of low entropy, we would still, if we posited random micro-states at both 1 ends, predict the 2 ends to be of higher entropy than the 1 ends and we would predict oppositely time-directed entropy changes for the two systems! The trick is to insist that if the posit of random micro-state is made at one time end of one system, it always be made at the same time end for all other systems.

Once again, such a posit is reasonable if we want to describe the world correctly. Systems do (in the probabilistic sense) have their entropies increase in a time direction parallel to each other and to the main system. That isn't the point. The point is that we wanted an *explanation* of that fact. We thought we were going to get one from the time direction of entropy increase of the main system. We didn't. We simply got a rule that generates a prediction of parallel behavior by positing it and avoids predicting anti-parallel behavior by prohibiting the other way of applying the posit of random micro-state. But parity of reasoning considerations would lead us to think that the posit is legitimate applied both ways. It isn't, of course, but we still don't know why.

To emphasize what has gone on here, consider Davies' argument to the effect that we are blocked from inferring a high-entropy past for a system created in low entropy by the fact that it has no past. Very good. But now consider a system in low entropy about to be reabsorbed into the main system. By parity of reasoning, we ought to assume that it had a high-entropy past. The inference to its future is blocked by the fact that it has no future as an isolated system. Such an inference would be mis-guided, of course. If it has low entropy in its final state, it almost certainly got there by coming from an even lower-entropy initial state when it was created. Or again, consider a low-entropy state in the middle of the lifetime of a branch system created and later reabsorbed. It has both a past and future from that midpoint. By the posit of random micro-state, we ought to infer that it had a higher-entropy past and will have a higher-entropy future. But, of course, the second inference is reason-able and the first isn't.

It is legitimate to posit randomness of the micro-state for low-entropy *initial* states of systems, but not for their final or intermediate states. Let me quote O. Penrose in *Foundations of Statistical Mechanics*:

The principle, which I shall call the *principle of causality*, is simply that the phase-space density at any time is completely determined by what happened to the system before that time, and is unaffected by what will happen to the system in the future. This principle, though usually taken for granted, is worth stating explicitly here because it contains the distinction between past and future in statistical mechanics.

Indeed, it is notions of causation that we are likely to find invoked at this point. The initial low entropy of a branch system is explained by its having been cut off from the main system. That is how it was "caused" to be brought into being. Because its state at that initial time then "causes" the future states of the temporarily isolated system, reference to the initial micro-states is in order. And if we are concerned with probable evolu-tion, reference to a probability distribution over these initial micro-states is in order in explaining future micro-states or probabilities of them. But the final state, even if of low entropy, can't be explained by refer-ence to the forthcoming reabsorption. Nor can the states that led to it be explained by its micro-states, because they caused the later states and not vice versa. So it would be unreasonable to infer the likely past behavior by a posit of probability over the later possible micro-states of the final low-entropy state of the system. This is indeed the right way to do sta-tistical mechanics.

But we were supposed to be avoiding bringing such obviously time-asymmetric posits into play here. We were supposed to be relying on the sole given asymmetry, the time direction of entropic increase of the main system, the universe as a whole. I have argued here that in Reichenbach's account to get parallelism of time direction of entropic change for branch systems, a posit has been made that is close to that stipulated by O. Penrose. It doesn't talk of past and future, but it does demand that the posit of "random micro-state" always be applied to states of two systems in the *same* time direction from the state whose nature we are trying to probabilistically infer. From it we can, in a sense, infer the parallelism of systems with each other, and hence with the main system as one system among many. But it is wrong to think that these arguments constitute in any way a derivation of parallelism from some time-neutral probabilistic assumption, such as that probabilities of "space ensembles" duplicate those of the time ensemble of an individual system, as Reichenbach intuitively characterizes his posit that the array is a lattice of mixture. The probabilities in the individual system are time-symmetric. Parallelism is gotten out only by a judicious and question-begging insistence that the probabilities of the individual system always be applied to the same time end of systems whose evolutions are to be comparatively inferred. And it is equally mistaken to think that parallelism can be derived in any explanatory way from the fact that systems are branch systems with only finite lifetimes. It is also wrong to think of these arguments as in any way explaining why the time direction in which entropy increases in the branch systems whose entropy is changing is always the same in the probabilistic sense. It is true that given the positing of parallelism for *all* systems, it follows that they have their entropy increase (probably) in the direction of entropy increase of the main system, because the main system is one of the systems. But in no way is the fact that that direction is the direction in which nearly all the systems do show parallel entropic increase in a certain time direction *explained* by the direction of entropic increase of the main system.

That elusive object of desire, the Second Law, has once again eluded our search for its explanatory ground. The overall entropic state of the universe does indeed provide a pool from which low-entropy systems can be "cut off" and obtained far more easily than by creating an equilibrium system and waiting for a fluctuation to a low-entropy state. But that is as far as it goes. Concerning parallel entropic increase, it seems, rather, that the universe is one *instance* of the general phenomenon, rather than its ground.

An additional note is in order here. Sometimes it is suggested that the origin or asymmetry for the local systems can be found in some combination of the idea that these systems are never really totally isolated from their surrounding environment and the cosmological facts. But this also seems implausible. We saw earlier how the usual "continuing intervention" thesis foundered on the fact that the plausible kind of intervention to posit would be random disturbance of the system by the outside environment. Positing intervention with low entropy for the system at the initial time might indeed lead to a justification of the inference to higher entropy at the later time. But positing low entropy at the final stage of the system, would, if we don't beg the question by introducing the asymmetric notion of causality that allows the outside world to intervene in the system so as to "effect" its future but not its past states, lead us, on the basis of a random intervention posit, to infer higher entropy at the earlier time.

Throwing cosmological considerations into the interventionist picture doesn't change the situation. The low entropy of the outside environment would not, for the situations we are considering, lead us to posit some time-asymmetric, rather than random, intervention of outside world into system. Nor does the fact that the global entropy of the universe is increasing in a time-asymmetric way – in one but not the other time direction – seem to help here. That is, indeed, an asymmetric fact, but how could that fact be used to account for some special asymmetry of intervention of the appropriate sort? I don't see how it could.

We could also just assume that the initial total state of the universe fully determines all its subsequent states. Then we could simply posit an initial state that gives rise to parallel entropic increase of branch systems with each other and with the main system. But to characterize the state in that way would, of course, not be offering us an explanation of the sort we expected. It would be one thing to be able to characterize the initial state in some simple way, say as having spatial uniformity at early times and a "probable" micro-state compatible with that macro-state, and be able to derive the Second Law from that. But to derive the Second Law from a bald assertion that "initial conditions were such that they would lead to Second Law behavior" hardly seems of much interest.

I believe also that a proposal by P. Horwich that attempts to get parallelism of entropic increase of branch systems with each other and with the main system also suffers from the defects noted here. It relies upon a dubious interventionist account of entropic increase for not-really-isolated branch systems. And it posits the appropriate non-correlation of the external world with the micro-states of the system in question in one time direction but not the other, a kind of interventionist *Stosszahlansatz*. But short of just demanding that this be true of the initial micro-state of the world, it is hard to see how such a posit, as opposed to a posit, say, of pure randomness for outside intervention applicable in both time directions, can be gotten from any simple, plausible characterization of the initial cosmic state that doesn't just postulate parallelism of branch systems with each other and with the main system.

It is also true, of course, that if the entropy of the whole universe is increasing, if we partition the universe into energetically isolated pieces that as a totality exhaust the whole universe, it can then be inferred by the sheer additivity of entropy for the systems (entropy of the whole being the sum of the entropy of the parts) that it must be the case that entropy increase dominates in the partitioning systems. But that trivial conclusion is quite different from the claim that it is the fact that the entropy of the universe is low and increasing in the future direction of time that explains why a specific sample of gas equilibrates in the future and not the past time direction.

IV. Further readings

A treatise on Machian ideas in dynamics is Barbour (1989). Volume 1 dealing with pre-relativistic physics has appeared, and a following volume on relativistic matters is forthcoming. Earman (1989) is useful on Machian ideas in dynamics and space-time, especially Sections 2–1; 4–8 and 5–8. Sklar (1974), Chapter III, Section D.2, is a brief summary of Machianism and some of its problems.

For Boltzmann's original cosmological ideas, see Brush (1965), Volume 2, Chapter 10.

A treatise on "anthropic arguments" in physics and cosmology is Barrow and Tipler (1986). Chapters 1–4, 6, and 7 are most relevant to the matters discussed here.

For basic treatments of modern cosmology, see Bondi (1961) and Sciama (1971).

A good survey of inflationary cosmologies is Linde (1984).

For the basic connections between cosmologies with its cosmic expansion and thermodynamic features of the universe, Tolman (1934) is a classic. The issues are also discussed in Sciama (1971). Davies (1974), Chapter 4, is a good summary of much of the material. See also Gold and Schumacher (1967).

On cyclic cosmologies and claims that entropics change is cyclic with the cosmic expansion, see Chapter 7 of Davies (1974).

Davies (1974) also provides a very good introduction to the alleged connections of radiation asymmetry and cosmology in Chapter 5.

A very subtle treatment of the need to posit initial low entropy and of the nature this must take is R. Penrose (1979). This piece also outlines Penrose's proposal that this initial low entropy is to be accounted for by a new law of nature.

On "teleological" or "design" arguments in general and on their connection to initial low entropy for the cosmos, see Barrow and Tipler (1986), Chapters 2 and 3.

The seminal exposition of the notion of branch systems and their use in characterizing entropic asymmetry is Reichenbach (1956). Branch systems are defended in Davies (1974), especially in Chapters 3 and 4. Grünbaum (1973), Part II, Chapter 8, and Mehlberg (1980) Volume 1, Part I, Chapter V, also discuss the place of branch systems.

For a critique of the role of branch systems, see Sklar (1987).

Another account of the role of branch systems and cosmology (using interventionism) can be found in Horwich (1987), Chapter 4.

On the "causal principle" in probabilistic inference, see O. Penrose (1979), Section 1.1.

9

◁═══════════════════════════════════════▷

The reduction of thermodynamics to statistical mechanics

I. Philosophical models of intertheoretic reduction

1. Positivist versus derivational models of reduction

The progress of science is marked by the continual success of attempts to unify a greater and greater range of phenomena in more and more comprehensive theoretical schemes. This unifying process often takes the form of a theory that encompasses the phenomena in one domain of experience being "reduced" to some other theory, the full range of phenomena handled by the reduced theory now being handled by the reducing theory. And the reducing theory continues to do justice to the phenomena for which it was originally designed as well.

Examples are manifold. We are told that Kepler's laws of planetary motion reduced to Newtonian mechanics, that Newtonian mechanics reduced to special relativity, and special relativity to general relativity. On the other hand, we are told that Newtonian mechanics reduced to quantum mechanics. The physical optical theory of light allegedly reduces to the theory of electromagnetism, a theory already obtained by the search for the unifying account to which the earlier separate theories of electricity and magnetism reduced. Nowadays we are told that the quantum theory of electromagnetism (quantum electrodynamics) reduces to the electro-weak quantum field theory and that there are hopes that theory will reduce, along with the quantum field theoretic account of strong interactions, to the grand unified theory. And that theory, along with the quantum theory of gravitation, may be reduced to the theory of super-symmetric strings.

Hopes abound that whole disciplines can be reduced to others. Isn't there at least the hope that biology can finally be reduced to chemistry and physics? And because chemistry can be reduced to physics, it is said, there is hope that an ultimate reduction of biology to physics is in view.

Other alleged possibilities of reductions of whole disciplines remain more dubious, of course. Just how conceivable should we take it to be that psychology, including in its domain the sensory direct contents of consciousness and the realm of the mentalistically intentional, can ever be reduced to the purely physical sciences?

Here, my aim will be only to sketch quite briefly a number of broad structural considerations that have caught the attention of philosophers of science concerned with unpacking just what we might be claiming when we claim that one theory reduces to another. As we shall see, the subject is one whose full richness and complexity have so far not been fully explored. One thing we shall note is that many important cases of alleged reductions seem to have such special features that none of the simpler general structural models can do them justice. The reduction of classical mechanics to quantum mechanics, of genetics to biochemistry, of mentalistic theory of one kind or another to physicalistic theory are all problematic in this way. And, as we shall see in Sections II and III, the alleged reduction of thermodynamics to statistical mechanics is another one of those cases where the more you explore the details of what actually goes on, the more convinced you become that no simple, general account of reduction can do justice to all the special cases in mind.

A look at a model of reduction once promoted by philosophers enamored of the positivist idea of theories as mere instrumentalist devices for generating lawlike relations among the observational data available to our "direct inspection" might be a good place to begin. J. Kemeny and P. Oppenheim, with this positivistic view of theories in mind, proposed a model of reduction that takes one theory to be reduced to another if the latter theory is able to generate all of the observational consequences of the former, and is superior to the former in some way. One way in which the reducing theory might be superior to the reduced could be the ability of the reducing theory to account for more phenomenal relations than the reduced could account for. That is, the reducing theory might be more general than the reduced theory. And the reducing theory might account for the same phenomena as the reduced theory but with less theoretical apparatus posited. That is, the reducing theory might be simpler than the reduced. Combining these considerations in some way, we might speak of the reducing theory as having "greater systematic power" than the theory reduced.

Someone who holds to a realist interpretation of theories will naturally be dissatisfied with this model. Surely, if an important part of the genuine cognitive content of a theory is its description of a posited real realm of theoretical structure, a theory can only be said properly to be reduced to another when that latter somehow accounts for the features of the world described by the former at the theoretical as well as at the observational level. Even those who think that ultimately a positivistic account of the role of theories in science is tenable are also likely to be dissatisfied with the Kemeny–Oppenheim account of reduction. If we look at actual cases of reduction in science, we find that there is always a close relationship between reduced and reducing theory at the level of theoretical structure. Indeed, given the great richness of observational consequences predicted by our sophisticated scientific theories, it would be astonishing to find a reducing theory that could account for all the phenomenal data accounted for by the reduced, but which bore no connection whatever to it at the level of the two theories' theoretical structures. Even if both theories are then nothing but instruments for generating observational correlation, we would expect that any real case of reduction would involve associating elements of the theoretical structure of the reduced theory with elements of the theoretical structure of the theory that does the reducing.

The simplest relation of reducibility of a kind stronger than the positivist would be a simple derivation or deduction of the reduced theory from the reducing. Here, the idea is that a limited lawlike generalization, or set of them, is found to be derivable from a general postulational scheme from which the earlier generalizations can be derived and also from which other generalizations covering a different range of phenomena can be derived. It is often claimed, for example, that Galileo's law of falling bodies and Kepler's laws of planetary motion were both shown by Newton to be derivable from his postulational dynamics and gravitational theory. Much more problematically it is sometimes claimed that Newtonian physics is derivable from special relativity and from quantum mechanics as well, and that special relativity is derivable from general relativity.

As is obvious from the latter cases cited, but as is clear even in the former cases noted, if close enough attention is paid, the simple idea of the reduced theory being deducible from the reducing cannot hold up. If Newton's theory is correct, Galileo's law is not true. Objects accelerate more the closer they are to the earth's center. And if Newton's laws hold, Kepler's laws are not true. The planet and sun travel around the center of mass as focus, not the center of the sun, perturbations disturb the elliptical nature of the orbits, and so on.

In the more modern cases, the situation is much more complex. Newtonian physics is derivable from special relativity only by a "limiting" process (the velocity of light going to infinity), not by any simple derivation at all. The exact details of the relationship between the space-time of special relativity and general relativity is a subtle one. The only obvious thing one can say is that Minkowski space-time of special relativity is a possible general relativistic space-time when the universe is empty of matter – which the actual universe surely is not. And even the formal relation of classical mechanics to quantum mechanics, the descendant of what was originally called the "correspondence principle," is a serious object of theoretical study, not a simple demonstration of a deduction in the narrow sense.

Once this has been noticed, a problem area is opened up that proves to be one of the slipperiest in philosophy. Simple assertions that the laws of the now refuted, reduced theory are good "approximations" to laws strictly deducible from the reducing theory won't do. First, the complex of "taking a limit" type of operations that characterize the interrelations of the theories in important cases – such as the limit of infinite velocity of light in the case of the reduction of Newtonian to special relativistic mechanics, and the limit of energy large relative to Planck's constant in the case of the reduction of classical to quantum physics – are far too subtle and formally complex for any simple notion of "approximation" to cover the relationship between the theories. Again, important questions of "limitation of domain," specification of the kinds of situations in which the reduced theory's laws can be expected to approximate the correct relations, are sometimes of importance and complexity.

But there is a yet deeper problem than these. Once we decide that the meaning of a term is fixed by its use in a language, the consequence that the meaning of a theoretical term in science is fixed by the totality of roles it plays in the network of assertions of the theory in which it appears seems almost inevitable. Then the consequence that any change in that network of assertions constitutes a change in the meaning of the terms of the theory as well also becomes almost a foregone conclusion. But then it becomes problematic to assert that the laws of the reduced theory can in any way be "approximations" to those derivable from the reducing, for if the theories are in any way incompatible with one another, mustn't we conclude that the very terms appearing in the theories have quite distinct meanings, making the comparison of the laws of one to those of the other, even in an approximative sense, impossible?

This seems pretty dubious if one is comparing, say, Galileo's or Kepler's laws with Newton's. Surely, it will be claimed, whatever meaning is, the kinematic terms referring to position, velocity, and so on mean the same thing in reducing and reduced theory. But assertions to the effect that "mass" just doesn't mean the same thing in special relativity as it does in Newtonian mechanics, or that space and time as projections of space-time in relativity theories simply don't count as the same features of the world as space and time did when they appeared in pre-relativistic physics, are all too plausible and, not surprisingly, frequently made.

Taken to its limit, the doctrine that every change of theoretical asser-tion generates a change of meaning of the theoretical terms involved leads to paradox, if not absurdity. If one person asserts "S" and someone else asserts "not S," they cannot be contradicting one another, for the terms in "S" and "not S" have, by the doctrine, different meanings, and the two assertions are not the negations of one another. Any assertion to the effect that the value of some physical parameter was slightly different than we thought it was would, again, have to count as a change of ontology! So the assertion that the charge on an electron was slightly greater than previously thought would have to count, because the new "electron" couldn't mean the same thing as the old "electron," not as attributing a changed attribution to some class of physical entities, but as denying that one class existed and positing something new altogether.

How to handle the "holistic" character of meaning, allowing meaning to be a "coarse enough grained net over truth" to do justice to what we want to say in the radical theory shifts and yet also to do justice to what we want to say in small changes of theory, is something we have yet to get sufficiently clear about in philosophy of language. As we shall see in Section II,2 in this chapter when we discuss the relationships between concepts in our particular reduced and reducing theory, thermodynamics, and statistical mechanics, much cruder distinctions in the ways the concepts function in the two theories will appear to force us to be cautious indeed in using the same term to try and deal with features of the world attributed to the world by the two theories. Exactly what "entropy" as used in statistical mechanics has to do with "entropy" as used in thermodynamics, for example, is a question replete with many problematic features of a much more specific kind than the general issues about meaning change as a consequence of theory change that we have noted here.

2. Concept-bridging and identification

Some of the most interesting cases of reduction are those where the reduced theory deals with a class of entities or properties apparently not mentioned at all by the reducing theory. Physical optics deals with light waves, but electromagnetism deals with electric and magnetic fields and their interactions. In the original electromagnetic theory, no mention of light waves seems to appear. How then could it be possible for the reduced theory to genuinely reduce to the reducing, in anything but the weak positivist sense?

An early attempt to deal with this problem by E. Nagel suggested that the reduction was constituted by the postulation of "bridge laws," lawlike propositions containing the concepts of both reduced and reducing theories that would serve, when conjoined with the laws of the reducing theory, to allow the derivation of the laws of the reduced theory with their "autonomous" concepts. But there is a serious objection to this account. If the reduced theory follows only from the reducing theory and the bridge laws, why say that a reduction to the reducing theory has occurred at all? Aren't the bridge laws themselves in need of some explanation in terms of the reducing theory alone if a genuine reduction can be said to have been accomplished? Doesn't the postulation of bridge laws as usable in reductions in fact trivialize the search for reductions? Can't we always derive the reduced theory from the conjunction of the reducing theory and a new bridge theory that simply posits that if the reducing theory holds, then so does the reduced?

One important response to this problem, proposed originally by H. Feigl and others, was to look to theoretical "identifications" as the core of these kinds of reductions. Just as the Evening Star and the Morning Star are one and the same object (also called the planet Venus), so light waves are nothing but electromagnetic waves of a particular frequency. But "The Morning Star" doesn't mean the same thing as "The Evening Star," and it is by no means a trivial proposition established by the "meanings" of the terms alone that the object referred to by the one is identical with that referred to by the other. The proposition "The Evening Star is the Morning Star" is not a priori or analytic, even it if is, according to some philosophers, a "necessary" truth. Similarly, Maxwell's discovery that light waves are nothing but electromagnetic waves – a bit misleading because what he actually said was that the optical aether was identical to the electromagnetic aether – was a genuine empirical discovery, not the exposition of a proposition knowable without reference to the data of experience.

Yet identities, even if established a posteriori, have a different status when explanation is the issue from that held by "bridging laws." If Maxwell had simply asserted that each electromagnetic wave was accompanied by a light wave, we would wonder *why* the two classes of entities were so associated. But taking light waves to be nothing but electromagnetic waves, there is no "lawlike correlation" of the presence of light waves and electromagnetic waves to be explained. There is a lot to be explained, of course. We want to know why, for example, electromagnetic waves revealed themselves in the world in the manner we were accustomed to consider the criterion of their presence. Why, for example, do electromagnetic waves stimulate our sight? Here the answer is, at least in part, the complex story that involves not only the identification of light waves with electromagnetic waves, but the embedding of that reduction in a general scientific context that includes other reductions as essential components. Another component of this overall reduction is the identification of the retina of our eyes to an array of components, many of which are subject to causal influence by electromagnetic fields.

It is frequently noted that such physical reductions by identification take place in a context in which some apparent features of the objects to be identified have already been stripped off these objects and put somewhere else, perhaps in the realm of the "subjective." The "sensed yellowness" of yellow light may be a puzzle for us, for example, for how could electromagnetic waves be said to have *that* perceptual feature? Of course, the context of the physical reduction accomplished by Maxwell is one in a scientific tradition that has already, for better or worse, stripped "sensed yellowness" from light, making it a "secondary quality in the mind" and leaving physical light with just the sorts of features (spatio-temporal locality, velocity, frequency, and so on) easily attributable to electromagnetic waves as well.

The problem of the ontology of the allegedly reduced theory having features implausibly attributable to the entities of the reducing theory is one that cannot be avoided by shunting off the problematic features into the realm of the "mental" if it is the plausibility of the reducibility of mentalist theories to physicalistic that is under debate. One hope of many of the explorers of identificatory reductions was that such reductions could serve to model the relationship of the mental to the physical. Could we reduce the psychological description of the mental phenomena of experience to, say, neurophysiology by, at least in part, an identification of some of the ontology of the former with some of that of the latter theory? The hope was that, for example, "having a yellowish sensation" could be identified with "having such-and-such a neural process going on in the brain." But numerous problems infect any such account. Some are the result of perplexities one gets into when it is events or properties or processes one wants to identify with other events or properties or processes, rather than identifying one class of *thing* with another. But the graver difficulty is the implausibility of finding any place for the "sensed yellowness," so clearly a component of sensing yellowly, in the comprehensive physical description of neural goings on.

Just as in the case of the simpler derivational reductions, in the alleged identificatory cases one again must confront issues generated out of the fact that it is rare indeed that the reduced theory suffers the reduction process unchanged. The very act of reducing one theory to another usually leads us to find flaws in the reduced theory as it was originally formulated, and to look for an alternative to it better suited to the reduction procedure. Because we have confidence in the correctness of the reducing theory and in the reduction, the modified theory is presumably better suited also as a theory of the phenomena the original reduced theory was generated to account for.

Once this has been acknowledged, a host of other considerations come to mind. If the modification of the reduced theory is severe enough, should we say that it has genuinely been reduced to the reducing theory? Or should we say, rather, that what we have discovered is that the reducing theory contains within its resources the appropriate substitute for a now refuted earlier theory of a range of phenomena? If what we say about the elements talked about by the reduced theory after the reduction is different enough from what we said about them before the reduction, should we say that we have "identified" them as elements spoken of by the reducing theory? Or should we say, instead, that what we have discovered is that the elements spoken of by the earlier theory don't exist, their explanatory role being taken over by new elements sufficiently like them to be called, misleadingly, by the same common noun terms?

So we can find those who will say that it is quite misleading to think of Mendelian genetics as having been reduced to modern molecular biology or to speak of Mendelian genes as being "identifiable" as DNA molecules. Rather, it is claimed, the new molecular biology provides a correct ontology of genetics showing us what was and was not correct in the older Mendelian account. Given the radical reinterpretation of thermal phenomena forced upon us by the joint theories of the atomic constitution of matter and statistical mechanics, it is not surprising that we will find much to say about just these issues when we discuss the details of the relationship of thermodynamics to statistical mechanics. So complex is the situation that we will be a bit surprised that people still say things like, "The identification of mental processes with brain processes is no more mysterious than the identification of temperature as mean kinetic energy of molecules," as if that latter "identification" (if that is indeed what it is) were as straightforward as all that.

One standard concern of those who deal with reduction by identification is how we could tell whether a class of entities at one level was identical with a sub-class of entities at the deeper level, as opposed to being merely correlated with that other class. In many cases, identity, as opposed to correlation, is forced upon us, unless we are willing to make substantive changes elsewhere in our theories. Once we attribute mass-energy to light waves and to electromagnetic waves, the conservation rules for that mass-energy won't leave room for distinct light waves and electromagnetic waves to be present. So if we are going to take light waves as something over and above the electromagnetic waves and merely correlated to them, we will have to so modify our theory to deprive the light waves of mass-energy, of other features (momentum, for ex-ample) as well. Once we say that the light waves are just the appropriate electromagnetic waves, the problem is gone.

Reduction by identification is plainly one of the primary ways in which our theories of the world are unified. A general methodological principle seems to be that we ought to identify, as opposed to positing a correlation, whenever we can. That is, we should identify whenever the assertion of such an identification is not blocked by some feature of the situation, typically by the reduced entity having some feature genuinely not attributable to the reducing.

As we shall see in Section II, the reduction of thermodynamics to statistical mechanics takes place in the context of a much broader scientific reduction by identification. This is the reduction of the theory of macroscopic matter to its micro-constituents by the identification of the macroscopic entities as structured out of microscopic entities. But whether the reduction also involves, in any simple way, the identification of the thermal features of things (especially temperature and entropy) with features of things characterized at the reducing level in non-thermodynamic terms is a subtler matter.

3. The problem of radically autonomous concepts

It is often alleged that there are cases where it would be inappropriate to speak of one theory being genuinely reduced to another, or one domain of nature reducing to the other, but where it is correct to say that everything that happens at one level is fixed or determined by what happens at the other. An account of the world at the second level, then, if complete, constitutes enough said about the world for all of the facts at the first level to be determined as well. The word "supervenience" has currently been employed to speak of this kind of relation among two domains.

But "supervenience" can hold for many different reasons. Moral philosophers and aestheticians who held that moral and aesthetic properties ("goodness," "beauty") were autonomous features of the world not identifiable in any way with the "natural" properties described by our scientific world picture, often held to the doctrine that these moral and aesthetic features were, however, "fixed" by the natural features of a situation. Fully describe a state of the world from the scientific point of view, and its degree of goodness was thereby determined. There couldn't, for example, be two naturalistically identical states of the world, one of which was morally good and one morally bad. Similarly, two works of art identical from the naturalistic point of view would have to be equally beautiful. Just why and how the naturalistic features fixed the non-natural moral and aesthetic features remained rather mysterious, as are "goodness" and "beauty" as "non-natural properties" themselves.

Faced with the apparent ineliminability of sensed yellowness and its apparent non-identifiability with a property of the brain (or, perhaps, of sensing yellowly and its ineliminability and non-identifiability with a brain process), but convinced that the physical facts when fully laid out would determine all of the mental facts about the world, philosophers of mind frequently became epiphenomenalists. Sensing yellowly they said, supervenes on the neural processes. What sensory states we have are fully determined by our neural condition, but the sensory states are autonomous qua states. Here, supervenience is grounded, apparently, in a causal relation, the mental phenomena being causally generated by the neural, at least on some traditional views of the matter.

A rather different set of issues, issues more relevant for our purposes, center around those cases where a simple reduction of one theory to another is blocked not by the appearance of some "irreducible" fundamental feature at the reduced level (like "goodness" or "yellowness"), but, rather, by the fact that the reduced-level description of the world is framed in terms of complex relational features of the world that fit into an "autonomous" schema not relatable to the way the world is described at the lower level in any simple way. The basic idea is that whereas the reducing level description of the world is "complete," the reduced level description, although not positing any fundamentally additional ontology of things or features absent from the lower level description, characterizes the order of the world in terms of relational features introduced for special purposes. This "recategorization" of the things and features of the world is such that no simple "definition" of the reduced level categories in terms of those of the reducing level can be given or ought to be expected.

A standard example of a trivial sort is frequently given by referring to the way we sort objects into household goods. Surely a chair is nothing but a complex array of the micro-components of the world described by physics. But any hope of defining "chair" in the standard fundamental vocabulary of physics founders on the complex, vague nature of that concept, a concept that takes its meaning from the place furniture plays in the complicated world of everyday life.

A much more interesting case might be that of evolutionary biology. Notions like that of genotype, phenotype, species, fitness, environmental stress, natural selection, and so on, form the fundamental vocabulary of this theory. The theory has its own explanatory principles that function well to offer us a coherent account of such phenomena as species diversification, extinction, modification, and so on. Any simple hope of defining these terms in terms of the concepts of fundamental physics seems out of the question. Yet the existence of the complex relational structure of world in which evolution takes place, and the usefulness of characterizing the world in the categories used by evolutionary biologists, doesn't imply (although some might claim that it does) the existence of some realm of being outside the realm treated by physics. Nor does it imply the existence of some new fundamental lawlike structure of the world that somehow puts the objects of its domain outside the realm of ordinary physical explanation.

It is often claimed that the part of our mentalistic discourse that deals with such notions as cognitive states (such as beliefs, desires, and so on), and the related linguistic-semantical discourse that deals with such things as reference and meaning are related to physics in some such way. A complex functional structure of the world exists, treated naturally in terms of its own appropriate discourse. Just as the totality of physical facts "fixes" all the evolutionary facts or facts about furniture, so too it fixes all the facts about mental cognitive contents, meanings of words, and so on. But that doesn't imply that there is any hope that one could somehow define the concepts of psychology and semantics in physical terms, or lay out some simple derivation from the laws of physics of the principles of explanation we invoke (for example, explaining behavior in terms of beliefs, desires, and the aim of the agent to maximize expected utility) familiar from these disciplines.

The phenomenon of "irreducibility" in the sense just described is closely connected with what has sometimes been called "emergence," the idea that if a physical structure becomes organized and structured to a certain degree of complexity, new phenomena "emerge" not present in simpler physical structures. Now, emergence can be (and has been) used to mean that what emerges is something genuinely ontologically novel (mental sensations, for example). But for others it is a kind of order and structure, demanding its own appropriate conceptual apparatus of concepts and lawlike regularities that is said to emerge. This is an ontologically much more modest claim.

A claim to the effect that humans (and some machines) are organized in such a way that their workings are best described by positing states whose interdependence is best characterized in terms of their logical (in the broad sense) relations to one another, rather than their physical, is hardly the claim that humans and machines are not still complex physical objects. The fact that a programmer describes and explains the operation of a computer by referring to its program states and their logical inter-relations to one another doesn't entail that he won't freely admit that the electrical engineer could, in the conceptual framework of physics and its causal relation of physical state to physical state, account for the behavior of the machine as well.

Given the special and quite idiosyncratic conceptualization that thermo-dynamics and statistical mechanics both impose on the phenomena, at least from the point of view of the conceptualization familiar from other physical theories, it won't come as a surprise that some such way of looking at the way these theories categorize the world and explain its phenomena so characterized, might prove useful when the relation of thermodynamics to statistical mechanics, and of the two together to the remaining body of fundamental physics in which they are embedded, becomes our object of study.

Once one has posited these autonomous concepts for the upper-level theory, the theory allegedly reduced, of course, and a group of laws (or, more generally, generalizations) that involve them and that we can use for predictive and explanatory purposes, the question immediately arises as to the relationship between this conceptual and generalization structure at the upper level and the concepts and laws of the lower-level theory, the alleged reducing theory. Simple derivation of the former from the latter seems out of the question, but other subtler relations surely hold. Those who posit "emergence" in some radical sense in which the new structure is somehow to be taken as incompatible with the underlying structure certainly have a problem on their hands. The less radical thesis that the upper-level structure is compatible with the lower level but somehow still "independent" of it also requires some understanding of how, if the lower-level structure offers a "complete" characterization of the world and its causal dependencies, such an independent level of organization can exist. Even when it is asserted that supervenience holds, so that all that goes on in the upper level is "fixed" by the lower-level pattern of causation, even if our description of the upper level behavior is not derivable from the framework in which we characterize the lower-level dependencies due to the "indefinability" of the upper-level concepts by the lower-level ones, there still is likely to be much more to say about how the upper-level conceptual scheme "fits in" to the general scientific framework characterized by the lower-level concepts and regularities.

The philosopher of biology wants to understand the place of the evolutionary scheme and its concepts in a biology that treats organisms as complex biochemical structures embedded in a physical world. The philosopher of psychology and of language wants to know where our conceptual scheme of behavior as governed by states having cognitive content and our scheme of language as governed by semantical rules fit into our picture of agents as biological beings in a physical world. And our concern is with the overall place of the schema of thermodynamics with its concepts of equilibrium, heat, temperature, and entropy, and its fundamental laws (especially the Second Law) in the world of micro-components evolving under dynamic evolution. As we shall see in Sections II and III, in this case we can get clear, to a degree still not obtained in the other two cases, on just what the place is of the thermodynamic view in the dynamical (or the dynamical supplemented by probabilistic posits). But being clear leads us to view simple-minded claims of "reducibility" with some skepticism.

II. The case of thermodynamics and statistical mechanics

1. The special nature of reduced and reducing theories

Given the peculiar place of thermodynamics in the body of physical theory, we will not be surprised to find that its concepts fit rather differently into our general description of the world than do the concepts of more ordinary physical theories. The theory is astonishingly universal. It is applicable to systems of radically different structural constitutions and to systems that are governed in their evolutions by radically different dynamics. It is even applicable to systems that have their evolutions characterized by distinct kinematical structures, indicating the way in which this theory "cuts across" the usual hierarchy of structural reduction and dynamical generalization that characterizes the rest of fundamental physics.

Whereas notions such as that of the equilibrium state, temperature, and entropy do introduce a kind of "dynamical" description governing the evolution of systems, the nature of this dynamical constraint on systems' evolutions strikes us as quite different in kind from the usual dynamical constraints that are the joint products of the structure of the system, including its internal and external forces, and the basic dynamical laws governing the response of systems to forces. The basic concepts of equilibrium states as attractor states, with its teleological flavor, and the basic rules of irreversible monotonic approach to these states for isolated systems, are quite unlike what we expect from ordinary dynamical considerations.

If we ask what gives the concepts of thermodynamics their meaning, the natural place to look is at the roles played by the concepts in the theory. Here, we discover three rather distinct kinds of propositions that are important. One class is the generalizations that connect the concepts to the measuring devices, described in a non-thermodynamic way, that are standardly used to attribute magnitudes of thermodynamic features to systems. These are most obvious in the early introductions of temperature into physics by the association of degree of temperature with readings on a thermometer brought into energetic contact with the system.

Next, there are the general thermodynamic laws that play such a crucial role in the axiomatic versions of the theory. In particular, the Zeroth, First, and Second Laws of Thermodynamics characterize the basic features of equilibrium (such as its transitivity as a relation among systems) and temperature. They also provide the ground for introducing the notion of entropy as a concept whose role is rationalized by the truth of the laws in the manner originated by Clausius. The place of a concept like entropy, being fixed fundamentally by its functional role as determined by its place in the specifically thermodynamic laws, may be suggestive to those familiar with functionalist accounts of concepts in other areas (logical states as functionally determined by the connection of input to output states of a computer – that is, by their place in programming descriptions; or intentional mental states as functionally determined in behavioral input-output theories of agent behavior, for example). A similar understanding of entropy (and perhaps temperature as well) would go some way in clarifying our understanding of these concepts. Here, the characterization of states of systems in terms of their features that are less specifically thermodynamic (say, of a gas by its volume and pressure and its temperature as thermometrically determined) would play the role analogous to the input-output states in the functional theories noted earlier, with the thermodynamic features of thermodynamic temperature and entropy introduced in a manner analogous to the functional states. When we look at the details of some of the specific concepts in the next section, we shall see a little more of this analogy appear.

Finally, there are the "constitutive laws," those principles connecting the thermodynamic notions to non-thermodynamic notions and that are highly dependent on the particular constitution of specific kinds of systems. A typical such law would be the ideal gas law, $PV = kT$ for certain kinds of idealized gases, or the other, similar equations of state describing the interrelations of thermodynamic and non-thermodynamic variables at equilibrium. Another set of examples are the more complex laws governing the hydrodynamic evolution toward equilibrium, or those applying in a non-equilibrium steady state. They, again, will vary with the specific system being described.

Evidence from within thermodynamics that its concepts are not of the usual physical sort is also obtained from a little reflection on "Gibbs' Paradox." Treat ortho- and para-hydrogen as just hydrogen, and the entropy of mixing of two volumes of the gases is zero. Treat them as discriminable by means of the different alignment within them of the two nuclear spins in a molecule, and the entropy of mixing is non-zero. So entropy change is relative to a "level of description" of the substances involved. Now it is clear that accusations of "subjectivity" for entropy and entropy change based on this fact are misguided. For a property to be relational is one thing, for it to be "subjective" (in any of the multitude of meanings of that notoriously ambiguous word) is quite another matter. Claims that entropy change is subjective because it is relative to characterization of the substances or to the level of fineness with which we are concerned in a given problem are as misguided as claims to the effect that length or time interval in relativity are subjective simply because they are, in that theory, relative to the inertial reference frame chosen.

That no entropy change is involved in mixing of the two kinds of hydrogen if all we are concerned with is the degree of uniformity of hydrogen distribution, but that entropy change is non-zero if it is uniformity of distribution of hydrogen with a specific alignment of nuclear spins in its molecules is perfectly understandable. In the former case, nothing need be done to restore the original non-uniformity from the final uniformity, for the original situation, like the final, is one of uniform distribution of hydrogen. In the latter case, work would need to be done from the outside to restore the original partitioned state of hydrogen-with-specified-alignment-of-nuclear-spins. No "subjectivity" is involved.

Nonetheless, the fact that entropy has this relation-to-detail of structure nature does tell us that it is unlike, in this way, such other attributions to the substance as its total energy, volume, and so on. Here, the role of entropy as fixed entirely by its functional place in the general thermodynamic laws (and, subsequently, by the constitutive equations) is important. For it is in exploring how those general laws are functioning in allowing us to predict and control that we begin to grasp the meaning of the thermodynamic concepts involved. Of course, these conceptual points become even clearer (if more complex) when the reduction of thermodynamics to statistical mechanics is achieved, but aspects of thermodynamics such as Gibbs' Paradox already begin to throw light on the special nature of the concepts.

If thermodynamics seems a bit out of place in the panoply of ordinary physical theories, statistical mechanics is plainly even more an unusual case of theory. Once again it shares with thermodynamics the features of being universally applicable to all manner of physical phenomena, even those radically differentiated from one another in other ways. And it is like thermodynamics in its mode of "cutting across" the usual theoretical hierarchy of constitution and general dynamical principles of evolution. But of course it is even odder from the point of view of most other physical theory in its aim of providing "probabilistic" accounts of phenomena rather than "deterministic" accounts.

For the first time, statistical mechanics introduced into physics the idea that the aim of a physical theory could be not to provide an account of what must happen, but of what might happen. Phenomena were now to be accounted for in probabilistic terms, events being accounted for as "overwhelmingly probable" or even "as predictable to occur with some probability" and macroscopic phenomena as being reflections of what happens "most probably," or sometimes "on the average," at the microscopic level.

The probabilistic theory differs from the earlier theories, as we have noted in detail, not only in its explanatory aims but in its basic posits. Postulates about the probability with which specific micro-states ought to be taken to occur, as well as other postulates introduced at least temporarily as independent probabilistic assertions (like the "rerandomization" posits of non-equilibrium theory), are fundamental to the theory. So this theory will differ in significant ways from most other theories of physics both in what it takes to be its fundamental explanatory goals and in the resources it will avail itself of to achieve those goals.

Furthermore, as we have repeatedly seen, the application of the probabilistic posits along with the structural facts and dynamical facts about the system requires that we work in an idealized context. All application of theory to the world requires some degree of idealization, of course. But, as we have seen, the idealizations utilized in statistical mechanics are sometimes much more radical in nature than those familiar from other fields. They are, for example, often "uncontrolled" in the sense of providing no insight into how deviant from the ideal a real system can be expected to be. This introduces one more factor that will complicate the analysis of the reductive relationship of thermodynamics to statistical mechanics.

2. Connecting the concepts of the two theories

Much of the problem concerning the reduction of thermodynamics to statistical mechanics can be localized in the exploration of the complex of relations that hold between the special concepts of thermodynamics and those concepts of statistical mechanics designed to do justice in that latter theory to the phenomena grasped by the concepts of the upper-level theory. How are we to "connect" the concepts of thermodynamics with statistical mechanics, or, on the object level of discourse, how are we to associate features of the world posited by the thermodynamic theory with features of the world posited by the reducing theory?

A first important point is that the reduction we are concerned with takes place in a general scientific context in which reduction is playing a crucial role in the background. Thus the thermodynamics of gases can only be reduced to statistical mechanics after we have already identified gases as collections of interacting molecules and reduced our gas theory to a theory of molecules and their interaction. So an identificatory reduction of the system plays here a necessary preliminary to a reductive account of the thermodynamic concepts predicated of it. Such background identifications as components of background reductions function in other ways as well to provide a context in which we can hope to reduce thermodynamics.

Some of the concepts that appear in our thermodynamic description of the phenomena seem to reappear unchanged at the reducing level. In thermodynamics, we talk of the volume of the container to which a gas is confined. In statistical mechanics, we deal with a posited volume of a container to which the moving molecules are restricted. Radical theories to the effect that any change of assertion is a change of meaning aside, it seems pretty clear that we ought to take "volume" (and other purely kinematic terms) to refer to the same physical feature of a system in both reduced and reducing theory. On the other hand, as we shall note once again in the next section, radical proposals have been made to the effect that even such a primitive notion as that of time takes on a new meaning in the statistical mechanical context from the simple parameter-of-change function that it played in thermodynamics and in our ordinary kinematical and dynamical descriptions of the world.

Concepts like heat and pressure in thermodynamics are closely tied within that theory to the dynamical concepts that appear in both the reduced and reducing level. As a consequence, we are directed un-equivocally toward the concept at the reducing level that we must asso-ciate with a specific concept at the reduced level. Once we accept the First Law of Thermodynamics and accept the model of a gas as consisting of moving molecules interacting by potential energy between them, the association of heat with a portion of the energy of the molecular assem-blage, and therefore with something constituted out of its molecules' kinetic energy of motion and potential energy of interaction, seems nec-essary. Or, again, once we view the gas as consisting of molecules that impinge on the walls of the container and bounce off it, the association of pressure with force per unit area leads us unequivocally to associating it with rate of transfer of momentum to the container wall per unit area and unit time due to molecular impact.

Although a direct argument, then, leads us from thermodynamic to statistical mechanical concept in these cases, it would still be quite mis-leading to baldly assert that the reduction proceeds by a simple-minded identification of the thermodynamic feature with some feature of the individual system characterized in mechanical terms. Take pressure, for example. There is, for a particular sample of gas at equilibrium, the actual momentum transferred by the molecules impinging on a wall of the box in a short time, and there is its average value per unit area per unit of that time. On the other hand, there is the quantity calculated for an ensemble of similarly constituted systems – say, by the usual method of taking ensemble averages of the appropriate function of the molecular dynamical quantities, or, alternatively by looking for the most probable value of the relevant quantity in the ensemble. Whereas the former sort of pressure, the feature of the individual system, will be expected to fluctuate, the latter kinds of ensemble quantities, quantities defined by the macroscopic characterization of the system and the chosen probabil-ity distribution over the ensemble, will, of course, not. Here, fluctuation

will show up as assimilated into the ensemble description by the calculation of averages or most probable values of quantities, but the averages
themselves are not the sort of things to fluctuate.

Just how, in detail, such ensemble quantities are connected to what we
expect to find as features of the individual systems has been the subject
matter of much of Chapters 5 through 7. The point being made here
is that we already see, even in the less problematic cases like that of
pressure, a ramifying of concepts at the statistical mechanical level,
and a situation sufficiently complex that any hope of doing justice to it
by simply saying that a thermodynamic feature is "identified" with some
feature of the system at the reducing level is misguided. It should not
surprise us that Gibbs, when he came to associate ensemble quantities
with thermodynamic quantities in Chapter XIV of his book, spoke of
the "thermodynamic analogies" when he outlined how thermodynamic
functional interrelations among quantities were reflected in structurally
similar functional relations among ensemble quantities. He carefully
avoided making any direct claim to have found what the thermodynamic
quantities "were" at the molecular dynamical level.

This contrast between a feature at the statistical mechanical level taken
to be, like the features posited at the thermodynamical level, a property
of the individual system, versus a feature thought of, rather, as a characteristic of the ensemble or of the probability distribution associated with
some fixed parameters of the system, is pervasive.

Take equilibrium, for example. In thermodynamics, it is posited primarily as a state that, in the axiomatic theory, obeys the Zeroth Law. In our
more generalized macroscopic picture, it is the end state of spontaneous
evolution. The constitutive equations characterize the interrelations among
the other posited characteristics of a system when it has the characteristic
of being in equilibrium. Within statistical mechanics we could think of
the equilibrium states as being those overwhelmingly most probable
distributions of microscopic conditions that lead to the usual equilibrium
constitutive equations. We would then expect the principle that systems
approach equilibrium and stay there to be replaced by a modified law
that allows for the fluctuational behavior summarized by Boltzmann's
picture of the system isolated for eternity.

Alternatively, we may think, rather, of equilibrium as a characteristic of an ensemble, justifying the choice of probability distribution for
this "equilibrium ensemble" in the ways described in Chapter 5. Of
course, there will be a variety of possible ensembles (such as the micro-
canonical and the canonical) dependent on the choice of "fixed" parameters for the system. The Zeroth Law will then hold exactly, but
now because it is no longer a proposition about the behavior of individual systems, but a proposition derivable from the construction of the

equilibrium ensembles. The approach to equilibrium and the fact that having arrived there one stays there, will also now not be a statistically true generalization, but (if we can somehow rationalize non-equilibrium statistical mechanics) a fully "lawlike" proposition. But, once again, the "exceptionlessness" of the "law" has been obtained by a reconstrual of the essential concepts appearing in it. They now no longer refer to properties of individual systems. Instead, they now refer to properties of a probability distribution constructed relative to a specified kind of system. Fluctuation from equilibrium is no longer part of the theory, but only because equilibrium itself is now construed as something that contains within its definition the fluctuational phenomena. The equilibrium probability distribution contains not only the overwhelmingly probable states of individual systems but the less probable states (and their probabilities) as well.

With the concept of temperature we begin to see some of the richness of conceptual variety that develops when we move from thermodynamics to statistical mechanics. As usual, we must distinguish at the reducing level the concept that is associated with thermodynamic temperature, and that denotes a feature of individual systems (temperature, for example, as the average kinetic energy of the molecules of a particular ideal gas at a particular time), from the concepts applicable primarily to ensembles (temperature as a defining constraint of a canonical ensemble or as a parameter of the micro-canonical distribution). As usual, the former concept functions in the new fluctuational surrogates to the thermodynamic laws, and the latter keeps the laws exact at the price of conceptual innovation.

The statistical mechanical surrogate for empirical temperature (as characterizing equivalence classes of systems in mutual equilibrium with each other) and absolute temperature (characterizing maximal efficiency of transformation of heat into work for ideal heat engines) is not an obvious function of the micro-dynamical characteristics of the system as is, for example, pressure. Here, the careful study of the circumstances in which ensembles will be in "equilibrium," as done by Gibbs, for example, in his study of joint ensembles, is necessary to find the right surrogate for the thermodynamic notion.

Although the primary structure fixing the meaning of "temperature" in thermodynamics is the fundamental laws, it is a nice feature of statistical mechanics that it can account for the place of temperature as fixed by standard measuring instruments as well. Again, this is a subtler matter than when one is dealing with pressure or volume. Here, the embedding of the thermodynamical reduction to statistical mechanics within the broader micro-reduction of matter to its constituent micro-components is crucial. Once we have said what, for example, a gas thermometer is, we can, within statistical mechanics, offer an explanation of why it is that gas thermometers – the devices by which temperature was "defined" in that stage intermediate between temperature as felt hotness and the full thermodynamic account of it – work as well as they do in measuring temperature values of systems. So the statistical mechanical reduction "completes" thermodynamics in the sense of providing us with a theory that describes the operation of the measuring instruments used to assign magnitudes of its quantities to systems. This is a virtue of theories much sought for. In space-time theories, it leads to attempts to frame general relativity in terms of the motion of free particles and light waves, these being more easily construable in their behavior within the theory than the usual rigid rods and clocks often posited as primitive measuring devices. And in quantum mechanics it lies at the root of "realistic" theories of measurement, in contrast to the accounts of measurement of Bohr and of the idealists, for example, that attempt to characterize the measurement apparatus and the measurement process within the world described by quantum theory itself.

Temperature shares with intentional cognitive states in functionalist theories of mind the property of "multiple realizability." Two systems can have the same temperature and be in equilibrium with one another, although radically distinct in their natures. Thus, radiation, for example, can be in equilibrium with a gas.

The microscopic feature of the system associated with its temperature can, then, vary from system to system depending on its constitution. It has been argued that this shows that temperature cannot be "identified" with any characteristic of a system in microscopic terms (such as the average kinetic energy of the molecules, the feature associated with temperature in the case of the ideal gas). Now we have seen a reason why a too naive application of the notion of identificatory reduction would be misleading, because "temperature" in many of its uses in statistical mechanics refers not to an instanced property of a particular system sample at a time, but, rather, to some feature of a probability distribution over systems of a specified type. But even sticking to "temperature" as referring to a property of an individual system, does the variation in microscopic correlate from system to system make an assertion of identity of temperature with any microscopic feature impossible?

It would seem not. To be sure, the temperature of one kind of system is "identical with" one feature for one kind of system and with a different feature for some other kind of system. But this hardly precludes our saying that in each case the temperature of the system is identical to its relevant microscopic feature. It is sometimes argued in philosophy of mind that whereas each mental process as a "token" is identical with some specific physical process, no general lawlike identification of mentalistic processes as kinds with kinds of physical processes is possible. This argument is, at least in part, grounded on the assertion that the functionalist states can be realized in a variety of physical processes, just as a logical state of a program can be realized in a computer whose workings are mechanical, electronic, or optical. In the case of thermodynamics and statistical mechanics, we do have a general procedure for determining, for a specific kind of system, the appropriate microscopic feature of the system with which to identify temperature. And insofar as it is temperature as the instanced property of a particular system that we have in mind, there seems nothing to block a strict assertion of the identity of that temperature for that particular system with the appropriate microscopically characterized feature of it instanced at the same time.

If we want to speak about the relation of temperature in general to the micro-features of a system, we can then, as S. Yablo has suggested in the mind-body context, think of the relation of temperature to the micro-feature on the model of a determinable property's relation to a determinate property falling under it. That is, philosophers think of yellow, say, as a determinate property falling under the determinable quality, color. So one could characterize the mean kinetic energy of an ideal gas as a property that is to temperature in general as a determinate feature is to the general determinable under which it falls.

Additional complexity arises out of the fact that the reducing level of description of the phenomena provides a context in which the concept associated at that level with some concept of the reduced level can have its theoretical role extended in useful ways that take its application beyond the realm dealt with at the upper level by the concept. Temperature provides a good example of this kind of concept extension applied after a reductive association of concepts has been achieved.

The ordinary notion of absolute temperature has all such temperatures non-negative and finite. Aren't we restricted in the same way in statistical mechanics? For many systems, the answer is positive. But the association of temperature with the measures of order and disorder in statistical mechanics leads to a natural extension of the absolute temperature concept in that theory. A typical situation in which the extension is applied is that of nuclear spins of atoms locked in a crystal. If they are aligned by a magnetic field in a direction and the direction of magnetization is then reversed, a slow process of transition ensues from a perfectly ordered state where the spins are all in the "wrong" direction, through a state of disordered spins, to an ordered state in which all of the spins are now in the "right" (lowest energy) direction. The natural statistical mechanical description of the situation, with temperature associated with internal energy and entropy in the usual ways, has us describe the situation as one in which the system goes from a temperature of "minus zero" degrees, through finite negative temperatures that go "down" to "minus infinity," and then from "plus infinity" (which is the same state as minus infinity) down to "plus zero," obtained when the ordered lowest energy spin state is obtained.

Here, the concept extension follows in a natural way from the formalism designed to handle the more usual cases (where energy is unbounded and maximal disorder is associated with highest energies, rather than here where the maximal energy state is a peculiarly ordered one). The reduction has directed us in the ways we have described to fix on an appropriate concept at the reducing level to associate with the concept of temperature at the reduced level. Once we fix on the formal surrogate for thermodynamic temperature in our statistical mechanical picture, we find that this concept can be extended to new situations for which the original thermodynamic concept was never intended. Such a natural extension of the reducing level concept is, of course, conceptually harmless. And it is scientifically effective in doing the additional work cut out for it. This phenomenon of concept extension – where a concept originally selected because of its association with a concept present at the reduced level is extended to realms outside the grasp of the original higher-level concept – is a familiar feature of the reductive process.

The concept of entropy is the most purely thermodynamic concept of all. Its primary meaning is fixed entirely by the role entropy plays in the characterization of systems that follows out the method introduced by Clausius of proving entropy's existence as a state function from the basic consequences of the Second Law. Given the abstractness of entropy and its place high up in the theory and unrelated to immediate sensory qualities or primitive measurements (as temperature is related to these), it is not surprising that in seeking the statistical mechanical correlate of thermodynamic entropy we have the least guidance from the surrounding embedding theory. It is that surrounding theory that guides us so quickly to the right surrogates for volume, heat, and pressure, for example. Nor do we have guidance from the connections of thermal characterization to measuring instrument construction, something that provides some guidance with respect to temperature. We need to find some feature of systems, characterized in our reducing theory, that will fill the theoretical role of entropy. As we have seen, the additional levels of complexity in the reducing theory may leave some openness in what to choose as the surrogate for entropy. This is exactly what we find – that is, a wide variety of "entropies" to correlate with the thermodynamic concept, each functioning well for the specific purposes for which it was introduced.

First, there is, as usual, entropy as a characteristic of an individual system at a time. Boltzmann's famous entropy, derived from the coarse-graining of the μ-space, the phase space of individual molecules, is the most well known of these entropies. For systems more complex than the ideal gas, deriving the right substitute for the original simple system entropy is sometimes a difficult task, but the general conception is always the same. Given the fact that the value of this entropy will depend upon the chosen partitioning of the molecular phase-space into finite "boxes," the value attributed will depend upon the coarse-graining chosen. As we saw in Chapter 2, Boltzmann already had good reason to pick the position-momentum partitioning he chose, even at its origin, rather than the a priori equally good position-energy partition. His choice was based upon his earlier work on the stationary solutions of his kinetic equation and their characterization as having the maximum value of the usual Boltzmannian entropy.

We have seen that the choice of the appropriate statistical mechanical correlate of this kind for thermodynamic entropy is fixed by a multitude of considerations, including the additivity of entropy for independent systems in thermodynamics and, most importantly, the demand that the function be maximized for the equilibrium configuration of the micro-components and that this configuration obey some kind of stationarity under the kinetic equation.

Constrasting with this entropy of individual systems is entropy as a characteristic of an ensemble or of a probability distribution relative to a system kind. What differentiates entropy from the other thermodynamic concepts is the wide variety of ensemble "entropies" suggested as the appropriate correlate with the thermodynamic feature.

Gibbs' "fine-grained entropy" is the primary notion. As we have seen, it fits nicely into the functional relation among ensemble features that Gibbs used in his discussion of the thermodynamic analogies in statistical mechanics where he found the appropriate analogue to the thermodynamic quantities. And it has the appropriate maximization properties for the equilibrium case, identifying the standard equilibrium distribution as that which maximized the ensemble entropy relative to the imposed constraints.

The relation of this entropy to the Boltzmannian is the familiar one of the relation of an ensemble feature to a feature of individual systems. The ensemble over which the average of the "index of probability" is taken – Gibbs' fine-grained entropy – contains systems with every possible value of Boltzmannian entropy predictable of them. Although Gibbs associated an ensemble quantity found by averaging over all of these systems with the thermodynamic entropy, Boltzmann associates that feature of individual systems that we know, from Khinchin type of results, to be overwhelmingly dominant (in the probabilistic sense) among all the systems in the ensemble with the thermodynamic quantity. This is the usual distinction between average and most probable values as representative of equilibrium that we discussed in Chapter 5. So long as we are clear on how the various concepts are invoked in our statistical explanatory structures, we will suffer no harm from the dangers of ambiguity latent here.

The usual alternative exists of sticking with a concept most like the thermodynamic concept and exchanging the law of thermodynamics for a statistical regularity, or of moving to an ensemble concept and keeping the unexceptionless law. We expect a system held in energetic isolation from its environment to have its Boltzmann entropy fluctuate, even after "equilibrium" has been obtained. But its Gibbs fine-grained entropy will, of course, remain constant, like the entropy of a system in equilibrium in thermodynamics, because that ensemble entropy will depend only on the constraints imposed on the system and will be indifferent to its fluctuational behavior. The Gibbs fine-grained entropy, by averaging over the ensemble of systems, has absorbed the fluctuations into the improbable but present low Boltzmann entropy members of the collection.

But, of course, making this distinction between individual system property and ensemble feature is only the beginning of the process of understanding the perplexing and complex role of entropy in statistical mechanics. We have seen how, in an attempt to characterize the non-equilibrium dynamical process, the idea of coarse-grained entropy was introduced. Ensemble fine-grained entropy is provably invariant over time. But the coarse-grained entropy can vary with time, and thus have its increase at least potentially represent the approach toward equilibrium of systems started away from the stationary state.

Fully understanding the role of coarse-grained entropy is a complex matter. First, there are the difficulties in connecting the change of coarse-grained characteristics with the observed changes in real macroscopic features whose dynamics we wished to understand in the first place. Then there are all the difficulties encountered in trying to show that the coarse-grained quantity, especially the coarse-grained entropy, will really have the behavior one wants – in the case of entropy, monotonic increase toward the equilibrium value. Getting this result out of the theory is a matter not fully resolved. The best results hold for certain kinds of ideal systems (C systems and the like). In an idealized sense, we get the right limit results as time goes to infinity or we get the right dynamical results for a coarse-graining proven to exist but not easily characterizable so as to be connectible to the macroscopic variables with which we are concerned. Nonetheless, it is plain that the invocation of these coarse-grained notions of entropy does hold some promise for grounding what might ultimately be a successful attack on the general problem of non-equilibrium dynamics.

The introduction of coarse-grained entropy can be understood as another case of "extending the concept" from the older narrow conception of the reduced theory to a broader conception allowed by the richer resources of the reducing theory. Already within thermodynamics, as we have seen, the concept of entropy had been extended in natural ways from the equilibrium case to non-equilibrium cases. When the non-equilibrium system had local equilibrium, for example, we could define a local entropy density and take the overall entropy to be that quantity integrated over the entire system.

In the coarse-grained case, entropy is being generalized in a different way. Here, the entropy is a Gibbsian ensemble quantity. For equilibrium ensembles, coarse-grained and fine-grained entropies will agree, because the ensemble points will uniformly occupy each cell of the coarse-graining. In the non-equilibrium cases, the fine-grained and coarse-grained entropies will generally disagree, the fine-grained remaining constant, and the coarse-grained, one hopes, evolving toward the entropy appropriate for the equilibrium state of the ensemble relative to the imposed constraints. Whatever the difficulties there are in applying coarse-grained entropy for statistical explanatory purposes, and whatever the difficulties there are in finding posits sufficient to get the coarse-grained entropy to behave as one wants it to behave, the process of concept extension here is clear.

It is worthwhile, at this point, to discuss the issue of the legitimacy of concepts such as coarse-grained entropy. Here, the issues I wish to discuss are not those centered around the suitability of using coarse-grained entropy to describe the non-equilibrium process, but rather accusations against the notion that it is too "subjective" to play a legitimate role in physics. The accusation of subjectivity comes from the fact that we can coarse-grain the phase space in an infinite number of different ways. For any given actual ensemble of systems, in any finite time, coarse-grained entropies can be found that will vary in any way one likes: increase, decrease, oscillate, whatever. How then could the increase of coarse-grained entropy (or its monotonic increase, if one can get that) in some coarse-graining represent the objective increase of entropy associated with approach to equilibrium?

One suggestion, that of A. Grünbaum, although ingenious, won't work. He suggests that for each coarse-graining, a majority of systems in the ensemble will behave in such a way as to have its entropy monotonically increase, even though it will be a different overwhelming majority for each coarse-graining. But for any coarse-graining there will be ensembles whose coarse-grained entropy relative to that coarse-graining will do whatever one likes, and, more importantly, for any specified ensemble there will plainly be coarse-grainings that make the ensembles' entropy do whatever one likes, at least for finite time intervals.

On the other hand, the results of the theory generalizing beyond the

Ergodic Theorem do provide some "objectivity" for coarse-grainings. Mixing results tell us, for example, that in the appropriate measure-theoretic sense, it doesn't matter what coarse-graining we use, for in the limit as time goes to infinity we will get the same equilibrium value for the coarse-grained entropy. Proofs that a system is a Bernoulli system, on the other hand, tell us that there exists some appropriate coarse-graining having the randomizing feature that we are looking for, and that is certainly strong enough to give us monotonic increase of coarse-grained entropy relative to that coarse-graining.

The much more important point is that a *relativity* of coarse-grained entropy value and behavior to a coarse-graining chosen is no mark whatever of "subjectivity." This is to repeat the points made earlier about the relativity of entropy in thermodynamics and the level of character-ization of the system. It is, of course, of vital importance that coarse-grained entropy does indeed behave the way it should with regard to certain "natural" partitionings of the phase-space, ones based on the usual breaking up of the phase-space region into cells small with regard to the overall phase space but large enough to contain a vast number of systems in an ensemble. These coarse grainings are based on cells of simple shape, and are based on cells whose definition is framed in the usual parameters such as position and momentum. For it is that kind of coarse-grained behavior that we can hope to connect in the right way with the observables whose dynamical evolution we wanted to com-prehend. And, of course, it is a vital question why systems are such that their coarse-grained entropy, so construed, does indeed behave in the expected manner. But none of this should lead us to characterize coarse-grained entropy as in any way "subjective." There are, to be sure, interesting objections to the legitimacy of the invocation of such partition-relativized notions into the theory, and proposals to avoid that relativity. An example of the latter is the proposal of Prigogine, closely connected to the results on systems that are mixing, K systems, or Bernoulli systems, to handle the problem of arbitrariness by proving the existence of a new ensemble representation that is Markovian in its evolution. But it is quite misleading to exclude the coarse-grained notions a priori on the basis of some claim that they are subjective.

The concept of entropy allows for a wide variety of other conceptual extensions, each suitable for a different purpose. Prigogine suggests a concept of entropy that is obtained by first making the transition to the new ensemble representation that obeys a Markovian equation of evo-lution, and that provably exists when the system is randomizing enough (being a K system). He then defines a new entropy from the new rep-resentation in a manner analogous to the Gibbsian fine-grained entropy of the original representation. We have seen how this entropy, which does vary in time, can be used to try and distinguish within Prigogine's program, the permitted singular states from those that are excluded and that would result in anti-thermodynamic behavior.

In another context, Prigogine suggests a different (but related) extension of entropy concept. We have seen how in the spin-echo experiment it is possible to act as a Loschmidt Demon, preparing the system in such a state that from the usual point of view, it spontaneously moves from a high- to a low-entropy condition. This can be construed as a violation of the Second Law more radical than that entailed by the existence of fluctuational behavior in any statistical mechanical account. In practice, the entropy generated in the external environment to generate the spin flip is vastly greater than any entropy decrease accomplished in reversing the motion of the spins.

Prigogine suggests that although the intervention into the system seems to generate a spontaneous decrease in entropy subsequent to the flip, this result is merely the consequence of an inadequacy in our usual notion of statistical mechanical entropy. The statistical mechanical entropies typically work by focusing on only a portion of the information contained in the full ensemble probability distribution. Boltzmann entropy, for example, utilizes only the one-particle distribution function, ignoring the correlational information contained in all the higher particle functions and contained entirely in f_n, the full n-particle distribution function. In the spin flip, this is what happens: A system originally of low entropy, where the entropy is calculated by using the one-particle distribution function or something like it, evolves into a higher entropy state. The information contained in the original parallelism of the spins is now distributed in all the correlational facts about the spins and hidden from the one-particle distribution function. The spin flip "reverses" all the correlations, so that the system evolves, once again, into a low-entropy state where entropy is calculated in the usual way. From this perspective, after the flip the system evolves from the high to the low entropy.

Prigogine suggests that a new entropy be defined that pays attention to the correlational information as well as to the one-particle information. He seeks such a notion that would have the following features: (1) Like the one-particle entropy, and unlike the entropy we would get using the full n-particle distribution function, the entropy can vary in time for an isolated system. (2) But like the n-particle entropy and unlike the one-particle entropy, this new entropy takes account of the information about correlations among the components of the system. (3) In the case of normal system evolution, where information dissipates from the one to the n-particle distribution, the entropy so defined will, as usual, increase. (4) But after something like a spin-flip, where anomolous anti-thermodynamic correlations among the components have been introduced by the Loschmidt reversal process, the entropy takes on a very *low* value, below even that of the original spins-parallel state.

With this new entropy so defined, we can hope to describe the spin-flipping process as one which, by intervention from the outside, radically decreases the entropy of the spin system. The spontaneous evolution that follows the flip will then be one in which entropy increases. So with this new definition of entropy the Second Law is "saved." But, of course, the anomalous nature of the evolution of the system after the flip is as surprising as it ever was, and the possibility of the Loschmidtian reversal process as important as it ever was. Once again we see the possibility of saving a regularity by modifying a concept. The new notion of entropy will, as usual, agree with the old in many important cases like that of ordinary equilibrium systems. In other circumstances it will differ from the old notion. And the benefit gained by moving to this novel concept will be the ability to save the old regularity from apparently refuting physical possibilities.

Summarizing, the situation is this: When we move from thermo-dynamics to statistical mechanics, a much richer set of conceptual pos-sibilities is opened to us. We need to find concepts at this new reducing level to associate with the concepts at the reduced level. We are directed in our search for the appropriate concepts by the place played by the concepts at the reduced level in the general thermodynamic laws, in the constituent equations, and by the standard measuring processes in-voked to determine magnitudes in the reduced-level theory. Further guidance is provided by the place of this particular reduction in the broader scientific reduction of macroscopic systems to arrays of constitu-ent micro-components, and by the place of this theory in the general kinematic and dynamical picture.

Concepts exist at the reducing level that designate properties like those posited in thermodynamics in that they are concrete features of the in-dividual systems. For these features it is reasonable to assert that the property referred to at the reduced thermodynamic level has been iden-tified with a feature referred to at the reducing, statistical mechanical, level. With this association of features at the two levels, the laws at the thermodynamic level become modified by the reduction into statistical regularities. Even restricting attention to this kind of concept at the re-ducing level, the process of extending the concept into realms out-side the scope of the associated concept at the reduced level can be important.

Concepts at the reducing level can also be used to designate features not of individual systems but of probability distributions relative to a system kind – that is, of ensembles. The regularities among these features at the reducing level will frequently be of the exceptionless kind, the fluctuational aspects of the world being absorbed into the ensemble structure. A wide variety of such ensemble concepts may correspond to one concept at the thermodynamic level. And the association of ensemble features to the macroscopic features of individual systems will be a subtle and complex one. The full understanding of the association will require the exegesis of the exact connections between the positing of facts about ensembles and the probabilistic assertions about the macroscopic features of individual systems that follow from these posits. Here, the notion that the thermodynamic feature is "identical" with the statistical mechanical feature misrepresents the interrelation of reduced and reducing theory.

Again at this ensemble level, a variety of extensions of the concepts are available, so that the concept naturally associated with the thermodynamical feature can be extended and modified in quite distinct ways. No harm is done by such conceptual modification as long as it is always kept clear about how the novel concept is utilized in some explanatory scheme, and as long as one avoids the pitfalls of confusion due to equivocation on the various meanings that words like "entropy" can have at the reducing level.

III. Problematic aspects of the reduction

1. Conservative versus radical ontological approaches

Scientific progress consists in part in a process of increasing the number of posited entities of the world and, simultaneously, in a process of decreasing the number of these posits. Novel theories introduce entities into our world picture previously unimagined. But theoretical novelty also consists in telling us that entities and features of the world we previously thought to exist do not exist (ghosts, phlogiston, and absolute simultaneity, for example). And scientific progress also consists in telling us that what we initially thought were two classes of entities in the world are but one class. Where we once thought there were both light waves and electromagnetic waves, we now know that there are only electromagnetic waves. Not because light waves don't exist, of course, but because the collection of light waves is a proper subset of the collection of electromagnetic waves.

Intertheoretic reduction sometimes encompasses a rejection of portions of the now reduced theory, and so a dismissal of some of its previously posited ontology. And intertheoretic reduction often consists in part in the identification of a class of entities or features posited at the reduced level with entities or features posited at the reducing level. The reduction of thermodynamics to statistical mechanics reveals both of these processes.

Within the development of thermodynamics itself we had the transition from the positing of heat as a conserved substance (caloric) to the energetic theory of heat as a quantity of "hidden" internal energy of the system. This elimination of a posited "heat stuff" was an essential preliminary to the later reduction.

Again within the context of scientific thermodynamics, we already had the separation from the physical system and its features of such things as "felt hotness and coldness." This was part of the familiar (if philosophically very perplexing) process by which sensorily apparent features of a system, originally used to characterize its physical nature, are declared "secondary qualities" and dumped into the realm of the "mind of the perceiver." The nature and rationale of this process remains problematic, but it certainly is a crucial preliminary to the reduction of thermodynamics to statistical mechanics. We need not worry how that latter theory is to encompass "felt hotness and coldness" because these have already been stripped off the physical system (and, perhaps, out of the physical world) by the implicit metaphysics of the reduced theory.

Whereas caloric is eliminated as an entity, heat, temperature, and entropy are, insofar as they are viewed as properties of individual systems at a time, identified with features of the system characterized by the reducing theory. The reduction is not a simple identificatory reduction, as we have seen, because the terms correlated with the thermodynamic terms are in many of their uses being used in a way quite distinct from being used to refer to "concrete features" of a system. But from the perspective of asking what happens to the heat of a system, its temperature, and its entropy, interpreted as concrete features of the individual system, it does seem plausible to say that they are not "eliminated" by the reduction, but rather assimilated by identification to the relevant kinematic-dynamic features of the system described at the level of the kinematics and dynamics of its micro-constituents.

Our grounds for picking some particular mechanical state of the system to serve as that identical with some thermodynamically characterized feature of the system may be statistical in nature, and may require reference to the system as a system of a specified kind (that is, as a member of a particular ensemble). Thus knowing which mechanical feature of the system is to be taken to be its temperature (in the concrete feature sense) requires putting the system into an ensemble and contemplating the features of that ensemble as being an equilibrium ensemble. Nonetheless, the feature of the individual system is not itself a statistical feature, but is definable in purely mechanical terms. Thus for an ideal gas, mean kinetic energy of the molecules is just the average of those purely mechanical kinetic energies for the molecules of the system as a time.

Objections to alleged reductions frequently rest upon the claim that some entity or feature posited at the reduced level can neither be eliminated from our picture of the world nor identified with anything in the total ontology of the reducing theory. Thus, we are sometimes told, it is wrong to think that the mental can be reduced to the physical. For sensations of yellow objects have a feature, perceived yellowness, that cannot be discarded from the world but that cannot sensibly be found as a feature of things, events, or processes in the physical theory that includes all the features of neurons and their processes in its domain.

It doesn't seem that the claim of the reducibility of thermodynamics to statistical mechanics can be objected to in this vein. As we have seen, thermodynamics has already eschewed such things as felt hotness as being outside its purview. And all of the entities and features of things it retains, having dismissed caloric, seem to be identifiable as mechanical features of the system construed as a structure of micro-constituents in the usual manner of identificatory reductions.

Yet it frequently is asserted that thermodynamics is *not* reducible to the mechanical world-view. We will examine the grounds of one such claim in the next section, a claim that rests on something other than a view about the ontological posits of the alleged reduced and reducing theory. But here it might be useful to say just a little about one such claim that does have an important ontological aspect – the claim of Prigogine and others, a claim noted briefly in Section IV,2 of Chapter 7, that the relation of thermodynamics to mechanics requires us to radically revise our posited ontology of the world at the mechanical level. Here it is not claimed that thermodynamics posits a need feature of the world not identifiable with any mechanical feature that allegedly blocks the reduction of thermodynamics to mechanics plus statistics. Instead, it is claimed that thermodynamics and the facts about mechanics we discover in an attempt to reduce thermodynamics to the latter theory show us that the mechanical theory we had as reducing theory had been misguided all along.

Although this radical proposal for revising our basic ontology has never been made fully and precisely enough for us to be able to explicate it in detail, some of its prominent features are clear enough. It is claimed that the existence of radical instability of trajectories, either in the measure-theoretic sense used for proving mixing and its generalizations or in the "Poincaré instability" sense in the unstable regions when the KAM Theorem holds and blocks measure-theoretic chaos over the whole phase space, makes the positing of exact micro-states for the system (represented by points in the phase space) a false idealization. If we cannot, in principle, "fix" the exact micro-state of a system, ought we not, on good positivist grounds, to deny reality of such a point micro-state? Instead, shouldn't we take the correct representative in our theory of the micro-state of a system to be some appropriate probability distribution over a region of phase space. In other words, shouldn't we take ensemble representations to characterize individual system states in the manner discussed in Chapter 7?

These general considerations are backed up by arguments that rely upon the use of the non-unitary transformations discussed in Section III,6 of Chapter 7. It is argued that the fact that initial probability distributions that are such that they are represented as singular concentrations of all the probability on a single phase point or other set of measure zero will still be such that their non-unitary transforms will evolve with time into "delocalized" probability distributions reveals formally the "false idealization" involved in assuming that exact point micro-state representatives correspond to any such genuine micro-state in reality.

On the basis of these considerations, the analogy is drawn with quantum mechanics, which, in nearly all accepted interpretations, denies the existence of an exact position-momentum state for a system. Instead, quantum mechanics attributes to the individual system the wave-function, which is a probability amplitude if not a probability distribution, as the "exact and complete" state of the individual system.

But, as we noted in Chapter 7,IV,2, this radical revision of our standard ontology is fraught with problems. The argument from non-precise-fixability of the pointlike micro-state to its non-existence is surely too quick. Even from a positivist perspective in which the postulation of unobservables is taken as at best instrumental, genuine reference being reserved for terms referring to the observables, the positing of the point-like micro-states could very well be a legitimate theoretical posit even if we could not fix the value of the micro-state for individual systems (at least where the radical instability holds of them). The positing of the exact but unknowable micro-state might still be justified on theoretical grounds. And it might very well be claimed that the very success of statistical mechanics, which posits flows on phase space as its basic object of concern, presupposes the existence of the exact micro-states of the individual systems as its underlying basis. Of course, the necessity for this presupposition is just what the radical reviser of our traditional ontology will deny.

The formal argument to the effect that the non-unitarily transformed dynamics takes initially singular probability distributions into distributions whose representatives are no longer totally concentrated upon sets of measure zero is also not entirely persuasive. Although such transformed representations may very well capture interesting aspects of the dynamics of the ensemble, the fact that the original phase-point representation still exists (and is, indeed, presupposed by the transformed way of representing the evolution of the statistics) seems to allow the defender of exact micro-states the option of arguing that the existence of this alternative representation, although revelatory indeed of the instability of trajectories that leads to the familiar coarse-grained ensemble spreading, in no way demands the rejection of underlying exact micro-states of the individual systems.

The analogy with quantum mechanics is dubious for several reasons. In quantum mechanics, the "no hidden variables" results militate in favor of the Bohrian interpretation of the wave function as being a "complete" description of the state of the system, and against earlier interpretations of uncertainty as the mere result of epistemic limitations on our abilities to know the values of existing but "hidden" exact parameters, such as the position and simultaneous momentum of a particle. But in statistical mechanics the existence of the "hidden variables," the states of the system represented in the theory by the points of the phase space, is presupposed by the construction of the phase space that provides the arena over which probability distributions are constructed and over which their dynamic evolution is followed.

In quantum mechanics, an additional reason exists for taking the wave function to be a characteristic of the individual system, and not a probability distribution in either the "ignorance" or "collective" sense that underlies the usual understandings of ensemble probabilities in statistical mechanics. This is the fact of superposition. The probability of an outcome that can be obtained by a multiplicity of possible but exclusive routes is, in quantum mechanics, not the sum of the probabilities of the outcome by each separate route. Rather, the probability amplitudes from each source (which, when squared would give the probability of the outcome due to that source) are added together. The total probability of the outcome is given by the square of the sum of the amplitudes. This gives rise to the familiar "interference of probabilities" in quantum theory. This behavior of probability amplitudes, like the additivity of physical waves so that the intensity of the whole is the square of the sums of the amplitudes of the component waves, allowing interference effects where amplitudes (unlike probabilities) can "cancel" one another, is one of the strongest pressures behind treating the wave-function in quantum mechanics as something more like a concrete feature of a system than like a probability construed as a measure of belief or of frequency in a collection.

But, of course, such interference phenomena are totally lacking in the ensemble probabilities of statistical mechanics. In quantum statistical mechanics, such interference must be taken into account, of course, but that is at the level of the wave function attributed to an individual system, not at the level of the ensemble probability measure over possible wave functions. Once again, a crucial argument that inclines us to take the wave function as a feature of the individual system and not as some "ignorance" or "statistical" measure fails to incline us to that attitude toward the probabilities of statistical mechanics.

The necessity of treating individual systems as if they had exact pointlike micro-states when the systems are simple enough to have stability of their possible trajectories, such as in the case of unperturbed planetary motion or perturbed motion that is still in the KAM regions of stability, would also seem to incline one to view the exact micro-states as real even when instability makes it impossible for us to determine which exact micro-state a system possesses. But the ontological radical can reply to this that the postulation of the exact micro-states in the stable cases is just a limit to the real ontology. This is taken by the ontological radical to be a very tightly focused probability distribution over initial states, all of whose trajectories are coherent with one another for at least some future time. Here the argument would be at least analogous to those who would argue in the quantum context that the postulation of "classical" systems is unnecessary, things behaving classically just being quantum systems whose interference features are "negligible."

A greater problem for the ontological radical is that posed by spin-echo-like results. Here, it would seem, the appropriate ensemble probability distribution to pick at the moment of flipping is the spread-out distribution, in some appropriate coarse-grained sense, at least, corresponding to the system's overt randomized nature. Yet the possibility of flipping the system so that it retreats to its initial overtly ordered configuration seems to imply that some exact, pointlike, micro-state, containing all of the subtle correlational facts among the spins, exists. Otherwise, how to explain the ability of the system after flipping to "find its way home?"

Added to these considerations is the fact that denying the existence of the exact pointlike micro-state is neither necessary nor sufficient to solve our ultimate theoretical problems. Positing the exact micro-states is compatible with the generation of an appropriate statistical structure at the level of statistical mechanics to give us a statistical explanatory account of the thermodynamic features of phenomena, even if getting that account requires some puzzling and problematic posits about the probabilities of initial micro-states of systems. Denying the existence of the exact micro-states, on the other hand, doesn't seem to help us at all in resolving the most crucially puzzling questions such as those centered around the origin of time-asymmetry. Whether we take the probability distributions as directly descriptive of concrete aspects of individual systems, as the ontological revisionist desires, or, instead, as probabilities over actual exact micro-states of systems, as the ontological conservative would have it, we are faced with exactly the same puzzles about the origin of the special initial distributions we must posit to get the non-equilibrium time-asymmetric results we want. It is far from clear why anything in the ontologically revisionist position helps with this fundamental problem.

2. The emergence of thermal features

If the arguments of the last section are correct, then thermodynamics requires no ontological posits about the world not already present in the posits of the theory of the micro-componential structure of matter and in posits of the kinematic and dynamic theories governing the behavior of these micro-constituents. There is no residue of unreduced entities or features of the world posited by thermodynamics yet not identifiable with systems or features of systems posited at the reducing level. Nor does the structure of the reducing theory require any radical ontological novelty in order that it do justice to the world described by thermodynamics. Yet the intuition persists among many who have thought carefully about these issues that it would be misleading to think of thermodynamics as a genuinely "reduced" theory. On what do these intuitions rest?

The thermodynamic picture of the world introduces a kind of structure on the world that doesn't find an easy place to reside in the general dynamical world picture. In particular, the idea of equilibrium states as attractor states for systems and of the time-asymmetric nature of the approach to equilibrium are the characteristic thermodynamical aspects of the world that require the greatest insight in constructing the reducing micro-theory that is intended to account for them. What do we need at the reducing level to do justice to this thermodynamic structure?

Mechanics by itself, along with the facts about the constitution of the macro-system out of its micro-components, won't do the job. And merely introducing "natural" probabilities over microscopic conditions won't, in one sense, do the job by itself either. What I mean here is that the arguments stemming from the first critical responses to Boltzmann's attempt to save the *H*-Theorem by making the interpretation of it probabilistic still carry strong weight. In order to get the results we want out of statistical mechanics, even taking the cosmological ways out stemming from late Boltzmann into account, it seems that we must asymmetrically apply our natural probabilistic assumptions to the dynamical state of a system at a time. We must apply them only to "initial" states and never to "final" states. Only thus can we get the correct thermodynamic predictions out of non-equilibrium statistical mechanics and block any anti-thermodynamic retrodictions that would seem to follow by parity of reasoning if we allowed ourselves the natural assumption that the probabilistic posits over states at a time could be indifferently applied at the beginning or at the end of a branch system's career.

Even if we put the time-asymmetry of the world described by thermodynamics to the side, although this asymmetry usually does lie at the heart of arguments for the "irreducibility" of thermodynamics to any lower-level theory, there remain the "natural" probability posits themselves. Rationalizing them as somehow mere instances of a priori inductive reasoning, and so immune to the demand for a physical account of why they are true, seems dubious. Except for the pure equilibrium case, where the ensemble probability distribution can be given its "transcendental" rationale as being the only stationary distribution constant in time and absolutely continuous with respect to the natural measure, the standard posited probability distributions over initial conditions of systems, no matter how "natural" they appear, constitute an independent posit of the theory, not derivable from mechanical considerations alone. Even in the pure equilibrium case, the assumption of absolute continuity with respect to the standard measure is a posit that cries out for further justification. And in the non-equilibrium case these posits are essential not only for time-asymmetry but for such crucial matters as getting the right finite relaxation times for systems.

One way in which claims of "irreducibility" for thermodynamics can be understood is in terms of a strong emphasis being put on the necessity of the basic posits about probabilities over initial conditions. The claim here would not be that there is no truth to the claims of reducibility of thermodynamics to statistical mechanics, but rather an assertion that the most crucial features of the thermodynamic aspects of the world, the features like approach to equilibrium and time-asymmetry that appear as a surprise from the purely mechanical viewpoint of the world, appear at the reducing level only because a basic posit, essentially thermodynamic in nature, is introduced at the reducing level in order to carry out the reduction. This is the posit of uniform probability over not too small and not too complex regions of phase space for the initial conditions of a system. Clearly this is what Krylov had in mind when he insisted so trenchantly that thermodynamics could not be derived either from classical or quantum mechanics alone.

We can summarize the situation by saying that statistical mechanics successfully reduces thermodynamics by replacing the structural constraints on the world imposed by the latter theory by its fundamental autonomous law – the Second Law – with a structural constraint on probabilities of the initial conditions of systems characterized at the micro-level.

What remains most disturbing here is the seeming absence of any place for such a constraint from the mechanical world-view. It is not that imposing the constraint is in any way inconsistent with the purely dynamical world-picture. As we have noted, although rerandomization might prove inconsistent with dynamics, imposing the distribution on the micro-states at one time cannot be. It is, rather, the more subtle point that the general context in which the mechanical world picture develops is usually presented as based on an implicit assumption that "initial conditions are freely choosable." The only constraint on the construction of a model of a possible world is taken to be the lawlike constraint that once an initial state is chosen states at other times be lawlike compatible with that choice. But now we seem to be saying that the choice of initial states of the world is constrained, at least in the peculiar probabilistic sense we require for statistical mechanics.

It is worth noting that different sorts of unexpected constraints on initial conditions appear in other portions of physics as well. It has been proposed that one can have a theory of special relativistic space-time (Minkowski space-time) and yet tolerate the existence of super-luminal causal signals – signals traveling faster than light in a vacuum, so-called tachyons. The theory seems to imply that what appear as signals going forward in time for one observer can transmit causal influence in a way that would be construed as transmission of a signal "into the past" from the point of view of some other observer moving with respect to the first. This seems to generate the possibility of a closed causal loop, in which the occurrence of an event causes something to happen in the past of the event that might prevent the first event from happening. That is, the theory seems to generate the possibility of a causal paradox. One proposed solution is just to insist that initial states of the world that are so causally incompatible with themselves cannot exist. Here again we see the suggestion that there might be constraints on initial states over and above the constraints on pairs of world-states imposed by the usual laws of the theory.

A similar argument arises in the theory of general relativity. There are world models of that theory that allow for closed timelike curves – that is, one-dimensional sets of events that are causally connected to one another and so arranged that, once again, an event can "cause itself" at a finite time interval along the curve. Here again, the possibility of causal paradox can be blocked by the demand that the world models have an internal self-consistency among all the events in them.

What kind of a constraint on initial conditions in the world is needed for statistical mechanics? It is only a probabilistic constraint, of course. Is it a new "law of nature?" That doesn't seem quite right. For one thing, it doesn't say that the improbable can't occur. Even construed as a new "statistical law" it seems a bit out of place. It is as if the full set of ordinary dynamic laws ought to be taken by us as exhausting all the genuine laws (exceptionless or statistical) there are. These additional constraints seem too anomalous to be counted among the fundamental laws of nature. That is why, of course, so many have insisted with Reichenbach, Mehlberg, Grünbaum, and others that it is a mere "de facto" condition of the world that the initial states are so distributed in the world. But, again referring to Krylov's persuasive arguments, that doesn't seem to do adequate justice to the situation either. The pervasiveness and generality of the Second Law phenomena seem so universal and fundamental that describing them all as the result of some "mere accident" about the configuration of initial conditions (which, on the model of the distance from the sun, the earth just happens to have) seems absurd. To go from the dynamic world picture to a world theory adequate to ground thermodynamic phenomena seems to require the positing of something *sui generis*, the fundamental probabilistic constraint on initial conditions that is the core of the statistical mechanical approach to non-equilibrium dynamics. I think it is this all important fact that lies at the heart of claims that the thermodynamic structure of the world is primitive, ineliminable, and irreducible to any other fact about the world.

A brief comparison of the reduction we have been concerned with with some other putative possible reductions will be enlightening here. Philosophers of biology often express the opinion that the ontology of genetics and evolutionary theory is exhausted by the usual biological entities and features such as DNA molecules, their cellular environments, the chemical, physical, and biological structure of organisms, their physical environments, and so on. But, they sometimes continue, it is unreasonable to expect a reduction of evolutionary theory, say, to molecular biology in any way that would provide a definitive correlate at the reducing level for such terms as "environment," "fitness," "niche competitor," and so on, found at the reduced levels of evolutionary theory. Again, some philosophers of the mind would argue that we can't expect to find neurological kinds that would capture the mentalistic kinds framed in our intentional vocabulary of cognitive states, but that this is no reason to deny the conclusion of physicalism that the realm of the physical (including the neurological) exhausts all there is in the world.

For these "supervenience" cases of reduction it is usually claimed that we can, at least in a preliminary way, understand why the reduced-level theory has the "irreducible" categorical kinds it has. We understand how organisms develop, the role of self-reproducing molecules in the genesis of their structure, biological facts about transmission of genetic character in the reproduction of organisms, the place of the environment in aiding or hindering the survival of an organism, and so on. This gives us at least some preliminary idea why such categories as those of fitness, species, and so on – the framework categories of the theory of natural selection

– work as well as they do in allowing us to organize the data of our "macro-level" experience of species differentiation and species dynamics in evolutionary history. From this perspective it can at least be argued that there is no need to posit anything over and above the usual physical entities and features of the world and the usual physical laws governing their behavior in order to find a place in nature for species and their evolution. Naturally, there are skeptics who still maintain that any such hope of accounting for the reduced-level structure in terms of that of the reducing level will founder, even if less than kind-to-kind identificatory reduction is all that is demanded.

A similar case can be made for the claim that one can at least offer a preliminary explanation, framed in biological-physical terms, for the successful place played by the "irreducible" structure of the language of cognitive states, content, intention, and so on in our "folk psychological" account of human behavior. Here, an account in terms of the place of organisms in their environment, the structure and evolution of the neural processing machinery of the brain, the behavioral strategies likely to lead to species survival, and so on will all be invoked to try to explain why humans (and animals) have the kind of inner-computational mechanism that results in behavior usefully describable and explainable in terms of the possession by the agent of beliefs, desires, and so on. Once again, there will be many skeptical that any such argument can really do the job. But at least the hope for a physicalist account of the place played by the intentional structure, even if that intentional structure is not reducible to a physical account by any reasonable type-to-type identificational reduction, is not unreasonable.

Seen in relation to these two cases, the problem of the relation of thermodynamics to ordinary dynamics provides an interesting contrast. Here there is no question of the irreducibility of the "non-physical" to the physical, of course. Rather it is the degree of autonomy that we must credit to such intrinsically thermodynamic notions as equilibrium, entropy, and irreversibility in our account of the world, as contrasted with the degree to which we can found these notions on the common notions of the remainder of physical theory that is in question.

Here, the connection between the thermodynamic structure and purely dynamic structure is much clearer than the relations speculated about between evolutionary and biochemical structures or intentional and neurological structures. We know pretty clearly what we must super-add to the usual dynamic description of systems to be able to generate the situation where the applicability of thermodynamic concepts in a useful descriptive and explanatory role becomes clear. It is the posits about the probabilities of initial micro-conditions of systems that are the additional elements. These, combined with elements from the purely dynamic picture (facts about instability of trajectories, for example), are what provide the characteristic structure of the world that makes the thermodynamic picture applicable.

So, the additional elements necessary to place thermodynamics into the general dynamic picture are clear, although, as we have seen, much debate remains about how, exactly, they are to be construed. What is striking is that this additional posit seems to be itself so autonomous and so out of place in the remainder of our world picture.

We do now understand how there can be an additional structure of the world that although not invoking any ontology not posited by the remaining structure of the reducing theory, can introduce novel elements into our structural picture of the world. The underlying compositional and dynamic theories leave open what the initial conditions of a system are to be at the micro-level. And these theories leave open the possibility of there being some interestingly characterizable distribution of those initial conditions in the probabilistic sense. It is into this "gap" in the remainder of physical theory that the basic additional posits of statistical mechanics fit. Once again, there is nothing in the remaining physical theory that blocks such posits of probabilities over initial micro-states. But there is nothing in the remaining theory that "grounds" these posits either. And there is a general methodological presupposition in most applications of that remaining theory – the presupposition that initial conditions for a system can be "freely chosen" – that makes the needed posits seem strangely out of place in the remaining body of physical theory. It is this that is at the core of the strongly held intuition of many scientists working in the fields of thermodynamics and statistical mechanics that the thermodynamic structures of the world are basic, primitive, and "irreducible" in our overall physical theory of nature.

It is important to note here also the degree to which macroscopic knowledge of the nature of a system, framed in the conceptual languages of thermodynamics, is utilized in choosing the basic probabilistic posits necessary for the underlying statistical mechanics. We have noted how, especially in the non-equilibrium case, it is essential to have some thermodynamic appreciation of the system in order to even begin to look for the correct initial probability distribution that must be posited in order to get the ensemble dynamics of statistical mechanics underway. Frequently, we have little guide as to how to choose this probabilistic posit other than our experience at the macro-level, conceptualized in thermodynamic-hydrodynamic terms, of what the appropriate "reduced descriptions" are for a system that will allow for the production of the dynamic laws of approach to equilibrium framed in macroscopic terms. When we come to undergirding this structure with a more complete statistical mechanical picture, it is this already existing macroscopic account that is used in framing that deeper picture.

Even when we already have the statistical-mechanical kinetic equation at our disposal, we generally cannot hope to find the general solutions of it. Instead, the search for solutions is guided, once again, as in the Chapman–Enskog mode described in Chapter 2, by the use of what we know, from the macroscopic thermodynamic-hydrodynamic level, about the structure of the solution we are seeking. The statistical-mechanical deeper account does provide us new insights of great importance, allowing us, for example, to derive the values of transport coefficients previously inserted into the hydrodynamic equations on the basis of their experimentally determined values. But it is vitally important to recognize that the very structure of the solutions found is guided by our antecedent knowledge at the empirical macroscopic level of the kind of solution we must look for.

The apparent ineliminability at certain crucial stages of previously obtained characterizations of systems, characterizations framed in thermodynamic terms and established by empirical observation, in formulating the correct statistical mechanical model posited to underlie the thermodynamic behavior in question once again makes it clear that if we wish to claim that thermodynamics is reducible to statistical mechanics, we must have a subtly contrived model of reduction in mind.

IV. Further readings

For the positivist view of theory reduction, see Kemeny and Oppenheim (1956). Nagel (1961), Chapter 11, discusses many aspects of reduction in science and posits lawlike "bridges" as essential. The many types of intertheoretic reduction are discussed in Sklar (1967). Hooker (1981) is an extensive review of philosophical models of reduction and their problems.

For general skepticism concerning the reductive relationship between theories and their successors, see Feyerabend (1962) and Kuhn (1970).

On the notion of supervenience, its formal structure, and its application, see Kim (1982), (1984), and (1985).

For the issues plaguing attempts to characterize the relation of mind to brain (or mental theory to biophysical theory), see the readings selected in Volume 1, Parts 2 and 3, of Block (1980).

Gibbs' Paradox is reviewed and studied in Denbigh and Denbigh (1985), Chapter 4.

For the use of the determinable-determinate relation to model the relation of a property to those properties that "realize" it, see Yablo (1992). For the determinable-determinate relation, see Johnson (1964) or Prior (1949).

On the extension of the temperature scale to "temperatures beyond infinity" and "negative absolute temperatures," see Landau and Lifschutz

(1979) or Ramsey (1970). A philosophical discussion of these physical concepts is Ehrlich (1982).

A discussion of the variety of entropies encountered in statistical mechanics is Grad (1961). Denbigh and Denbigh (1985), Chapters 1–3, covers many of the issues, focusing primarily on allegations of subjectivity for entropy in many of its guises.

Grünbaum's way out of the relativity of entropy change to choice of coarse-graining is in Grünbaum (1973), Chapter 19.

For non-equilibrium entropy designed to take account of of correlation (as in the spin-echo experiment), see Prigogine (1970) or (1973). For non-equilibrium entropy based on a change of ensemble representation see Prigogine (1980), Chapter 9, or Misra (1978). Another treatment of non-equilibrium entropy is Goldstein and Penrose (1981). O. Penrose (1979), Section 3.5, surveys a number of proposals for non-equilibrium entropies.

"Ontological radicalism" is suggested in Prigogine (1980), Chapter 9. See also Prigogine (1974), Goldstein, Misra, and Courbage (1981), and Misra and Prigogine (1983b). For a detailed critique, see Batterman (1991).

On issues of biology and reduction, see Hull (1974) and Schaffner (1967) and (1969).

10

◁═══▷

The direction of time

I. The Boltzmann thesis

Let us review once more the basic components of Boltzmann's final account of the asymmetry of the world described by the Second Law of Thermodynamics. In his final picture of statistical mechanics, we would expect to find an isolated system almost always at or near equilibrium. Excursions to states of very low entropy ought to be rare, and we should expect to have higher-entropy states immediately after and immediately before improbable excursions from a close to equilibrium condition. How can we reconcile this account of the probabilities of micro-states in the world with what we actually find? What we find is a universe apparently quite far from equilibrium. It seems to be approaching equilibrium in the future direction of time, but, as far as we can tell, seems to be ever further from equilibrium as we move back in time into the past direction. In addition, this entropic asymmetry of the universe as a whole is matched by the parallelism of entropic increase of branch systems temporarily isolated from the main system.

Boltzmann, the reader will remember, offers a multi-faceted story about the world to reconcile his probability attributions with the observed facts. First, it is posited that the universe available to our inspection is only a tiny fragment of the whole universe. Both in terms of spatial extent and temporal duration, it is assumed that the universe as a whole is vastly larger than the universe we can observe at the farthest reaches of our astronomical investigation of the cosmos. Next, it is posited that this vaster universe is, in fact, overwhelmingly at or near equilibrium in its spatial regions at any one time and overwhelmingly most often at or near equilibrium for any one spatial region over most of that region's temporal duration.

But, the argument continues, one will, given the standard probabilities attributed to micro-states, expect to find "small" regions of this vast universe in conditions far from equilibrium for "brief" periods of time. But "small" and "brief" are implicitly relative notions, the overall size and duration of the universe constituting the standard of size. So we would expect to find regions as vast as that of the universe that we can observe in conditions as far from equilibrium as ours is for periods of time of the order of the billions of years revealed to us by the evolutionary lifetimes of the stars.

Next, Boltzmann applies his version of what has come to be called the "weak anthropic" argument. Although a point picked at random in the universe in space and time is likely to be in a region of the cosmos at or near equilibrium for a long period of time, a place characterized as inhabited by sentient observers will have a probability, given that condition, of being very far from equilibrium. For only in a quite large portion of the cosmos that for quite a long period of time is very far from equilibrium could the stable structures necessary for sentience and intelligence survive, supported as they are in their integrity and functioning by continual energy flows. When we add that these creatures arose by the process of evolution by natural selection, the need for an even vaster and longer lived excursion from equilibrium becomes clear. Only such a vast fluctuation could provide the conditions under which the evolution of stable sentient-intelligent creatures could take place.

Let us assume, then, that the universe is overwhelmingly close to equilibrium in space and in time but that we are in a "small" region of entropic fluctuation. This region is absolutely improbable but highly probable relative to the presence in it of living creatures, although, as usual, not as probable as the much smaller region of fluctuation that would be sufficient for life. But, finding ourselves on a region of the local entropy curve with low and high entropies in the two different time directions, why is it that we find the high-entropy end to be in the future time direction and the low-entropy end in the past, and not the other way around? It is at this point, of course, that the claim is made that the direction of time in which entropy is increasing is "constitutive" of what *counts* as the future time direction. What makes the "future" time direction the future direction is its being the time direction in which the local universe is heading toward equilibrium.

Here, the illustrative claim is also made that if there is more than one region of the universe far from equilibrium at a given time, some of the regions will be going toward equilibrium in our future time direction and some moving away from equilibrium in that time direction. In these latter regions, Boltzmann alleges, the inhabitants will have as their local future direction of time the direction of time "parallel" to our past time direction. And if creatures could live in the regions of the universe more or less at equilibrium, which is impossible, they would have no future (or past) direction of time at all. There would, to be sure, still be two oppositely directed time directions, but neither could appropriately be labeled past or future.

This Boltzmann picture has now been modified by our new understanding of the cosmological structure of the universe. The new orthodox account is of the entire universe in a state of disequilibrium. Starting from the Big Bang in a state of very low entropy, the universe is currently expanding. Its entropy as a whole is constantly increasing, the increase coming both in the form of an initially uniform space-time becoming progressively more "clumpy" and with the matter so clumped dissipating its now concentrated energy into the now cold reaches of empty space as the stars radiate light and material entropy increases. The assumption is, then, that the universe as a whole is on one of those slopes of the entropy curve, with one direction of time pointing toward the higher-entropy end of the slope and one pointing toward the lower-entropy end. This direction of entropic change for the universe as a whole parallels the direction of entropic increase for the dominant class of branch systems, although the reason for this parallelism of branch systems with each other and with the main system is far from clear.

Once again, we note the possibility that this picture might be embedded in a larger Boltzmannian framework if one tolerates the possibility of "multiple universes." In some versions of this idea, the multiplicity of universes are all expanding "bubbles" in a more encompassing manifold than the space-time that constitutes the entire world of each bubble universe. In other versions, the multiple universes are the "parallel" worlds, each of which represents one possible evolution from the initial state according to a quantum mechanical dynamic of evolution. In this account, each individual universe represents one of the allowed possible evolutions, and the collection of them constitutes the entire "wave function" of the equation of evolution. These multiple universe views may once again be supplemented by an "anthropic" argument to the effect that although universes with initial low-entropy states are rare, those with sentient inhabitants will necessarily have a low-entropy endpoint in order that they can, at some point in their history, support sentient life. Again, the problem of why our universe has such an "excess" of low entropy will arise.

But, in any case, the view will be supplemented once again with the final Boltzmannian argument. Why is it that in the universe as we find it, with its radical variance in entropy from one end to the other, we find the high-entropy end in the future? Once again, one is likely to encounter the claim that the very constitution of what distinguishes future from past time direction is to be found in the entropic asymmetry of the universe and its branch systems.

The motivations behind such a claim are clear. If the claim were true, it would obviate the need for finding some explanation as to why entropy increases in the future as opposed to the past time direction. For that fact would follow automatically and trivially from the "definitional" status of what counts as future. Of course, the existence of a time direction of parallel entropic increase of isolated systems would remain non-definitional and in need of an explanatory account. Further, a broad naturalism would hold that all of the facts of the world ultimately require an explanation in terms of the fundamental physical facts about the universe. If it is truly the case that all of the physical asymmetry in time of the world resides in the asymmetric temporal facts about entropic behavior of systems, then any asymmetries implicit in the notion that the future direction of time is distinguished from the past direction would, somehow or other, have to supervene on the basic physical asymmetry that is fully encompassed in the entropic asymmetries.

But many questions remain. First, there are general issues about just what kind of claim of "definition," "dependence," "constitution," or "super-venience" the claim that the future-past asymmetry as a whole is "defined by," "dependent on," "constituted by," or "supervenient upon" the entropic asymmetry is supposed to be. It is not a trivial matter to get entirely straight just what kind of dependence relation is had in mind by those who maintain the dependence thesis. Without knowing exactly what the claim comes down to, it is hard to know what kinds of arguments can support it and what kinds can confute it. Once one has become clearer concerning the nature of the claim, the soundness of both the support-ing and confuting arguments can be evaluated more clearly. I shall be exploring the first sort of question in some detail but will, alas, have much less to say about the plausibility of the claims to the truth of the "entropic theory of temporal asymmetry" than I will about what the structure of such a claim comes down to.

II. Asymmetry of time or asymmetries in time?

There can be no question that the entropic features of the world reveal a pervasive asymmetry of the behavior of physical systems in time. Both for the cosmos as a whole, and, in a probabilistic sense for the tem-porarily isolated branch systems, later states are higher-entropy states. The main questions in the issues of the "direction of time" problem are concerned with whether or not we can ground the familiar "intuitive" distinctions between past and future on the entropic asymmetry. But there is a preliminary topic to which we first ought to devote a little attention. Ought we to say that the entropic asymmetries in the world reveal an asymmetry of "time itself" or ought we to say, rather, that they only represent an asymmetry of the world in time? Indeed, what would be the difference between maintaining one of those positions against the other? As we shall see, neither question has an obvious answer.

1. Symmetries of laws and symmetries of space-time

Symmetries of the observational phenomena in the world play a vitally important role in contemporary physics. The observation that specific location in space is irrelevant to the evolutionary dynamics of a system (at least if the general theory of relativity is ignored) underpins the law of conservation of momentum. That the spatial orientation of a system is irrelevant to its behavior underlies the conservation law for angular momentum. And that the specific time at which a process is allowed to evolve is irrelevant to its evolution grounds the principle of conservation of energy.

Other symmetries abound as well. For example, Galileo's crucial observation that phenomena in one uniformly moving reference frame were identical to those in any other led, in the context of Newtonian mechanics, to the principle of Galilean relativity. Generalized to encompass electromagnetic phenomena, and with a now necessarily modified mechanics taking the place of the Galilean–Newtonian mechanics, the uniformity of phenomena with respect to uniformly moving reference frames becomes the relativistic principle of Lorentz invariance.

In quantum field theory, the invariance of the behavior of systems under the combined operations of going from a system to its mirror image, replacing particles with anti-particles and examining the evolution of the new system from the "time reverse" of what was the final state of the original system to the time reverse of what was the original system's initial state, follows from the fundamental assumptions of the theory. This invariance shows up in the identity of the original quantum-mechanical transition probabilities with the new "backward" probabilities. In many cases the various transformations, taken one at a time, constitute symmetries of systems as well, although, for example, invariance under mirror imaging is certainly violated by weak interaction processes. Time-reversal invariance is probably violated for some special systems as well. Other symmetries, that, like symmetry under replacement of particle by anti-particle, are not directly connected to space-time structure, such as the gauge symmetries of modern field theory, are important as well.

Failure of symmetry can be as important as the existence of symmetry. Indeed, it is, of course, the failure of symmetry under time reflection that we find in thermodynamics in the form of the time-asymmetry of the Second Law that is our main concern. The realization that weak quantum field processes failed to be symmetrical under the exchange of a system for a mirror image of that system was a crucial breakthrough in the understanding of the interaction of the elementary particles with one another. Much earlier, the fact that although systems behaved in exactly the same manner irrespective of which uniform motion they were in, but that accelerated motion, in the absolute sense, had demonstrable physical effects on systems making their behavior quite different from that of similarly constructed systems that were in uniform motion, was crucial to the Galilean–Newtonian mechanics and to the theory of space and time posited to underlie it.

Many asymmetries can be summarized as asymmetries in the fundamental laws of nature. In these cases, we are given a posited foundational law that itself is such that it informs us that two systems related by some appropriate transformation of the one into the other (the first a mirror image of the second, for example, or the first in accelerated motion with respect to the second) will behave in different ways as determined by the structure of the fundamental law.

In some cases, the asymmetry of the law may ultimately be derived by having that law subsumed under some higher order more general law that is itself asymmetrical in nature. But in other cases, the asymmetrical nature of the law is taken as fundamental or foundational, not to be derived from some more basic law. How then can its asymmetrical nature be explained? In some cases, it becomes plausible to blame the underlying asymmetry of the law, and of the phenomena it describes, on an asymmetry of "space-time itself."

Thus, Newton took the differential behavior of accelerated systems to be due to their acceleration with respect to "space itself." It was the existence of space as a single reference frame of motion, and causal effects due to accelerated motion with respect to that space itself, that explained why objects in genuinely accelerated motion displayed effects never found in any system in uniform motion. A general proposal is to introduce just enough asymmetry in the space-time to account for the asymmetric phenomena summed up in the asymmetric law and no more. The critical examination of Newton's notion of absolute space, and the transition from Newton's space-time to so-called neo-Newtonian space-time illustrates this point.

Once a fixed reference frame, at rest in "space itself," is posited, the notion of the absolute velocity of a system, its velocity with respect to space itself, becomes well defined. Yet by Newton's own theory, the absolute velocity of a system will have no discernible effects, all phenomena being symmetrical with respect to the transformation of a system into a system in uniform motion with respect to the first system. So the asymmetry of space-time posited by Newton seems excessive. One can, indeed, move to a space-time in which the distinction is retained between those systems that are in uniform motion and those that are absolutely accelerated, but in which no notion of absolute rest or of absolute uniform velocity is defined. This is the so-called Galilean or neo-Newtonian space-time. In it, the inertial reference frames – those in uniform motion – are distinguished from the other reference frames, but no one of the inertial frames is taken to be "at rest," with the others possessing some uniform absolute velocity. Similar refinements of the space-time lie behind the move in relativistic theories from the Minkowski space-time of special relativity that retains the global inertial frames of neo-Newtonian space-time to the curved space-time of general relativity in which only local free-fall reference frames are well defined and no global frames of uniform motion can be characterized.

So, the general line here is to look for symmetries and for asymmetries in the foundational laws of physics. Some of these may have no obvious relation to space-time, such as the symmetry of processes that captures the fact that for many processes physics is invariant under the exchange of particles for their anti-particles. Others, such as the dynamic symmetries and asymmetries we have just noted among systems in various kinds of motion with respect to one another, can be connected to the structure of the underlying space-time. When such space-time symmetries and asymmetries among the phenomena exist and are captured by such foundational symmetric or asymmetric laws, there is a general proposal to "explain" the results by positing just enough asymmetry of the underlying space-time as to account for the observed lawlike asymmetries and no more.

Before going on we ought to note, however, that many methodological, epistemological, and metaphysical puzzles are encountered in trying to fully understand the structure of such attributions of symmetry or asymmetry to "space-time itself." For the relationist, the alleged explanatory virtue obtained by positing "space-time itself" with its various symmetries and asymmetries, and then using that posit to explain the observed symmetries and asymmetries among the phenomena of the world, is a spurious virtue. For reasons of a familiar reductionist or eliminationist nature, the relationist is generally happier to simply assert that some symmetries and asymmetries exist among the phenomena, and that some of these are space-time-like in their nature having to do with the symmetries and asymmetries among systems related to one another in space-time-like ways. The systems may be related by relative position or state of motion or mirror imaging, for example. The relationists further maintain that any additional positing of "space-time itself" with its underlying structure is theoretically otiose. For the relationist, the foundational laws are genuinely foundational, and attempts to explain their structure by postulation of a structure's "space-time itself" is futile. The relationist has other arguments as well to try and persuade us that the positing of the space-time itself is also fraught with other unnecessary methodological and metaphysical difficulties. For the "substantivalist," the positing is a genuine explanatory posit, no worse than the positing of, say, molecules, atoms, and elementary particles with their structure as explanatory of the observable behavior of macroscopic objects.

Pursuing the debate here, one that is hard even to formulate in a way that makes it very clear just how those on opposite sides of the issues really differ from one another, is something we cannot do. Instead, we will focus on a much narrower issue central to our concerns. This is the issue of whether or not it is plausible to attribute the asymmetry of phenomena we have been exploring – the asymmetry summed up in the Second Law of Thermodynamics or one of its statistical surrogates – to an asymmetry of space-time itself, in something of the manner in which the dynamic asymmetries we have noted are so "explained." Or, instead, are we forced to think of this particular asymmetry of phenomena as not indicative of an asymmetry of "time itself"? The issues here are also far from clear.

2. Entropic asymmetry and the asymmetry of time

Everyone will agree that the universe we experience is one in which physical systems show a remarkable asymmetry in time, the asymmetry summed up by the slogan that in an isolated system, entropy increases, at least in the statistical sense, in the future time direction. But should we view the world as one in which systems show a remarkable asymmetry in time or, more "profoundly," as one in which "time itself" is asymmetric? Ultimately, I think, it just isn't clear what the issue in this debate really comes down to. But let us start with an initial idea that has frequently proved attractive.

Over the years, a number of thinkers, from H. Mehlberg to P. Horwich, have argued that an asymmetric nature can be attributed to time itself only if the asymmetry displayed by physical systems is one that is grounded in a temporally asymmetric physical law. Only then, they have argued, is the argument from the temporal asymmetry of processes to the asymmetry of time as explanatory of the fundamental lawlike asymmetry possible.

But, they continue, the asymmetry summed up in the Second Law is now known to be a mere "de facto" asymmetry not grounded in a fundamentally asymmetric foundational law. Its origin is to be found, rather, in the asymmetry in the distribution of initial conditions of ensembles of systems that is represented by the differing appropriate representing statistical ensembles for initial and final microscopic states of systems that have common initial and final macroscopic conditions attributed to them. But if the physical asymmetry is merely an asymmetry of the way in which microscopic initial and final conditions are distributed over the possibilities allowed by macroscopic conditions, how could the asymmetry be attributed to "time itself" as the ultimate explanatory factor?

Perhaps some of the intuitive plausibility of this argument comes from considering asymmetries of systems in the world that are grounded in "mere de facto initial conditions" and that are much less "grand" than the entropic asymmetry that is our primary concern. Consider, for example, the fact that in living creatures it is frequently only one of the two possible stereo-isomers (mirror image molecules) that plays an organic role. Why is it that the DNA of living creatures is overwhelmingly found to twist in one spiral direction and not in the other, equally possible, direction?

There have been, to be sure, attempts at deriving this asymmetry from the deeper underlying parity asymmetry of the weak interactions of elementary particles, although none of the attempts to so explain the biological asymmetry have been very convincing. Usually, the asymmetry is credited to the "accident" that in some initial life forms the one stereo-isomer and not the other happened to play the crucial biological role. Once started in this asymmetric way, the asymmetry was passed on to the manifold descendants of the initial life form by molecular reproduction. To be sure, this account also faces a number of puzzles, but let us assume, for the sake of argument, that it is the correct one.

Would, then, the existence of this molecular, biological asymmetry be grounds for attributing some asymmetry to space under mirror imaging? Surely not. It is a mere "accident" that the asymmetry appears with the universality it does in the life forms on the earth. We can fully account for this asymmetry of the systems without any recourse to something as dramatic as an underlying asymmetrical nature to space-time itself. That seems correct.

But does a similar argument do justice to the claim that, likewise, the entropic asymmetry of physical systems, being "merely de facto" in nature, should not be thought of as resting on any asymmetry of "time itself?" Alas, we are hard pressed to know what to say here, if for no other reason than the fact that we really have no sound account of the explanatory grounds we ought to take as established that would explain the physical asymmetry. But, as N. Krylov so forcefully argued, any attempt to credit the distribution of initial microscopic conditions to "mere happenstance" or "mere accident" lacks plausibility. Nothing as pervasive and as universal as the entropic asymmetry can be thought of in a similar vein to the sort of "accident" that might, indeed, ground the asymmetry of stereo-isomers in molecular biology. If it is not a "law of nature" that the initial conditions of similarly prepared systems are distributed as they are – that is, in just the appropriate manner to generate not only the Second Law asymmetry but the "lawlike" approach to equilibrium that macroscopic systems invariably show – it is not a "mere matter of fact" either. Indeed, the very distinction between what is forced by the laws of nature and what is a merely contingent initial state seems to dissolve in the light of statistical mechanics.

Perhaps a cosmological solution to the physical asymmetry of entropy increase would give us an explanatory picture more like that which accounts for the molecular biological asymmetry as a merely propagated initial "accident." But that doesn't seem too likely. Even the usual characterization of the cosmological picture – with the entropy of the space-time at the Big Bang very low to provide the reservoir of low entropy from which low-entropy branch systems are subsequently selected – doesn't' by itself account for the branch system asymmetry. And that very condition of the Big Bang is itself, in at least some accounts like that of Penrose, to be explained in terms of a fundamental lawlike asymmetry. In the light of all that we have seen so far, it seems a misrepresentation to think of the entropic asymmetry as "merely de facto" in any sense that would reduce it to some sense of pure contingency without a shred of lawlikeness to it. This is so whether that view is held by Mehlberg and Horwich, who would use it to deny that the physical asymmetry is something that ought to lead us to attribute an asymmetry to time itself, or by H. Reichenbach and A. Grünbaum, who think of it as constituting a sufficient basis to speak of "time's asymmetry."

Finally, we must return, once again, to the question of whether any really coherent sense can be given to the distinction between the two positions in the debate. There is an overwhelmingly pervasive and universal asymmetry in the distribution of initial conditions and final conditions of micro-states of systems characterized in similar ways from a macroscopic point of view, the initial distribution lacking the "uniformity" appropriate to a close to equilibrium state and the final distribution having more of that characteristic. If this asymmetry of distributions of microscopic conditions is that which in the world represents at the microscopic level the grand asymmetry of entropic increase, isn't that enough to say that by itself it *constitutes* the asymmetry of "time itself?"

When one reflects upon the difficulties encountered in trying to posit symmetries and asymmetries of space-time as some "independent fact" that "explains" dynamical symmetries and asymmetries, and sees how much plausibility there is to the claim that attribution of the symmetric or asymmetric structures to the space-time really amounts to little more than a convenient way of summing up the lawlike dynamic symmetries and asymmetries observed of systems in the world, it becomes not implausible to say that the admittedly still mysterious asymmetry of initial and final microscopic conditions that is the entropic asymmetry of systems itself is what the asymmetry in time of space-time comes down to.

III. What is the structure of the Boltzmann thesis?

1. The intuitive asymmetries

The issue just discussed – whether we are to think of the entropic asymmetry of phenomena in time as reflecting some underlying asymmetry of time itself – is not, of course, the crucial one for the Boltzmann thesis. His thesis requires, as its final step, the justification of the assertion that "what we mean by the future, as opposed to past, time direction, just is that direction of time in which the entropy change is an increase."

The argument begins by trying to itemize for us some of the features of the world that characterize the asymmetry of future and past. Just what makes us say that the future is very "unlike" the past? If we can get a little clearer on that, then we will be able to get a little clearer on just what it might be for the entropic asymmetry to underpin the more intuitive asymmetries by which we characterize the future-past distinction.

To begin with, we think that we have some direct, non-inferential, immediate awareness that the later-than relation of one event to another just isn't like the earlier-than relation. Getting straight on just what this might mean is not easy, but it certainly includes the idea that the later-than (or earlier-than) relationship is asymmetric. It is unlike, say, the relation two things bear to one another when they are "next to" one another, so that if a is next to b, it follows that b is next to a. If a is an event later than b, surely b is not later than a (peculiar universes with pathological global time topologies left to the side). And, we feel, we have some sort of "direct awareness" of two non-simultaneous events that we experience, an awareness that tells us which of the two events is later than the other. We need not, if the events are both given to us in our direct experience, make any inference from some other fact to determine which of the two events is the later, although of two events about which we have only indirect information such an inferential process might be necessary.

It is a little easier to come to grips with the intuitive asymmetry between past and future than is summarized in the very different kind of epistemic access we have to the events that occur in these two time domains. It is, of course, wrong to say that we can know what happened in the past but not what will happen in the future. We can have very certain knowledge, indeed, about what some future events will be, and we can be limited in a variety of ways from ever knowing, of some sorts of past events, which ones actually took place. Yet there is, clearly, an asymmetry of *method* by which we can know what events have taken place versus what events will take place. We have records of the past.

Vast amounts of information about what took place in the past are available to us from data about the events present in records of the event. Yet we have no records of future events yet to take place. It seems plausible to say that a record of an event is some present event causally related to the original event in such a way as to allow us to infer the occurrence of the original event in some especially simple or trivial way. There are also present events causally connected to future events that also allow us to infer that the future events will occur. But there is some difference in the relation of present to past events and the relation of present to future events that makes the special kind of correlation that generates a "record" of a past event unavailable for future events.

Indeed, there is that very special record of past events that we call memory. We can remember the past, yet we cannot remember the future. And, surely, this is not a mere formal triviality in our confining the word "memory" to our knowledge of past events and refusing to apply it to our inferential knowledge of what will occur. There is something about the past and its relation to the present that makes it possible for there to be something of the sort we call memory of the past, but that makes it impossible for us to have any similar epistemic access to the future. It is even a profound question as to whether or not memory is just one species of record akin to physical records of past events, although general physicalist considerations make us think that something like that must be the case, and we speculate about "memory traces" in the brain corresponding to memories in the mind. Why can we remember the past but never the future?

We have, as well, the intuitive idea that the direction of causal influence is always from past to future. Once again, it seems far too shallow to simply claim that of events lawlike connected to one another we choose, by stipulation, to always call the earlier event the cause and the later the effect. The idea that the former makes or determines the latter to occur, but not vice versa, is a pervasive one, even if it is hard for us to pin down why we hold to such intuitions with such firmness. Surely the asymmetry in time of the causing – the making or determining – relation among events is one of the crucial features about the world in time that, along with the epistemic asymmetries just noted, constitutes a ground for our intuition that future and past are radically unalike in a way that has no parallel in, say, some alleged asymmetry of space.

Human concerns are also radically asymmetric in time. We have a very different attitude toward past and future events. Past events may be regretted or remembered with nostalgic pleasure. But they cannot be planned for, anticipated, or worked for or against. The idea that the future, being a realm yet to come into being and causally effected by the present, is one for which we should take prudential steps, but that such activity toward what has already occurred is pointless, there being no use crying over spilt milk, is again one that pervades our everyday intuitive distinction between how we think and act vis-à-vis future and past.

Those metaphysically inclined find even deeper grounds for the distinctiveness of past and future. The past, some have asserted, is "fixed." Past events have occurred and propositions about what has occurred are either true or false. But the future is the realm of a reality that is at best "indeterminate." This is not a denial of determinism in the sense of past events lawlike necessitating what will follow, but a denial of any kind of "determinate reality" to future events at all. There have always been those who would deny reality to both past and future, reserving it for that which exists in the present. But another strand of intuition, at least as early as the thoughts of Aristotle, is that although the past can be said to be a reality, the future, or at least that part of it that is at present still contingent, cannot be said to have any determinate reality at all.

Exploring any one of these alleged asymmetries at length would be a massive project, and surely one that would have to be undertaken as a preliminary to an exploration of the question as to whether Boltzmann was right in claiming, as he would have to in order to establish his thesis, that the "ground" of each such intuitive asymmetry could be found in the pervasive entropic asymmetry of physical systems. Rather than pursuing this grand task, though, I will instead focus, first, on some rather broad preliminary issues that need to be faced by anyone who espouses the claim that entropy provides the basis for our intuitive distinction between the nature of future and past. That is, I will explore some general issues about just what such a claim of "grounding" the intuitive distinctions in the entropic asymmetry would come down to. After that, I will note only briefly a few important suggestions about how the tasks set out in this preliminary outline have been taken on in at least a beginning fashion by some who have thought about these matters.

2. What is the nature of the proposed entropic theory of the intuitive asymmetries?

In order to carry out the last stage of the Boltzmann argument, we would have to show that the intuitive asymmetries by which we distinguish the future from the past in their special natures are "grounded" in the entropic asymmetry of physical systems in time. But what does "grounded" mean here? The very nature of the alleged "reduction" is sometimes, I believe, misconstrued. Only when we have become a little clearer on the *intention* of the entropic thesis can we begin to explore whether it can be successfully defended.

Defenders of the entropic thesis sometimes begin with the proposal that one consider watching a film of a physical process. How can one tell if the film is being run in the proper direction – that is, with earlier projected images being the images of earlier events – and not in the improper, reversed direction that is, with the initial images being of the final events? Unless entropic features play a role, it is argued, one simply cannot tell. This argument assumes, of course, that the fundamental laws of nature are time-reversal invariant and that the film is not of some process among the elementary particles where time-reversal invariance doesn't hold.

But if there are entropic features of the processes involved, then one can easily tell if the film is being run in the proper direction. The film of the diver being projected out of the water and onto the diving board by the spontaneous confluence of small dispersed agitations of the water in the pool is one of those processes declared so improbable as to be ruled out of court by the usual Second Law considerations. It is argued, then, that entropic features are our guide to which events in a process come later than which other events. The higher entropy events can be inferred to be the later stages of the process. Is this an adequate argument for accepting the entropic thesis?

No, it isn't. It may very well be true that in trying to decide of a record of a process like a film which is the temporally beginning stage and which the temporally ending stage of the process recorded that we must rely on entropic considerations. But that doesn't mean that it is by noting the entropic features of states of the world itself that we decide which of those states are the earlier and which the later. It has been argued, justly I think, that we determine of events that are within our experience which are earlier and which later by something that can be called, if anything can be so-called, a totally non-inferential process of acquiring knowledge. Watching a ball rolling across the floor in a frictionless way provides us with no entropic clues as to which position of the ball is later and which is earlier. But we have no trouble in knowing which way the ball is rolling, and need not seek some subtly associated states that do have entropic characterizations and that are simultaneous with the states of the ball in order to determine by some inference which states of the ball were first and which later.

There is a kind of proposed reduction of B phenomena to A phenomena that works by trying to convince us that all our knowledge of B phenomena is summed up in an evidential basis that is totally characterizable in terms of A phenomena. It then uses a verificationist account of meaning to try and convince us that this evidential relationship between the phenomena ought to convince us that all assertions about B phenomena "reduce by meaning" to assertions about A phenomena. Thus, for example, the phenomenalist argues that all assertions about material things "reduce" to assertions about the immediate contents of awareness: But such an account of reduction will not do to convince us that the phenomena of temporal asymmetry reduce to entropic phenomena.

A much more persuasive case for the entropic account can be found in Boltzmann's original analogy of his entropic theory of the future-past distinction with the gravitational theory of up and down. We have an intuition that space, at least in our vicinity, is asymmetric in that there are distinguishable directions of up and down. The two directions in space are associated with a variety of asymmetric physical phenomena – rocks fall down, flames go upward, and so on. Indeed, some early theories of space, such as Aristotle's, proposed that this asymmetry was a fundamental constitutive structure of the space of the cosmos. And we can even tell which direction is down in a non-inferential way. Even locked in a darkened room we know, immediately and without inference, which way is down.

But now we understand that the up-down distinction is one grounded in gravitation. The local direction of the gradient of the gravitational potential is that which determines, at any point of space, the direction that is the downward direction. The theory explains why things behave as they do with respect to the up and down directions. It tells us why, for example, unsupported rocks accelerate in the downward direction. The theory even explains how it is that we can have our "direct and non-inferential" knowledge of which direction is the down direction, an explanation that adverts to the semicircular canals of our inner ears that serve as our orientational sense organs, and the behavior of the fluid in them. Of course, we don't infer from the state of the fluid in the canals to the downward direction. We don't even know, usually, that the canals exist. But it is the state of those sense organs determined by the local gravitational field that gives us our awareness of the direction that is downward.

Understanding all of this, we now realize that it would be quite parochial on our part to take the direction in space that is down for us and declare that direction to be the downward direction everywhere. It is much more reasonable to declare at each point "down" for that point to be the direction of the gravitational force at that point, admitting that in doing so "down" for someone else can be in a different direction than "down" is for us. What we mean by sameness of direction is determined by parallel transport of directions from one point to another. And, we then realize, one consequence of thinking of "down" in this way is that there may very well be places in the universe where there is no up-down distinction at all – that is, in those places where there is no direction of gravitational force.

So it is, Boltzmann alleges, with entropy and the intuitive future-past distinction. If the direction of time in which entropy increases is the direction of time that we label the future on the basis of those intuitive asymmetries we have discussed, and if the existence of the intuitive asymmetries is totally accounted for explanatorily by the existence of the entropic asymmetry, then it becomes just as plausible to say that the future direction of time is that direction of time in which entropy increases as it is to say that "down" just is the direction of space that is the direction of the local gravitational force. And there would be no more point to asking why entropy increases in the future time direction than there would be in asking why the gravitational force points in the downward direction. Surely it is this idea of entropic asymmetry as the explanation behind the intuitive asymmetries that is the intended force of the entropic account of the future-past distinction.

Just as the gravitational theory of up and down provides an explanatory account of how we can have non-inferential knowledge of which direction of space in our vicinity is the down direction, so we should expect from the entropic account of time-asymmetry some account of how it is that we can know non-inferentially of events which is later and which earlier, given that we know that there is some time interval between them. Just as the gravitational theory of up and down allows for there being places in the world at which down varies from its direction here, so the entropic theory would allow for at least the possibility of there being regions of the universe in which the future in that region is counterdirected to our direction of the future in time. And just as the gravitational theory allows for regions of space in which there is no up-down distinction at all, so the entropic theory would allow for regions of the cosmos in which, although there were still two directions of time, it would be incorrect to refer to either of the directions as the past or as the future. The counterdirected regions would be those, of course, that are far from equilibrium and in which the statistical preponderance of entropy increase of small systems was in the opposite time direction from the time direction in which systems have their entropy increase in our vicinity. And the regions of the universe at equilibrium, for Boltzmann in his speculation the overwhelming bulk of the cosmos in space and time, would be those with no past or future time direction at all in the proper sense.

Surely the entropic theory of the future-past distinction thought of in this way is at least initially coherent. But to explore its plausibility it might be helpful to first look at another distinction in the world, one that, unlike the up-down distinction and its association with the direction of gravitational force, provides a sort of "counter-analogy" to the Boltzmann thesis.

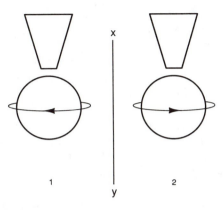

Figure 10-1. Weak interactions not invariant under spatial reflection. A nucleus with angular momentum decays by liberating electrons. There is a preferential direction in which the decay particles are emitted that is along the axis of rotation of the nucleus but in the direction of that axis from which the nucleus appears to spin in the clockwise direction. The mirror image of the physically possible situation seen in Fig. 1 is shown in Fig. 2. But this physical situation is in violation of the laws of decay with their preferential orientation.

The failure of invariance under mirror imaging in what are now taken to be the fundamental laws of particle interactions tells us that a system and its mirror image will not, in general, both be physically possible systems. If a system shows a certain spatially distributed decay pattern governed by the weak interactions, the mirror image of that system will be one showing a decay pattern spatially arranged in a way that is never observed to occur in nature. We could use this lawlike correlation between orientation and other, not intrinsically orientationally defined, processes, to explain, for example, to someone what we take a right-handed system or a clockwise rotation to be (as distinguished from a left-handed system or counterclockwise rotation) without even showing the person an example of the system of the kind in question. We could do this by having the other person construct a physically possible system and then viewing, say, its rotation from direction fixed by the emission of a beam of decay particles, explaining that the rotation so observed has the appropriate handedness. (See Figure 10-1.)

But surely a claim that handedness is somehow "reducible" to behavior under weak interactions is absurd. We certainly don't determine which of two gloves is left-handed by first examining weak interaction decay processes and then using them to fix appropriate orientations. But a theory that somehow or another the behavior of systems under weak interactions reduces the left-right distinction in the sense of telling us

what that distinction "is," as the theory of gravitational force tells us what
the up-down distinction was "all along," whether we knew it or not, and
as the entropic theory alleges the entropic asymmetry tells us what, "all
along," the future-past distinction was, whether we knew it or not, also
seems manifestly absurd. Granted that there are many puzzles about the
left-right or orientational distinction, some connected with the fact that a
local distinction between left- and right-handed systems may fail to be a
global distinction if the space of the world is non-orientable and that the
distinction is also connected with the dimension of the space into which
the objects are embedded, there just doesn't seem to be any plausibility
to the claim that somehow or other the theory of weak interactions will
explain what "underlies" the left-right distinction or explain how it was
that we were able to discriminate between a system in one orientation
and one in the other all along.

The theory that seems plausible is that the orientational distinction is
fundamental and irreducible, and that it is our perceptual awareness of
the world that gives us direct awareness into the orientational structure
of a perceptually presented object. Although it seems plausible to claim
that without the existence of gravity there would be no up-down distinction in the world, and possibly correct that without the entropic asymmetry there would be no future-past distinction of the standard kind, it
just seems absurd to say that had weak interactions turned out to be
invariant under mirror imaging there would be no distinction at all between a left-handed and a right-handed glove.

One thing reflection on this case shows us is that we ought to be
dubious of claims to the effect that if no other phenomenon were connected with a given relation, either by a law of nature or by something
quasi-lawlike, like the connection between entropic increase and time
order, the given relation would have to be taken as "unreal." There
would be nothing "unreal" about the distinction between a left- and
right-handed object even if there were no foundational laws of physics,
nor quasi-laws, that were non-invariant under mirror imaging. Similarly,
claims to the effect that were there no such associated phenomenon, a
world with a given structure of the relation in question and one that had
all the relations reversed by some symmetry operation would be "one
and the same possible world" cannot be glibly accepted. If the relation
in question is one that is itself fundamental and irreducible, it can be
quite misleading to make such a claim. Even if all the fundamental laws
were invariant under mirror imaging, there is no immediate reason to
assume that a claim to the effect that the world and its entire mirror
image would be the same possible world would be correct. Even claims
that one could not tell things related in one way from things related in
the reverse way were there no such lawlike or quasi-lawlike associated

phenomenon seem misleading. If a relation is available to us by perception, say, as the distinction between a left- and a right-oriented object seems to be, we could indeed distinguish the one from the other without there being any associated phenomenon, such as a differential decay pattern, to provide us the means for the distinction.

Nor is the existence of such a lawlike or quasi-lawlike association by itself in any way sufficient to make a grounding or reduction claim. What we seem to have in this situation is the existence of a distinction between two kinds of spatial systems, left- and right-handed ones, and the discovered existence of a lawlike correlation between some other phenomena (weak interaction patterns) with handedness in the world. But the existence of this correlation, even if it has a full lawlike status, in no way tells us what handedness "is." Even if the laws of particle interactions had been completely symmetric under mirror imaging, the distinction between the two kinds of handedness would remain the same.

The situation is quite different in the up-and-down case. Here we have discovered that the reason the direction we call down in our vicinity has the interesting distinguishing features we have long known it to have is because that direction at our location happens to be the direction of the gravitational force at that point. Had the gravitational force been in some other direction, that other direction would have been the one that was the down direction. And had there been no gravitational force at all in our vicinity, there would have been no up-down distinction selecting out two directions (the "up" and the "down") as distinguished at all.

Plainly, it is this model that the entropic theorist of the future-past distinction is proposing. At any event location, there are two directions of time. We could arbitrarily label them, say "plus" and "minus." Under the circumstances of the world being time-orientable, we could extend this labeling in a consistent way so that there would be a global "plus" time direction and a global "minus" time direction. But, the entropic theorist claims, all of the interesting features that characterize one of these time directions as future at an event location are features existing at that location solely because there is, in that region of space-time, an entropic asymmetry in the behavior of physical systems. Just as in the gravitational case, such an account will allow for one of the directions of time (where global time orientation is possible) to be the future time direction at one event location but the past time direction at some other. And at those event locations where no entropic asymmetry exists, it will allow for the possibility of neither time direction counting as future or past.

The question is, of course, whether entropy is really connected to the intuitive past-future distinction in the same way that gravity is connected to the intuitive up-down distinction, or whether, instead, the ground of the future-past distinction is to be found elsewhere, and the correlation of entropic increase with the future direction of time is to be viewed as a contingent fact about the world. If the entropic theorist is correct, it is not a contingent fact that entropy increases into the future, for the very notion of what the future *is* is given by the stipulation that it is the direction of time in which entropy increases. But is the entropic theorist correct?

Before looking at a few arguments that try to make the entropic account plausible, it will be helpful to notice some other features of the proposed account that make it somewhat different in its detailed nature from the gravitational account of the up-down distinction. Although increase of entropy is taken to determine which direction of time is the future direction, it is certainly not claimed that for any isolated system the higher-entropy end state of that system will be later than the lower-entropy end state. For many isolated systems – say, a single particle acted upon by a force in a frictionless and otherwise non-dissipative environment – there will be no entropies distinguishing states of the system. In other cases, we will expect the final end state of even an isolated system to have lower entropy than its initial end state, because we know that the increase of entropy into the future is only a statistical probability and not an inevitability.

Consider, similarly, the role played by the notion of a "common cause" in arguments about the source of our intuitive notion of the direction of time. We know that there will be processes in which a group of events at one time will display a higher probabilistic correlation of occurrence than we would expect given their independent probabilities. That is, if A and B are two such events, the joint probability of A and B will be higher than the product of their independent probabilities. Such a correlation is often explainable by the existence of a single common past cause of both events. Thus, a pebble tossed into a pond will result in a correlated spatially dispersed wave in the future. We have noted the close connection between such "spreading out of consequences of a common cause" with the thermodynamic principle of entropy increase when we discussed Einstein's claim that the asymmetry of radiation (correlated outbound radiation from an accelerated charge, but no converging correlated radiation onto a charge that accelerates the charge) was not, as Ritz proposed, the source of entropic asymmetry or independent of it, but a consequence of it.

Whereas the events A and B are not be probabilistically independent, if we conditionalize all probabilities on the common cause event, C, probabilistic independence will be restored. That is, while

$$P(A \ \& \ B) \neq P(A) \times P(B), \ P(A \ \& \ B/C) = P(A/C) \times P(B/C)$$

Event C, the common cause, is said to "screen off" the correlation of A with B. Associated with the view that earlier stages of a process are the lower entropy stages is the view that the direction of time is determined by the fact that such screening off common causes is earlier in time than the pairs of events whose correlations they screen off by their serving as the common cause of the events otherwise correlated.

Yet, as B. van Fraassen clearly has pointed out, we can find in such statistical theories as quantum mechanics cases where a common cause, in the intuitive sense, fails to screen off the probabilistic correlation between the two caused events. Other cases can be constructed where a screening off common cause can be found subsequent to the corre-lated events. Once again, no simple, uniform association of temporal direction with the associated probabilistic relation that is supposed to "ground" the temporal asymmetry can be constructed.

One needs to take a somewhat subtler view of the relationship be-tween the future-past distinction and the alleged grounding phenom-enon. The basic strategy to employ was made clear by H. Reichenbach. Independently of the problem of characterizing the ground of temporal asymmetry, we have the time ordering of events that allows us to say of two pairs of events, A and B and C and D, that D is the same time direction from C as B is from A. That is, we have a notion of time par-allelism. It is not our concern here how that temporal relation is grounded nor how we can gain epistemic access to it. Given this parallelism, if we can establish which direction of time is the future direction for some pairs of events, utilizing some asymmetric relation between them – for example, one event being a record or memory of the other or one event being the effect of the other as cause, where the relation is in the cases in question founded upon some existing entropic asymmetry applicable to the cases in question – we can then "project" that asymmetry onto all other cases of pairs of events by using the parallelism of direction in time.

We might, then, claim of a pair of events that one is the cause of the other as effect, even when no entropic consideration characterizes the events at all, such as in the case of the initial and final position and velocity of a point particle. Here, we could simply decree that we take the earlier of two such lawlike connected events to be the cause and the later the effect, in this case building temporal priority into the very meaning of "cause" and "effect." The claim is, though, that the deeper reason for so building temporal order into our causal notion is that in the entropic cases, some other reason exists for distinguishing the cause event from the effect event, some reason grounded in entropic considerations. The claim then continues that it is the parallelism in time of the entropic order that tells us that we can think of one direction of time (the past) as the direction in which to look for causes and the other direction of time (the future) as the one in which to seek effects. Once an asymmetry of future and past has been established for these special cases, where entropic considerations play a role, we then project the notion of asymmetric causation outward by time parallelism so that even events symmetrically related to one another by a lawlike connection and for which no entropic features can be found to characterize directly an asymmetric relation between them can still be said to be asymmetrically related by cause and effect. The asymmetry of the causal relation is now grounded in the mere fact that one event is earlier in our asymmetric time ordering than the other.

Similarly, it may be argued that there is something special about those cases where events in the direction of time we think of as the past do screen-off probabilistic correlations of pairs of events future to them and lawlike connected to them. This, it can be argued, gives us the ground of a notion of an asymmetric causal connection and an ensuing notion of an asymmetric past-future asymmetry. If in other cases we find that the past event fails to screen off the correlated events, or that some event future to the correlated pair does the screening-off, we can still call the earlier event the cause and the later the effect, now relying upon a general notion of causation as having an asymmetry dependent on our already established temporal asymmetry of past and future. This is so even if that asymmetry is itself grounded at a deeper level on a notion of causal asymmetry itself grounded on some probabilistic asymmetry related to entropic dissipation of order into the future.

It can be seen, then, that an entropic theory of the asymmetry of future and past is neither an absurdity nor a hypothesis that can be rejected out of hand by some immediate and short argument. On the other hand, if the theory is to be believed it will require a good deal of argument to back it up. As we have seen, the account will not be a simple one. Basic to it will be some argument that parallels the arguments we give in the case of the gravitational theory of up and down. That is, we need some arguments to the effect that all of the intuitive features of the world that we ordinarily take as distinguishing past and future are features whose origin and nature can be shown to arise out of the entropic asymmetry of systems in time.

3. Sketches of some entropic accounts

Although the appeal of the Boltzmannian account of our intuitive time-asymmetries is clear, there really has not been very much done in the way of trying to "fill out" the account with a plausible, detailed, demonstration that all of the intuitive asymmetries we have noted can be grounded – explanatorily accounted for – in the entropic asymmetries we have studied in the previous chapters.

To begin with, there is the issue of which intuitive asymmetry ought to be thought of as "fundamental," or, indeed, if one of them should be so privileged. One might try first to account for the asymmetry of knowledge by means of the entropic asymmetry, trying to explain why we can remember the past and not the future and why we have records of the past and not the future. One might then argue that our intuitive idea that causes precede their effects in time is a reflection of this original asymmetry of knowledge. The idea might be, perhaps, that we are often in a situation of knowing what has gone before and wondering what will come next. Our desire to infer the future from the past knowable by memory and records would then be argued to underlie our idea that the direction of inference, and hence of explanation, goes from the past to future. The explaining events, the events from which we infer – that is, causes – would then be taken to precede in time the events inferred – that is, explained, that is, the effects. Again a plausible case might be made out to explain how our ideas of past as having determinate reality and of the future as being a "mere realm of possibility" are to be accounted for by the fact that we have definite knowledge of the past through memory and record and only, at best, a more explicitly inferential knowledge of what the future will be like.

Alternatively, one might try to get at the causal asymmetry directly, arguing that our intuition of causes preceding effects can be accounted for in terms of an analysis of what we have it in mind for the causal relation to be and how facts about entropic asymmetry then fit in with this analysis of causation. This analysis would be used to explain why it is that anything we take to be related to something else as cause is to effect must display the asymmetry in time parallel to that of the entropic increase of systems. Once having established the entropic basis of the cause-effect asymmetry, one might try to account for the others by arguing, for example, that our very meaning of the record or trace of some happening is some other feature of the world for which the original happening was causally responsible, thereby building on the initial cause-effect asymmetry in time a derivative asymmetry of record, trace, or even memory. We will look, although only very sketchily, at approaches of both these kinds.

One important attempt at grounding the epistemological asymmetry on the entropic is that of H. Reichenbach, an approach looked upon with favor by A. Grünbaum and J. Smart among others. Here we are to consider the famous "footprint in the sand." Walking along a beach we see a footprint, and we immediately take this to be the record, or trace, of a person having walked along the beach in the past. Why? It is, says Reichenbach, a matter of the footprint being a "low entropy" state of the beach system.

Now we must be careful here, for, as Reichenbach acknowledges, it isn't entropy in the sense of statistical mechanics, determined by the micro-distribution of the basic molecular constituents of the system in the usual phase-space way, that we must have in mind. Here, "entropy" is being used in an analogous way and is referring to what Reichenbach calls "macro-entropy." It is the distribution of the sand grains themselves that determines the appropriate macro-entropy in this situation. The idea is that the uniform, disordered arrangement of sand as we should find it more or less smoothly arrayed on an untrodden beach is a high macro-entropy state. But the ordered state of the sand with the footlike pattern in it, analogous to the gas on the left-hand side of the box only, is to be taken as a low macro-entropy state.

It is then argued that isolated systems left to themselves will usually evolve into, and remain in, a high macro-entropy state. If we find such a system in a state of low macro-entropy, what must we infer? By analogy with the behavior of the micro-entropy of systems, we can take it as highly improbable that such a state, like the footprint, is one of those rare spontaneous fluctuations of the system that are possible but too improbable to realistically countenance. Rather we ought to infer that with high probability, the surprising low macro-entropy is the result of the past causal interaction of the system (the beach) with some external system, an interaction that generated the transient, low macro-entropy. This is just like the account given of low micro-entropy branch systems that can be causally split off from the low-entropy main system in the Reichenbachian account of thermodynamic parallelism. It is, these theorists claim, such temporary low macro-entropy states of the systems that are indicators of their past, but not of their future. They are the result of causal interaction with external systems that make them records or traces of that past causal interaction.

But, as J. Earman has noted, there is an obvious objection to this account. What we take as records or traces are often items of unexpectedly high macro-entropy. We expect to find the beach uniform and treat this non-uniformity of the footprint as an anomaly to be explained by referring to a past causal interaction. But in cases where we expect the order and coherence corresponding to low macro-entropy to be the norm, it may very well be a high macro-entropy, disordered state that requires our explanatory skills. Walking in the supermarket it is the chaos of cat food cans all over the floor that makes us wonder about what past causal interaction with an expected, orderly stack of cans resulted in the mess.

A more plausible account of what a record or trace is would then look for states of the world, whatever their macro-entropy might be, that would be otherwise unexpected unless some quite specific past causal interaction could be adduced as their source. There is certainly no doubt that we make such inferences and that we make them legitimately in some cases. Nor can there be much doubt that it is inferences of this kind that play a crucial role in some of the things we call records or traces of the past. When I read of yesterday's Dow Jones average in the paper today, I take it that this record provides me with reliable information about an event in the past. And I take it so because I believe that an appropriate chain of causation does connect the average's being calculated with the numbers appearing in the paper that I read that reveal what that average was.

That this account is true of memory is less clear. But it does seem at least plausible to maintain that memories provide a record of the past because an event now – my seeming to remember a fact about the past – is something that would not have occurred had not an event in the past been relevantly connected to me in such a way as to causally generate the current seeming memory. And that it is therefore legitimate to infer from what I seem to remember to the event actually having occurred (even if, of course, the inferences can sometimes go wrong).

But, of course, we infer as well from present events that would otherwise be unlikely to events in the future that given their occurrence, make the current event more probable as well. Seeing the Scud on the radar screen, the operator infers that an explosion will occur. The appearance of the image on the screen is, everything else being equal, quite an unlikely event. But given the nature of the setup, it might be very likely, indeed conditional on the future event taking place. Now, of course, we don't take the radar image to be a record or trace of the future event. But, it is plausible to argue, that is because we think of records as being the causal result of what they are a record, and we think of causes as going from past to future. The trouble is that we don't want to presuppose any of that when we are trying to ground the very notion of causal asymmetry on that of the epistemological asymmetry, and we are trying to ground that latter only on some appropriate relation among events itself grounded on the entropic asymmetry.

The natural way to proceed at this point, it would seem, would be to first explore in some depth the nature of the probabilistic relations among events that form the basis of our inferences from events we know to have occurred to those we wish to infer, be they events in the past or future, or indeed in the present. Here, all the issues familiar to those who worry about such matters concerning the role of one event in making another more probable or highly probable relative to background material would have to be discussed in detail. And one would then have to go on to explain what it is about the probabilistic relationship among events that makes some events traces or records of events in the past, and never of events in the future. And we would have to be clear how it is that these probabilistic relations take their source from those probabilistic relations among micro-events summed up in the statistical mechanical principles underlying the entropic asymmetry of thermodynamics.

Such an account remains to be given, but let me suggest one factor that will indicate just how difficult it will be to provide such an account that will be satisfying. We know from our earlier explorations in this book that probabilistic assertions are always quite sensitive to the way events are categorized and classified. Principles of a priori probability, for example, such as the Principle of Indifference, are empty without something being provided to provide the right or allowable "partitions" of the event space. I think similar issues are bound to surface here. For any probabilistic relationship among the macro-events that one prescribes as being the essence of one event being a record of another, earlier, event, it seems likely that with sufficient ingenuity another way of classifying events into sorts will be found that will generate the same probabilistic relationship between the record event and some event in its future. I think that this will be true no matter how many facts about the probabilistic relationships among the usual classes of events at the micro-level generated out of the usual statistical mechanical entropic asymmetries are brought into the picture.

The problem here is that we have not yet specified any limits to how events are to be sorted, classified, or partitioned at the macro-level. Yet this must be done, for it doesn't seem plausible that the entropic asymmetry in its pure "micro" form will provide the clarification that we want of that relationship between record and that recorded that directly explains the asymmetry of knowledge. At least in the case of classifying events for statistical mechanical purposes we did have some grounds, however slender, for thinking of some categorizations as privileged – in particular, those that used the standard measures over the appropriate phase spaces. But that resource won't be available to us at the macro-level. Perhaps some other will.

Presumably there is some macro-fact about the world today that is correlated in the right way with tomorrow's Dow Jones average, so that I could, were I to know this present macro-fact, infer tomorrow's average. Would that I knew what this macro-fact is! But such a macro-fact will, presumably, be sufficiently "unnatural" that unlike the "natural" fact of the appropriate numbers appearing in the appropriate place on the financial pages of a newspaper, knowing that it does, in principle, exist will not give us any way of accessing it. "Natural" facts, facts somehow accessible to our comprehension or grasp, may be probabilistically related to one another in a manner appropriate for our being able to have records of the past but not of the future. And it may very well be that this fact is itself grounded in the standard entropic asymmetry of systems in nature. After all, it is a quite grand fact about the world that we do have memories and records of past and not future. And what other time-asymmetric fact about the world other than the great entropic asymmetry would be big enough to do the job? But it remains an unfinished task to fill in the appropriate details to make such a story of a possible reduction of the knowledge asymmetry to the entropic asymmetry a convincing tale.

Another important suggestion for grounding an intuitive asymmetry on an asymmetry that might at least be thought to be connected to the entropic asymmetry focuses on the intuitive asymmetry in time of the causal relation. If we can show that our sense that causation always proceeds from past to future is founded on the entropic asymmetry, we might hope to ground the other intuitive temporal asymmetries upon the causal asymmetry as basic. That, for example, we feel that action can influence the future but not the past might follow from the fact that actions influence by means of causation. One might even try to ground the epistemological asymmetry on the causal. Here one would argue that traces or records, or even memories, are, perhaps by definition, the effects of the events of which they are the traces, records, or memories. It certainly seems plausible to claim at least that a record of an event could be a record of that event only if that event was somehow causally instrumental in forming the record, whether this is definitional of what it is for the record to be a record of that event or not. There is, therefore, at least plausibility to the claim that if we can show the causal relation temporally asymmetric and ground that asymmetry on the entropic asymmetry, we would have gone some way to explaining why we can have records of the past and not of the future as well.

One very clear and very suggestive attack on this problem is due to D. Lewis. Lewis wants to characterize the causal relationship by the use of counterfactual conditionals, thinking of the cause of an event as an event such that had it not occurred, the given effect event would not have occurred. Such an analysis requires tidying up to avoid counterexamples from a variety of different directions, including issues of preemptive causes and of non-causal dependencies that still ground appropriate counter-factuals. But the simple version will do for our purposes. When we think of what would have happened had some events in the world been other than as they actually are, we generally think that the world future to those events would have been otherwise than it was when they actually occurred. But, Lewis insists, even if we sometimes initially think that had an event been otherwise, the past relative to that event would have been otherwise as well, on reflection we usually drop our confidence in such "backtracking" counterfactual conditionals. If this is so, then, we can think of the temporal asymmetry of causation as itself being grounded on our intuition that had things been otherwise at a time, the future to that time would have been other than it actually was. Had the causal event not occurred, its future effect events would not have either. But the absence of that actual event would not have changed what happened in its past, so that causal influence would be propagated only one way, into the future.

But why do we think of future-directed counterfactuals as legitimate and of back-tracking counterfactuals as illegitimate? At this point, the familiar "macro-entropic" facts are noted about how localized distinctive events lead, in the future, to correlated macro-events at separations from one another, without there being any such parallel association of a single event with a set of correlated separate events in the past. The wave spreading out from the stone tossed into the pond or the electromagnetic wave spreading out from the oscillated electron are, once again, the familiar examples.

Now it does seem plausible to argue that had the stone not been tossed in the pond, any one of the local pieces of a circular wave propagated out from it would not have occurred. Yet we are more reluctant to say that had not one of those bits of the ripple appeared, the stone would not have been tossed. Why? Well one answer might be that given the remaining disturbances caused by that initial pebble-tossing event, their existence, even if one little bit of the wave is thought out of existence, is still good reason to believe that the stone tossing occurred in their past. One single event at a time, the stone tossing, gives rise to a multitude of later events correlated with one another and whose correlation with one another is, as previously noted, "screened off" probabilistically by the occurrence of the single past event.

Now, the argument goes, think of how we evaluate counterfactual conditionals. We think of a world like ours, except that in this new "possible world" the event taken not to have occurred in our counterfactual is thought away. What is such a possible world like in other respects? Here, following a suggestion originally made by R. Stalnaker and developed richly by Lewis, we can think of possible worlds as "closer" or "farther" from one another. Think of the world in which a given event, A, which actually does occur, has not occurred. What is that world like? To decide this we should ask ourselves what the possible world is like that of all possible worlds, is one in which A does not occur yet is closest to our actual world.

Of course, there are lots of technicalities involved in these notions. For example, is it clear that one and only one distinctive possible world will, in general, be closest? Again, if our aim is to decide which counterfactuals are true, our possible worlds picture isn't yet much help. We need to know which possible worlds are closer and which farther from our actual world, and that is, of course, just knowing which counterfactuals are true and which false. But, Lewis suggests, we can think of some of the criteria we do in fact use in thinking about possible world nearness. We generally think that worlds with "big, widespread, diverse" violations of the laws of nature are very far from the actual world. Next, he argues, our intuitive evaluations of the truth of counterfactuals leads us to maximize the spatio-temporal region throughout which the other possible world and ours match exactly in matters of particular fact. Less stringently, we then seek to avoid even local, small, violations of the laws of nature in our world. Finally, he argues, it is of little or no importance that we secure approximate agreement with regard to matters of particular fact even if these facts are of great importance to us. As Lewis demonstrates, such rules for evaluating closeness of possible worlds will lead, for example, to the conclusion that in a world in which the President has a button connected to a Doomsday machine, the counterfactual we take to be true comes out true if we imagine him pushing the button. Doomsday occurs rather than, say, his pushing the button and the signal down the wire violating the laws of nature on the way to the machine or any of the other possibilities occurring.

The net result of these conditions is such that only a small miracle is needed to allow an event that did happen not to have happened, keeping the past of that event constant. But a gigantic one would be needed to allow that event to have happened keeping the future exactly the same. Why? Because events are usually connected in the past to only a "small" determining event (certain neurons firing in the President's brain, say, that determined his decision to press or not to press the button). But events leave an enormous multitude of traces or records of their occurrence into the future. For A to have not occurred when it actually did would require a vast array of miracles to break the connection of A with all of the multitudinous other future events that were, as a matter of fact, its causal consequents. Thus, only a little would have to change between past and future for me not to have tossed the stone. But, the stone being tossed, there are all the bits of spatially dispersed ripples from its landing in the pond into the future, plus all of the light waves emitted from them into the universe, and so on. So, here, the familiar story that a bit of spatially compact low macro-entropy at one time is associated with a vast set of correlated but dispersed low macro-entropies in the future, but not in the past, is being used, along with our intuitions about the closeness of possible worlds and hence about the truth conditions for counterfactuals, to get us time-asymmetry. First, there will be a time-asymmetry of counterfactual dependence, and then, from the characterization of causation in terms of counterfactual dependence, an asymmetry of cause.

Of course, there are many questions to be asked. One is whether this diagnosis of our intuitions for evaluating the truth of counterfactuals really is the source of our felt time-asymmetries about causation. As usual, the theory is that such macro-entropic spreading situations give us our basic intuitions about the time-asymmetry of counterfactual and causal dependence, and that this time-asymmetry is then projected onto the cases where no such spreading of effect is present. Once again, there is the important issue of the criteria by which events are grouped into classes at the macro-level. If we are perverse enough we can reassort the events of the world to generate, even in the stone-in-the-pond cases that are paradigmatic for Lewis' argument, macro-spreading into the past and not into the future. So, once again, much insight into why some ways of thinking of the world as composed of macro-events are natural and others are not is needed here. Finally, as Lewis himself asserts, "I regret that I do not know how to connect the several asymmetries I have discussed with the famous asymmetry of entropy."

Well, suggestions can be made. As we noted in Chapters 2 and 8, Einstein and others have made serious efforts to connect the outbound wave condition for electromagnetic waves, allowing them to diverge in a correlated way from an accelerated charge but never to be found converging on it in perfect correlation, with more purely thermodynamic/statistical mechanical assumptions about emitters and absorbers in the universe. It is certainly quite conceivable that with sufficient insight we could construct an argument that starts with the standard temporal asymmetry of temporarily isolated systems in a time parallel process of entropic increase in one time direction and not the other, and end with a plausible generation of the conditions in the natural world that ground our intuition of causation going from past to future or of records being only of the past and not the future. Perhaps we could even explain the one-way efficacy in time of the process we call memory. It remains to be seen if this is so. This final stage of the Boltzmann thesis is neither proven nor disproven at present.

4. Our inner awareness of time order

Let us suppose that we can in fact provide the filling-in of the Boltzmann thesis that would convince us that indeed the ground of all of our intuitive asymmetries with respect to time was the entropic increase of temporarily isolated systems in the world. Let us suppose, that is, that we can explain, in the manner characterized in the last section, why it is that we have records and memories in one, parallel, time direction and not the other, and why we intuitively take it that causation goes from the direction of time remembered to that anticipated. Suppose also that the other intuitions about the openness of the future or its not having determinate reality can also be explained away, directly or indirectly, by the entropic asymmetry. Were that so, how would it be best to describe the relationship between the "after" relation and the relation one state of a system bears to another when the former state is in the time direction from the latter in which entropy, in general, increases? Let "*aLb*" mean "*a* is later than *b*." Let "*aEb*" mean "*a* is in the time direction from *b* in which higher entropy states of isolated systems are, generally, from the lower entropy states of those systems." What is the relation of the *L*-relation to the *E*-relation?

One suggestion that has at least initial plausibility is that there is only one relation involved here, a single relation that can properly be called the *L*-relation or can properly be called the *E*-relation. That is, that the *L*-relation is *identical* to the *E*-relation. This idea derives by analogy with the role substance identification plays in science, as described in Chapter 9,I,2, in the context of theory reduction. We discover that salt crystals simply *are* arrays of sodium and chlorine ions. Such micro-reductions are empirical discoveries. Yet they are discoveries that certain identities are truths, that each salt crystal is the same thing as a specifiable array of ions of a particular kind. Some empirically discovered substantive iden-tifications are not *micro*-reductive in nature. There is the important discovery, for example, that light waves are nothing but electromagnetic waves whose frequencies are within a certain specified range. It isn't that light waves are associated with or correlated to the presence of electromagnetic waves (as smoke might be to fire, say), even in a lawlike way. It is that the light waves just *are* certain kinds of electromagnetic waves.

Property identification can be thought of analogously. Once again we can think of the case of "down" and the "local direction of the gradient of the gravitational field." It might be argued that we can understand the relation of the one feature of the world (which direction is down) with the other (which direction is the direction in which the gravitational potential is deepening) as just the relation of identity. Being the "down" direction *is* just being the direction in which the gravitational gradient is pointing and "downness" *is* just the "gradient of the gravitational potential pointingness." Of course, all of this has now become "localized" as noted before, the direction of the gravitational gradient varying from point to point (and even at some points vanishing) and carrying with it a variation in the direction that is the downward direction.

Recent philosophers, following the work of S. Kripke, have insisted that a careful distinction must be drawn between an identification's being an empirical discovery, being a posteriori in Kant's sense, and the iden-tification's modal status. It is plausible to suggest that a genuine identity statement like that asserting the identity of light waves with electromagnetic waves is, if true, necessarily true. If a light wave is an electromagnetic wave, it is argued, it could not be the case that there could be a light wave that was not an electromagnetic wave.

Now none of this is meant to deny that there might be things in the world that behaved like light waves in their detectable aspects – stimulated retinas, reflection off metals, and so on – but were not electromagnetic waves. But, it is argued, the semantics of our substance terms is such that the right thing to say in such cases is that these would not be light waves at all, because they are not electromagnetic waves, and that is what light waves are. Instead, they are things other than light waves that behaved as light waves do. In some other possible worlds, there might be a substance that filled oceans, that was used to make lemonade along with lemon juice and sugar, and so on, but that was not H_2O. But, it is argued, that it wouldn't be water (no matter how water-like), because "water" is the word we use to refer to that substance that has the appropriate identifying features in this world, and in this world that stuff is H_2O. Nor, of course, is any claim being made that denies the fact that it is an empirical discovery on our part that light waves are, in fact, electromagnetic waves and that water is H_2O. From this perspective, if it is true that "down" is the local direction of the gradient of the gravitational field, then it is necessarily the case that that is what "down" is.

In this context, how plausible does the claim look that "after" is just "in the time direction in which entropy generally increases," construed as a thesis about the identification of one relation with another? Put this way, the thesis has some curious difficulties, difficulties seen some time ago by the physicist-philosopher A. Eddington. Interestingly, these difficulties parallel those made of an important identificatory claim made by philosophers of mind. Once again, as in our investigation of the reductive relation between thermodynamics and statistical mechanics, issues from the philosophy of the mind and the philosophy of statistical mechanics turn out to be mutually illuminating.

An important part of the process of substance identification as it appears in science is the "secondarizing" of properties. "Look," it might be argued, "how can we think of light waves as being identical to electromagnetic waves? Light waves have *color*, but there is no place in our physical theory for "blueness," say, to be a property of an electromagnetic wave." A frequent response is to deny that "blueness," properly so-called, really is a property of the physical object, the light wave, at all. Indeed, it is often, since Boyle and Locke, denied to be a property of any physical object. Instead, it is a quality "in the mind," a secondary quality. In physical matter, we may of course find some physical property that is in some way correlated with these mental sensations. For example, the wavelength of light impinging on the retina is closely associated with the sensed color qualia. But the color itself is not genuinely predictable of any physical object. We have seen how such a stripping away of "secondary qualities" from the physical system is instrumental in the identificatory account of the reduction of thermodynamics to statistical mechanics.

But suppose the entities or properties to be identified with others are themselves "mentalistic?" Such a possibility arises, for example, in those philosophical theories of the relation of mind to brain that argue that mental processes – for example, having a pain – can be simply *identified* with appropriate physical processes in the brain. In this case there is no place to which to remove the "secondary quality," so that if we are to take "feeling a pain" to be some process in the brain, we must wonder if we are willing to attribute to the brain itself (or its process?) the "felt quality" of the pain.

The implausibility of doing this has led many, in different ways, to be skeptical that the identificatory reduction of mind to brain can be carried out, at least where such mental processes as the experience of qualia (visual sensations, for example) or feelings (pains, for example) are to be accounted for. One version of such skepticism, also due to Kripke, in-vokes modal intuitions in its formulation. If pain really is a stimulation of neural fibers, then this is a necessary truth. But it seems to us that we can imagine pain without neural stimulation or neural stimulation without pain, and the possibility of such an imagining is usually taken as good evidence that such a connection is not a necessary one but only one that is contingent.

Now the mental-physical identity theorist can reply that we might think that we can imagine water that is not H_2O, but we would be wrong. Water being H_2O, it is necessarily H_2O, and our vaunted imagination misleads us as to the apparent contingency between being water and having the appropriate molecular structure. But, Kripke argues, this isn't a plausible way out for the identity theorist of mind and brain. In the water/H_2O case, we can explain why we are deceived into thinking that we can imagine water that is not H_2O because we can imagine something other than water having those "accidental" or "non-essential" properties by which we originally identified stuff as water (such as filling oceans, being drinkable, and so on). What we can really imagine is something other than water having those features. But our imagination of pain without neural stimulation or neural stimulation without pain isn't of that sort. Our "direct epistemic access" to "what pain is like" tells us what the essential features of pain are and what they are like. And knowing what we know of these essential features, we see that an explanation of our sense that the pain/neural stimulation connection is only contingent cannot be "explained away" as an error, as in the water H_2O case. We can imagine that something feels to us like heat but isn't molecular motion. This explains why, although heat is molecular motion and therefore necessarily molecular motion, we mistakenly think we can imagine heat that is not molecular motion. But we can imagine pain that is not accompanied by neural stimulation, and there is no "accidental" feature associated with pain that is the feature by which we recognize it as pain, because feeling the way it does is essential to being pain. The conclusion of this argument (one that is challenged, to be sure, and one that requires much further thought) is that where the identity of the mental with the physical is concerned, our imagination that tells us that pain could not be associated with a specific neural process is enough to tell us, as Descartes believed, that feeling pain is *not* identical with that neural process. However, it might be the case that without the appropriate neural process going on no pain would be felt.

Why is all of that relevant to issues concerning the possible identity of "afterwardness" and "being in the direction in time of higher entropy?" To see this, one needs to consider some arguments of A. Eddington, formulated in his Gifford Lectures of 1927 and published as Chapter V of his *The Nature of the Physical World* in 1929. In this chapter, Eddington is at pains to emphasize to his audience the fact that temporal asymmetry reveals itself in the physical world in the form of the entropic asymmetry and in no other guise. But he is also at pains to indicate his grave doubts that one could identify temporal asymmetry "as we experience it" with entropic asymmetry, say in the manner one identifies light waves as electromagnetic waves.

The crucial issues here have to do with deep questions about time and its nature that go far beyond the issues of the grounding of temporal asymmetry. Eddington holds to a view of science in which we are presented with items of conscious awareness and construct our picture of the physical world in such a way as to have an explanatory account of our phenomenal experience. But our direct knowledge is limited, he thinks, in the usual empiricist way, to the contents of direct awareness. At least that is so for most aspects of the world. Light, he claims, is just a "something" identifiable by its ultimate phenomenal causal effects (hitting our retinas it causes us to have visual flashes). Electromagnetic waves are similarly "unknown" in their inner nature, identifiable only by the role they play in a causal structure that has phenomenal consequences. There is nothing whatever incoherent in our discovery that the "somethings" identified through one such causal chain are the same things as the "somethings" identified through a quite distinctively characterized causal process.

But when we know what something is like – indeed, what its "essential" features are – we may be blocked from any such identificatory claim with some other kind of entity by knowing something about the essential features of that other sort of entity and knowing that two such radically different kinds of things could not be identical with one another. At least, this is the Eddington version of an argument quite closely related to that used by Kripke to defend a kind of Cartesian argument against the mind/brain identity thesis.

Now we might want to argue that virtually all of the features we ascribe to the physical world are of the kind known only indirectly to us through their causal consequences in our world of phenomenal experience. Certainly we will want to say this is true of such things as electric charge or strangeness or electromagnetic field potential. Someone like Eddington would argue that this is also true of those features of objects that we normally invoke to characterize even the familiar middle-sized things of ordinary life. Even the *spatial* aspects of things, he would say, are only indirectly known. Even shape, size, and other geometric features, as meant to characterize the realm of the physical, are only "inferred features" that are, in Eddingtonian terms, "only symbolically known." There is, of course, a problem at this stage, for we also usually feel that our immediate phenomenal experience has a spatial aspect. Don't even the components of our visual awareness have shape and size? Aren't they spatially related to one another in position?

Here, the Eddingtonian would argue that there has been an equivocation on the notion of "spatial feature." He would argue that one must not confuse spatiality as predicted of the physical world with the perceptual spatiality of immediate experience. Although the latter may be a guide to postulating the nature of the former, the two are not to be identified. After all, how our visual experience goes is evidence to us concerning the nature of physical space, but when one considers what physics now says about the nature of physical space (say, in the general relativistic context), it is argued, one immediately realizes that the "space" of objects and the "space" of perception must not be claimed to be the same thing, any more than "blueness" of visual experience should be thought of as anything like the real physical properties of the light that stimulates those blue sensations (such as its wavelength or electric field intensity).

Such a theory of the relation of perceptual experience to the world is one fraught with the gravest epistemological difficulties. But its appeal remains in that it seems to give us a framework for talking about perception and reality with which the various alternatives suggested to take its place have never adequately dealt. Suppose this account is plausible? Can we then go on to argue for a radical disidentification of *all* features of the realm of the perceived from all features of the physical realm?

Eddington thinks not. The problem is time. Our experience of time, he believes, provides "a true mental insight into the physical condition that determines it." Time in the physical world cannot be, he thinks, a mere "something," some relation I-know-not-what, identifiable by us only "symbolically" by its place in the explanatory structure we posit to account for our directly experienced awareness. If we are to have any understanding of the world at all, he seems to think, we must take it to be the case that physical events are temporal, that they participate, as he puts it, in "becoming," in just the same sense as we directly experience temporality in the temporal relations our inner experiences have to one another.

Then comes the argument that we know what "later-than-ness" is like, directly and non-inferentially. We know it because the fact of one event's being later than another is an immediate part of our direct experience of our inner mental life, the temporal part of that experience. And knowing what this is like, we know that it is *not* a relation like that of one state of a system being more disordered than some other state and separated from that other state in time. So we know that whatever the relation between temporal asymmetry and entropic asymmetry may be, it is not plausible to *identify* the "later-than" relation with the "has greater entropy" relation.

Perhaps it would be worth ending this discussion with an extended quote from Eddington, one that despite its obscurity, is rich with suggestiveness as to just why the issues connecting time itself with related physical features of the world in time, like asymmetric entropy increase, remain so puzzling on a philosophical level as well as on a physical level:

In any attempt to bridge the domains of experience belonging to the spiritual and physical sides of our nature, Time occupies the key position. I have already referred to its dual entry into our consciousness – through the sense organs which relate it to the other entities of the physical world, and directly through a kind of private door into the mind . . . While the physicist would generally say that the matter of this familiar table is *really* a curvature of space, and its color is really electromagnetic wavelength, I do not think he would say that the familiar moving on of time is *really* an entropy-gradient . . . Our trouble is that we have to associate two things, both of which we more or less understand, and, so as we understand them, they are utterly different. It is absurd to pretend that we are ignorant of the nature of organization in the external world in the same way that we are ignorant of the nature of potential. It is absurd to pretend that we have no justifiable conception of "becoming" in the external world. That dynamic quality – that significance that makes a development from the future to past farcical – has to do much more than pull a trigger of a nerve. It is so welded into our consciousness that a moving on of time is a condition of consciousness. We have direct insight into "becoming" which sweeps aside all symbolic knowledge as on an inferior plane. If I grasp the notion of existence because I myself become. It is the innermost Ego of all which *is* and *becomes*.

The amount of presupposed philosophy that is tendentious in all of this is great, and the issues far too deep to explore here. What we can say, though, is this. That although the notion of property identification might serve very well to let us say what we want to about the relationship between down and the direction of the gravitational gradient, it will be a problematic reading of the relation between temporal asymmetry and entropic asymmetry even in the best of all circumstances – that is, even if the Boltzmann thesis is made plausible to the degree that we become convinced that all of the intuitive temporal asymmetries can be explained on the basis of the entropic asymmetry of systems in the world. On the other hand, given the role memory plays in our immediate awareness of what it is for one event to be after another, it isn't clear that one could not frame the Boltzmann claim as an identity thesis even given Eddingtonian scruples about the inability to lift off perceived time from the actual physical time of the world.

IV. Further readings

Boltzmann's proposal that the direction of entropic increase defines the direction of the future can be found in Brush (1966), Chapter 10.

For introductions to the notion of symmetry of laws and its relation to symmetry of space-time, see Sklar (1974), Chapter V, Section B, and Earman (1989), Section 3.4 and 3.5.

That "de facto" asymmetry is enough for asymmetry of time is defended by Grünbaum in Grünbaum, (1973), Part II, Chapter 8. The argument that only lawlike asymmetry would be sufficient for space-time asymmetry is given in Mehlberg (1980) and in Horwich (1987), Chapter 3.

For an outline of the intuitive time asymmetries, see Sklar (1974), Chapter V, Section A, Part I.

For the origins of a detailed attempt to generate the intuitive asymmetries out of a temporal asymmetry of macro-entropy increase, see Reichenbach (1956), Part IV. See also Grünbaum (1973), Part II, Chapter 8, and Horwich (1987), Chapters 5, 6, and 8. See also Sklar (1984), Chapter 12.

On "the principle of the common cause" and "screening off," see Reichenbach (1956), Part IV, Chapter 19. See also Salmon (1984), Chapter 8, and Horwich (1987), Chapter 4, Section 8. For quantum mechanical example of the violation of the screening-off principle, see van Fraassen (1982).

Criticisms of the entropic account of the origin of the intuitive asymmetries can be found in Mehlberg (1980), Volume 1, Part 1, Chapter V, and Earman (1974). See also Sklar (1974), Chapter V, Section F.

For the time asymmetry of counterfactual conditionals, see Lewis (1986), Volume II, Chapter 17.

On the issue of our "direct awareness" of temporal asymmetry, see Eddington (1929), Chapter 5, "Becoming." See also Sklar (1984), Chapter 12. For the related argument concerning mind and brain, see Kripke (1972), pp. 144–155.

11

◁========================▷

The current state of major questions

At this point, it might be useful to take a retrospective look at some of the major questions posed throughout this book, asking ourselves to what extent the questions have been answered and to what extent important puzzling issues still remain to be resolved. The reader will not be surprised by now, I expect, to discover that it is the author's view that many of the most important questions still remain unanswered in very fundamental and important ways.

A reasonable understanding of the role played by probabilistic assertions in statistical mechanics requires, I believe, some version of an interpretation of probability that views it as frequency or proportion in the physical world. Although "subjectivist" or "logicist" interpretations certainly are suggested by the role played by symmetry principles and principles of indifference in generating the posited probabilities, and although, as we have seen, the probabilities can sometimes be generated out of the lawlike structure of the underlying dynamics alone, still an understanding of how these probabilities are then used to describe the evolution of systems and to explain that behavior requires that they be understood in the manner of proportions.

But, as we have also seen, this understanding of probability is fraught with subtle difficulties. First, there are the problems of idealization. In equilibrium theory, for example, the proportions of time spent in various micro-states that are actually derivable from the theory are limiting proportions as time goes on to infinity. Only with care can such ideals be associated with actual proportions over finite times when we come to explain the real behavior of systems. There are also those serious problems that lead us to puzzle over just what reference class is the appropriate one to pick when we are trying to say just what actual proportion a theoretically derived probability is supposed to describe. Beginning with the early confusions over such matters as ensemble averages and time averages, and in such puzzles as those Boltzmann confronted when he tried to reconcile equilibrium as overwhelmingly probable with the prevalent non-equilibrium nature of the world by means of cosmological speculation, there has been a continuing necessity in understanding the theory to carefully explore at each stage just how the "probabilities" derived by some theoretical schematism or other are to be appropriately associated with actual and experienced proportion in the world.

As we have seen, one of the most crucial problems in the foundations of statistical mechanics is the justification or rationalization or explanation of why the probabilities interjected into the theory are the correct probabilities to use. Work on this problem has clearly met with its greatest success in the equilibrium portion of the theory. Here, the results of Ergodic Theory do go quite a way toward showing us, on the basis of the constitution of the system and the laws governing its micro-dynamics, that the standard probability measures have a distinctly privileged status over alternative probabilistic choices. But even here, as we have seen, important problems remain. Thus, for example, despite resort to topological arguments, the neglect of sets of measure zero remains an interesting problem for the theory, a problem not resolvable by resort to structure and micro-dynamics alone. Again, the fact that most realistic systems pretty clearly do not meet the conditions for a system to be genuinely ergodic leaves the current theory rationalizing the standard probability choices in a shaky state, even given the successes of ergodic theory.

The rationalization of posited probabilities is certainly much less clear in the non-equilibrium case. Here, as we have seen, at least some progress has been made in eliminating the need for the most dubious probabilistic posits of non-equilibrium statistical mechanics. These are one or another of the descendants of the Hypothesis of Molecular Chaos. They get us the equation of evolution we want, but only at the cost of imposing on the deterministic evolution of the ensemble a continual rerandomization that is impossible to justify, given the fact that the evolution is deterministically fixed by the initial ensemble distribution.

But, as we have clearly seen, a remaining probabilistic posit over the initial conditions of the system in question – an initial ensemble distribution – is essential to get the non-equilibrium results we want. These include the appropriate finite relaxation times and the appropriate kinetic behavior in the approach to equilibrium. Whether we model approach to equilibrium by one of the mixing approaches or by the alternative Lanford approach, such an initial probabilistic assumption is essential to the account. And, as we have seen, although much imagination has gone into concocting proposals that would explain why the familiar probabilistic posits are justified, and why the symmetric posits that would generate anti-thermodynamic behavior are not, no convincing rationale yet exists for why the standard probability assumptions over initial conditions work as well as they do.

When we ask how much we understand about the nature of statistical explanation in statistical mechanics, it is once again in the equilibrium case that matters are most clear. Here, as we have seen, Ergodic Theory, combined with elements that rely upon the thermodynamic limit to connect average values with overwhelmingly most probable values, gives us a framework in which many questions can be answered. It is important to remember, however, that the notion of statistical explanation ultimately arrived at in this case is rather different from what the standard philosophical accounts of statistical explanation would have led us to expect. In the equilibrium case, it is neither subsumption of events under statistical generalities nor probabilistic-causal explanations that we derive. Instead, it is the "transcendental deductions" of the equilibrium distribution that we described in Chapter 5 that are the most unproblematic result of equilibrium theory. By this I mean such results as the demonstration that the standard probability measure over a phase space is the unique time-invariant measure absolutely continuous with the standard measure. We have seen how such results can be applied to try to ground arguments of the sort, "If equilibrium exists and has a certain statistical characterization, then here is what it must be like." As we noted, such arguments certainly don't give us an account of why systems go to equilibrium nor why they get there by the route they actually follow.

It is also important to remember how many puzzles still remain even in the equilibrium case and even when the modest notion of statistical explanation is the one had in mind. The measure zero problem and the fact that realistic systems are not ergodic remain problems, for example.

When we consider what we understand about statistical explanation in the non-equilibrium situation, the picture is much less clear. I have argued that some approaches to accounting for non-equilibrium behavior, for a variety of reasons, seem less plausible than the more orthodox accounts. Thus, I have argued, neither Jaynes' subjectivism, nor interventionism, nor the positing of time-asymmetric basic dynamical laws are plausible routes to follow in seeking an account of the approach to equilibrium.

But, as we have seen, there is not a single "orthodox" approach to the problem but, instead, a variety of different approaches. We have also seen that in some cases, as in the opposition between the mixing approach and that using the Boltzmann–Grad limit to construct a rigorous derivation of the Boltzmann equation, there are deep conceptual irreconcilabilities between the approaches taken. Although both models rely on initial probability assumptions over the micro-conditions of a system and then follow out the evolution of the initial ensemble by using the micro-dynamical laws, the models offered of approach to equilibrium are quite different in the two cases. Mixing follows out the ideas implicit in the Ehrenfests' critique of Boltzmann and in Gibbs' coarse-graining, identifying the approach to equilibrium with the path followed by most probable states of systems as the ensemble evolves. The Lanford approach, on the other hand, derives the kinetic equation as the description of "most probable" evolution, even if this claim is incompatible with recurrence.

It is very important to note how the two approaches rely on distinct "idealizations" of real physical systems. Mixing and its variants rely, for example, on such things as hard-sphere models for molecules, and often resort to infinite time limits. The Lanford approach, on the other hand, relies upon utilization, crucial utilization, of the Boltzmann–Grad limit. The role played by the many degrees of freedom of the system is also, as we have noted, quite different in the two approaches. One can learn from this. Although idealization in physics is often quite innocuous and quite "tractable," in statistical mechanics, and perhaps in other areas of fundamental physics, idealization plays a subtler role. Whereas in simple cases, which idealization is the right one to pick is clear, here it is not. Where in simple cases what errors will be introduced when one goes to the idealization may be calculable, here the way an idealization may differ from a realistic system may not be clear at all. Because all the approaches to non-equilibrium theory that are viable rely crucially upon such idealizations, their success in answering our original "why?' questions always remains in doubt.

Within any of the orthodox approaches, however, the pattern of statistical explanation is that of probabilistic causal explanation familiar to philosophers. The idea is to argue that given certain initial probabilities over initial conditions, the probabilities of evolutions of various kinds then follow from the assumed deterministic laws. Of course, what we can say about those trajectories of interest to us varies from one model to the other. But all will suffer the problem of accounting for the initial probability distribution. As we have seen, this is the most central and most intractable problem for the non-equilibrium theory. Although suggestions abound (facts about preparation of ensembles, "mere de facto" proportions in initial states, the separation of the system from the cosmological main system, singular initial states, among others) it seems quite clear that there is no single account of why initial states of systems are distributed as they are that is generally taken to be satisfactory. Of all the conceptual puzzles about statistical mechanics, this seems the most profound.

So great is the disagreement in this area that the opponents will often not even agree about the appropriate ontology to presuppose. Although most of the orthodox adhere to exact micro-states of systems, there are, as we have seen, radicals who will seek the solution to the initial probability problem in a denial of the very existence of an exact micro-state. Along with this, they will generally try to trade in the initial probability as a proportion over exact micro-states for some kind of initial probability of a dispositionalist sort for the new statistical states that have been substituted for the rejected exact micro-states. But, as we have seen, such proposals need a long way to go before we even understand them well enough to judge their plausibility.

The most notorious unsolved mystery of statistical mechanics is, of course, the problem of irreversibility or the time-asymmetry of entropic increase. From the orthodox perspective, the problem of irreversibility is just a sub-problem of the problem of picking the initial probability distribution. It must be emphasized that the bigger problem, picking the right distribution and rationalizing it, would still exist even if the asymmetry problem did not. But, naturally, the problem of asymmetry does serve marvelously to focus the mind on just how difficult the broader problem is. For now we must find a rationale for the initial probability distribution that does not fall prey to one of the parity-of-reasoning arguments that informs us that anti-thermodynamic behavior is just as inferable as is thermodynamic behavior on the basis of the rationale offered. As I have emphasized throughout, I don't see any resolution of that problem at hand.

It is to resolve the problem of temporal asymmetry that cosmology has been introduced into the picture, or at least that is the primary aim of introducing it. We have also noted how cosmology has been invoked since Boltzmann's time in attempts to reconcile the apparent conflict between the conclusion of statistical mechanics that equilibrium ought to be pervasive with the clear pervasiveness in the real world of systems very far from equilibrium. As we have seen, cosmology cannot be counted fully successful even in accomplishing this latter task. Nor, as we have also seen, does the invocation of cosmological facts, now in the form of one version or another of a Big Bang cosmology, fully resolve the issue of temporal asymmetry.

Most of the attention in the literature on this subject has been devoted to the quite fascinating problem of trying to account for the temporally asymmetric increase of entropy of the universe taken as a whole. As we have seen, there is no clear solution to that problem that has been universally accepted. Indeed, it isn't completely clear just what kind of explanatory account is being asked for, or is legitimately asked for, in this context. I have tried to focus attention, however, on another issue, one that I believe has not received its due in the previous discussions of cosmology and irreversibility. I have argued that it is very far from clear how one can derive the kind of temporal asymmetry one desires from the cosmological entropic asymmetry – that is to say, the statistically parallel increase of entropy in the same time direction for those systems temporarily isolated from the remainder of the universe from the cosmic entropic asymmetry. I have argued that the existing arguments attempting to derive this branch system asymmetry from the temporal asymmetry of cosmic entropy increase are all fatally flawed. I have also argued that this remains true even when the derivation, like that of Reichenbach, is based on its own additional statistical posits, at least so long as those posits are themselves time symmetric in form. I have argued that Reichenbach has put the rabbit of time-asymmetry into the hat by his formulation of his so-called Lattice of Mixtures, and that it is in this trick, not in the time direction of entropic increase in the cosmos, that the time-asymmetry derived in his proof of parallelism for branch systems is to be found. I think that such hidden asymmetric positing goes on in the other derivations as well. Although cosmology can provide us with a vast reservoir of low entropy, making low-entropic systems more likely to be pieces cut off from the main system than spontaneous fluctuations from isolated high-entropy states, the direction of entropy change of the cosmos cannot by itself, I have argued, provide the full reason for parallelism in entropic increase of the branch systems.

We have seen that there are those who would argue that in certain deep senses thermodynamics is an irreducible discipline. Perhaps this is better put by saying that they would argue that what thermodynamics and its underlying statistical mechanics shows us is that one cannot hope to understand these theories so long as one remains committed to the standard ontology of the traditional micro-theory. I am thinking here of those who would argue that the results on instability of trajectories that we discover in our search for the grounds for the approach to equilibrium show us that the very notion of an exact micro-state for the system, thought of in the traditional mechanical vein, is a false idealization. But, I have argued, this is too radical a conclusion to draw from the results. As far as things look now, nothing in statistical mechanics forces us to abandon the appropriate exact micro-states posited by the underlying dynamical theory, be it classical or quantum in nature. The best ontological picture that seems satisfactory overall is the orthodox one, in which the exact micro-states genuinely exist, instability and all.

But even from this orthodox perspective, the issue of reduction is one that remains complex and subtle. From the orthodox standpoint, we need not seek any radical revision in our basic ontology for physical systems. But we do discover that both the reduced thermodynamic theory and the reducing statistical mechanical theory are, in their conceptual structure, quite unlike the more usual theories of physics. And for this reason, the reductive relationship in question is more complex than the usual reductive relationship grounded on identificatory reduction at its core. The interrelationships that can hold between subtle concepts like entropy at the reduced level and the statistical concepts of the reducing

lead, as we have seen, to a multiplication of concepts and a very illuminating sorting out of the variety of notions implicit in the original naive
versions of the theories.

It is also important to remember the one clear sense in which it does
seem so far that thermodynamics remains an "autonomous" theory. Even
if reducible to statistical mechanics, that latter theory seems, as far as we
can tell, to require the fundamental posits about the proportions of initial
conditions in the world that are essential to derive its basic non-equilibrium results. It remains very hard to see how such posits can somehow
be eliminated or made innocuous. To this extent, thermodynamics and
statistical mechanics do require the positing of fundamental facts about
the world whose origin seems to remain outside the provenance of the
underlying theories of dynamics or the underlying constitutive facts about
the systems to which statistical mechanics and thermodynamics are
applied.

Finally, we come to that aspect of the discussion that has seemed of
most interest to philosophers – the question of whether our intuitive
notion of the asymmetry of time can itself be founded upon the entropic
asymmetry. I have argued that some aspects of the discussion – for
example, the debate over whether entropic asymmetry constitutes an
asymmetry of time or merely an asymmetry in time – may not be real
issues at all. But the question of whether an explanatory account of the
origin of all of our intuitive temporally asymmetric concepts can be found
in the facts about the asymmetry of entropic increase is a real one. It is
that question that must be answered in order to judge the correctness of
the final Boltzmann thesis. I have argued that although that final thesis
is intelligible and far from rejectable as absurd on its face, it is also far
from being established.

The difficulties encountered in trying to ground all of our intuitive
asymmetries of time on the entropic asymmetry have led many to argue
that the "grounding" is actually the other way around. There is, they
claim, a basic asymmetry to time itself, and it is this asymmetry that
accounts both for our intuitive asymmetries of knowledge, causation,
and concern, and for the entropic asymmetry as well. I have not explored these claims in this book. Suffice it to say at this point that although some imaginative "metaphysical" characterizations of time have
been offered that try to make sense of the past having determinate reality
and the future being merely a realm of possibility, we are far from having
any account that would be accepted by most as even intelligible. Even
if we had one, it is very far from clear how the intuitive asymmetries
would really be accounted for on the basis of the metaphysical model.
And, I think, it is quite implausible to hope that any such view of the
primitive asymmetry of time in some metaphysical sense would really

give us the needed explanation of the entropic asymmetry. That, I'm afraid, would still remain a mystery that needed solving on its own terms.

Anyone who has followed the debate from the days of Maxwell and Boltzmann to the present cannot but help be struck by the way in which the fundamental problems of the theory – the problems posed by its original discoverers and by the brilliant early critics – have remained as deep puzzles for over a century. Attempts at solving the profound quandaries at the foundations of statistical mechanics have led to some of the most innovative conceptual developments in physics. Furthermore, whole rich branches of mathematics, such as Ergodic Theory in all its present very general glory, have been inspired by the need to find the right language and the right basic postulates to deal with the fundamental issues that arise when probabilistic reasoning is applied to the dynamics of systems. Yet despite the richness of the resources that have been developed, and despite the immense clarification of the issues that has been obtained, the most basic questions of the explanatory accounts to be offered for the fundamental probabilistic posits of the theory and for the appearance of statistical temporal asymmetry in the world remain.

It is impossible to resist ending with a quote from an important worker in the field, J. Lebowitz:

I do not know if anyone is making bets on the eventual resolution of the apparent paradoxes relating to the coexistence, in the description of the same phenomena, of both determinism and randomness, reversibility and time asymmetry, and so on. If there are people betting, however, I would be very happy to be the banker and keep the money until everyone has agreed on the matter.

References

Material demanding familiarity with mathematics or physics at the intermediate level is marked (*). Material demanding comprehension of more advanced mathematics or physics is marked (**).

Arnold, V. and Avez, A. 1968. *Ergodic Problems of Classical Mechanics.* New York: Benjamin.(**)

Baker, G. and Gollub, J. 1990. *Chaotic Dynamics.* Cambridge: Cambridge University Press.

Balescu, R. 1975. *Equilibrium and Nonequilibrium Statistical Mechanics.* New York: Wiley.

Barbour, J. 1989. *Absolute or Relative Motion,* Volume 1, *The Discovery of Dynamics.* Cambridge: Cambridge University Press.

Barrow, J. and Tipler, F. 1986. *The Anthropic Cosmological Principle.* Oxford: Oxford University Press.

Batterman, R. 1990. "Irreversibility and Statistical Mechanics: A New Approach?" *Philosophy of Science* 57: 395–419.

Batterman, R. 1991. "Randomness and Probability in Dynamical Theories: On the Proposals of the Prigogine School." *Philosophy of Science* 58: 241–263.

Berry, M. 1978. "Regular and Irregular Motion." In S. Jorna, ed., *Topics in Nonlinear Dynamics.* New York: American Institute of Physics.(**)

Blatt, J. 1959. "An Alternative Approach to the Ergodic Problem." *Progress in Theoretical Physics* 22: 745–755.

Block, N., ed., 1980. *Readings in Philosophy and Psychology.* Cambridge, MA: Harvard University Press.

Bogolyubov, N. 1962. "Problems of Dynamical Theory in Statistical Physics." In J. deBoer and G. Uhlenbeck, eds., *Studies in Statistical Mechanics.* Amsterdam: North-Holland.(**)

Bondi, H. 1961. *Cosmology.* Cambridge: Cambridge University Press.

Brener, R. and Hahn, E. 1984. "Atomic Memory." *Scientific American* 251: 50–57.

Brush, S., ed., 1965. *Kinetic Theory.* Oxford: Pergamon Press.

Brush, S. 1976. *The Kind of Motion That We Call Heat.* Amsterdam: North-Holland.

Brush, S. 1983. *Statistical Physics and the Atomic Theory of Matter.* Princeton: Princeton University Press.

Buchdahl, H. 1966. *The Concepts of Classical Thermodynamics.* Cambridge: Cambridge University Press.(*)

Cardwell, D. 1971. *From Watt to Clausius*. Ithaca, NY: Cornell University Press.

Carnap, R. 1950. *Logical Foundations of Probability*. Chicago: University of Chicago Press.

Cornfeld, I., Fomin, S., and Sinai, Ya. 1982. *Ergodic Theory*. New York: Springer Verlag.(**)

Courbage, M. 1983. "Intrinsic Irreversibility of Kolmogorov Dynamical Systems." *Physica A* 122: 459–482.(**)

Cox, R. 1961. *The Algebra of Probable Inference*. Baltimore: Johns Hopkins Press.

Cramér, H. 1955. *The Elements of Probability Theory and Some of its Applications*. New York: Wiley.

Davies, P. 1974. *The Physics of Time Asymmetry*. Berkeley: University of California Press.

de Groot, S. and Mazur, P. 1984. *Nonequilibrium Thermodynamics*. New York: Dover.

Denbigh, K. and Denbigh, J. 1985. *Entropy in Relation to Incomplete Knowledge*. Cambridge: Cambridge University Press.

Devaney, R. 1986. *An Introduction to Chaotic Dynamical Systems*. Menlo Park, CA: Benjamin/Cummings.

Dias, P. and Shimony, A. 1981. "A Critique of Jaynes' Maximum Entropy Principle." *Advances in Applied Mathematics* 2: 172–211.

Earman, J. 1974. "An Attempt to Add a Little Direction to 'The Problem of the Direction of Time'." *Philosophy of Science* 41: 15–47.

Earman, J. 1986. *A Primer of Determinism*. Dordrecht: Reidel.

Earman, J. 1989. *World Enough and Space-Time*. Cambridge, MA: MIT Press.

Eddington, A. 1929. *The Nature of the Physical World*. Cambridge: Cambridge University Press.

Eels, E. 1991. *Probabilistic Causality*. Cambridge: Cambridge University Press.

Ehrenfest, P. and Ehrenfest, T. 1959. *The Conceptual Foundations of the Statistical Approach in Mechanics*. Ithaca, NY: Cornell University Press.(*)

Ehrlich, P. 1982. "Negative, Infinite and Hotter than Infinite Temperatures." *Synthese* 50: 233–277.

Farquhar, I. 1964. *Ergodic Theory in Statistical Mechanics*. New York: Wiley.(*)

Feller, W. 1950. *An Introduction to Probability Theory and its Applications*. New York: Wiley.(*)

Feyerabend, P. 1962. "Explanation, Reduction and Empiricism." In H. Feigl and G. Maxwell, eds., *Minnesota Studies in the Philosophy of Science*, Volume III. Minneapolis: University of Minnesota Press.

Fine, T. 1973. *Theories of Probability*. New York: Academic Press.(**)

Friedman, K. 1976. "A Partial Vindication of Ergodic Theory." *Philosophy of Science* 43: 151–162.

Friedman, K. and Shimony, A. 1971. "Jaynes' Maximum Entropy Prescription and Probability Theory." *Journal of Statistical Physics* 3: 381–384.

Gibbs, J. 1960. *Elementary Principles in Statistical Mechanics*. New York: Dover.(*)

Gleick, J. 1987. *Chaos*. New York: Penguin Books.

Gold, T. and Schumacher, D., eds. 1967. *The Nature of Time*. Ithaca, NY: Cornell University Press.

Goldstein, S., Misra, B. and Courbage, M. 1981. "On Intrinsic Randomness of Physical Systems." *Journal of Statistical Physics* 25: 111–126.

Goldstein, S. and Penrose, O. 1981. "A Nonequilibrium Entropy for Dynamical Systems." *Journal of Statistical Physics.* 24: 325–343.(**)

Grad, H. 1961. "The Many Faces of Entropy." *Communications in Pure and Applied Mathematics.* 14: 323–354.(*)

Grünbaum, A. 1973. *Philosophical Problems of Space and Time*, 2nd. ed. Dordrecht: Reidel.

Guttman, Y. 1991. *The Foundations of Probability in the Context of Statistical Mechanics.* Ph.D. dissertation. Columbia University.

Hahn, E. 1953. "Free Nuclear Induction." *Physics Today* 6 number 11: 4–9.

Hempel, C. and Oppenheim, P. 1948. "Studies in the Logic of Explanation." *Philosophy of Science* 15: 135–175.

Hempel, C. "Aspects of Scientific Explanation." In his *Aspects of Scientific Explanation and Other Essays.* New York: Free Press.

Hobson, A. 1971. *Concepts in Statistical Mechanics.* New York: Gordon and Breach.

Hollinger, H. and Zenzen, M. 1985. *The Nature of Irreversibility.* Dordrecht: Reidel.

Hooker, C. 1981. "Towards a General Theory of Reduction." *Dialogue* 20: 38–60, 201–235, 496–529.

Horwich, P. 1987. *Asymmetries in Time.* Cambridge, MA: MIT Press.

Hull, D. 1974. *Philosophy of Biological Science.* Englewood Cliffs, NJ: Prentice-Hall.

Hume, D. 1888. *A Treatise on Human Nature.* Oxford: Oxford University Press.

Hume, D. 1955. *An Inquiry Concerning Human Understanding.* New York: The Library of Liberal Arts.

Humphreys, P. 1989. *The Chances of Explanation.* Princeton: Princeton University Press.

Jancel, R. 1969. *Foundations of Classical and Quantum Statistical Mechanics.* Oxford: Pergamon Press.(*)

Jaynes, E. 1983. *Papers on Probability, Statistics and Statistical Physics*, Dordrecht: Reidel.

Jeffreys, H. 1931. *Scientific Inference.* Cambridge: Cambridge University Press.

Jeffreys, H. 1967. *Theory of Probability.* Oxford: Oxford University Press.

Johnson, W. 1964. *Logic*, volume 1. New York: Dover.

Katz, A. 1967. *Principles of Statistical Mechanics.* San Francisco: Freeman.(*)

Kemeny, J. and Oppenheim, P. 1956. "On Reduction." *Philosophical Studies* 7: 10–19.

Khinchin, A. 1949. *Mathematical Foundations of Statistical Mechanics.* New York: Dover.(**)

Kim, J. 1982. "Psychophysical Supervenience." *Philosophical Studies* 41: 51–70.

Kim, J. 1984. "Concepts of Supervenience." *Phil. and Phenomenological Research* 65: 153–176.

Kim, J. 1985. "Supervenience, Determination and Reduction," *Journal of Philosophy* 82: 616–618.

Kolmogorov, A. 1950. *Foundations of the Theory of Probability.* New York: Chelsea.(**)

Kreuzer, H. 1981. *Nonequilibrium Thermodynamics and its Statistical Founda-tions*. Oxford: Oxford University Press.(*)

Kripke, S. 1972. *Naming and Necessity*. Cambridge, MA: Harvard University Press.

Krylov, N. 1979. *Works on the Foundations of Statistical Physics*. Princeton: Princeton University Press.(*)

Kubo, R., Toda, M. and Hasitsume, R. 1978. *Statistical Physics-II*. Berlin: Springer-Verlag.(**)

Kuhn, T. 1970. *The Structure of Scientific Revolutions*. Chicago: University of Chicago Press.

Kyburg, H. 1970. *Probability and Inductive Logic*. London: Macmillan.

Kyburg, H. 1974. "Propensities and Probabilities." *British Journal for the Philo-sophy of Science* 25: 358–375.

Kyburg, H. and Smokler, H. 1964. *Studies in Subjective Probability*. New York: Wiley.

Landau, L. and Lifschitz, E. 1979. *Statistical Physics*. London: Pergamon.(*)

Lanford, O. 1983. "On a Derivation of the Boltzmann Equation." In J. Lebowitz and E. Montroll, eds., *Nonequilibrium Phenomena I: The Boltzmann Equa-tion*. Amsterdam: North-Holland.(**)

Lavis, D. 1977. "The Role of Statistical Mechanics in Classical Physics." *British Journal for the Philosophy of Science* 28: 255–279.

Lebowitz, J. 1983. "Microscopic Dynamics and Macroscopic Laws." In C. Horton, L. Reichl and V. Szebehely, eds., *Long-Time Prediction in Dynamics*. New York: Wiley.(**)

Lebowitz, J. and Penrose, O. 1973. "Modern Ergodic Theory." *Physics Today* 26: number 2: 23–29.

Leff, H. and Rex, A. 1990. *Maxwell's Demon: Entropy Information and Comput-ing*. Princeton: Princeton University Press.

Levi, I. 1967. *Gambling With Truth*. New York: Knopf.

Levi, I. 1980. *The Enterprise of Knowledge*. Cambridge, MA: MIT Press.

Lewis, D. 1986, *Philosophical Papers*, Volume II. Oxford: Oxford University Press.

Lichtenberg, A. and Lieberman, M. 1983. *Regular and Stochastic Motion*. New York: Springer-Verlag.(**)

Linde, A. 1984. "The Inflationary Universe." *Reports on the Progress of Physics* 47: 925–986.(*)

Malament, D. and Zabell, S. 1980. "Why Gibbs Phase Averages Work – The Role of Ergodic Theory." *Philosophy of Science* 47: 339–349.

Mayer, J. 1961. "Approach to Thermodynamic Equilibrium." *Journal of Chemical Physics* 34: 1207–1220.(*)

Mehlberg, H. 1980. *Time, Causality and the Quantum Theory*. Dordrecht: Reidel.

Mellor, D. 1971. *The Matter of Chance*. Cambridge: Cambridge University Press.

Misra, B. 1978. "Nonequilibrium Entropy, Lyapunov Variables and Ergodic Prop-erties of Classical Systems." *Proceedings of the National Academy of Sciences USA* 75: 1627–1731.(**)

Misra, B. and Prigogine, I. 1983a. "Time, Probability and Dynamics." In C. Horton, L. Reichel, and V. Szebehely, eds., *Long Time Prediction in Dynamics*. New York: Wiley.(**)

Misra, B. and Prigogine, I. 1983b. "Irreversibility and Nonlocality." *Letters in Mathematical Physics* 7: 421–429.

Misra, B., Prigogine, I. and Courbage, M. 1979. "From Deterministic Dynamics to Probabilistic Descriptions." *Physica A* 98: 1–26.(**)

Moser, J. 1973. *Stable and Random Motions in Dynamical Systems.* Princeton: Princeton University Press.

Munster, A. 1969. *Statistical Thermodynamics.* Berlin: Springer-Verlag.(**)

Nagel, E. 1961. *The Structure of Science.* New York: Harcourt, Brace and World.

Nakajima, S. 1958. "On Quantum Theory of Transport Phenomena." *Progress of Theoretical Physics* 20: 948–959.(**)

Newton-Smith, W. 1980. *The Structure of Time.* London: Routledge and Kegan-Paul.

Ornstein, D. 1974. *Ergodic Theory, Randomness and Dynamical Systems.* New Haven: Yale University Press.(**)

Oxtoby, J. 1971. *Measure and Category.* New York: Springer-Verlag.(**)

Pauli, W. 1928. In *Problemen der Modern Physik: Festschrift zum 60 Geburtstage A. Summerfelds.* Leipzig: Huzel, p. 30.

Penrose, O. 1970. *Foundations of Statistical Mechanics.* Oxford: Pergamon Press.(*)

Penrose, O. 1979. "Foundations of Statistical Mechanics." *Reports on the Progress of Physics* 42: 1937–2006.(*)

Penrose, R. 1979. "Singularities and Time Aymmetry." In S. Hawking and W. Israel, eds., *General Relativity: An Einstein Centenary Survey.* Cambridge: Cambridge University Press.

Petrina, D. Gerasimenko, V., and Malyshev, P. 1989. *Mathematical Foundations of Classical Statistical Mechanics.* New York: Gordon and Breach.(**)

Pippard, A. 1961. *The Elements of Classical Thermodynamics.* Cambridge: Cambridge University Press.(*)

Plancherel, M. 1913. "Beweis der Unmöglichkeit Ergödischer Mechanischer Systeme." *Annalen der Physik* 42: 1061–1163.(*)

Prigogine, I. 1970. "Dynamical Foundations of Thermodynamics and Statistical Mechanics." In E. Stuart, B. Gal-Or, and A. Brainard, eds., *A Critical Review of Thermodynamics.* Baltimore: Mono Books.(**)

Prigogine, I. 1973. "A Unified Foundation of Dynamics and Thermodynamics." *Chemica Scripta* 4: 5–32.(**)

Prigogine, I. 1974, "Physics and Metaphysics." *Bulletin de l'Academie Royal de Belgique.* Special Issue for the Proceedings of the Colloquium "Connaissance Scientifique et Philosophie."

Prigogine, I. 1980. *From Being to Becoming.* San Francisco: Freeman.

Prigogine, I. 1984. *Order Out of Chaos.* New York: Bantam Books.

Prigogine, I. and Grecos, A. 1979. "Topics in Nonequilibrium Statistical Mechanics." In *Problems in the Foundations of Physics: Varenna International School of Physics "Enrico Fermi."*(**)

Prior, A. 1949. "Determinables, Determinates and Determinants." *Mind* 58: 1–20, 178–194.

Quay, P. 1978. "A Philosophical Explanation of the Explanatory Function of Ergodic Theory." *Philosophy of Science* 45: 47–59.

Railton, P. 1978. "A Deductive-Nomological Model of Probabilistic Explanation." *Philosophy of Science* 45: 206–226.

Ramsey, N. 1970. "Spin Temperatures and Negative Absolute Temperature." In E. Stuart, B. Gal-Or, and A. Brainard, eds., *A Critical Review of Thermodynamics*. Baltimore: Mono Books.

Reichenbach, H. 1956. *The Direction of Time*. Berkeley: University of California Press.

Rhim, W., Pines, A. and Waugh, J. 1971. "Time-Reversal Experiments in Dipolar Coupled Spin Systems." *Physical Review B* 3: 684–696.(*)

Rosenkrantz, R. 1981. *Foundations and Applications of Inductive Probability*. Atascadero, CA: Ridgeview.

Rosenthal, A. 1913. "Beweis der Unmöglichkeit Ergodischer Gasssystemse." *Annalen der Physik* 42: 796–806.(*)

Ruelle, D. 1969. *Statistical Mechanics*. New York: Benjamin.(**)

Salmon, W. 1984. *Scientific Explanation and the Causal Structure of the World*. Princeton: Princeton University Press.

Schaffner, K. 1967. "Antireductionism and Molecular Biology." *Science* 157: 644–647.

Schaffner, K. 1969. "Theories and Explanations in Biology." *Journal of the History of Biology* 2: 19–33.

Schroeder, M. 1991. *Fractals, Chaos, Power Laws*. New York: Freeman.

Sciama, D. 1971. *Modern Cosmology*. Cambridge: Cambridge University Press.(*)

Shannon, C. 1949. *Mathematical Theory of Communication*. Urbana, IL: University of Illinois Press.(*)

Shimony, A. 1970. "Scientific Inference." In R. Colodny, ed., *The Nature of Scientific Theories*. Pittsburgh: University of Pittsburgh Press.

Shimony, A. 1985. "The Status of the Principle of Maximum Entropy." *Synthese* 63: 35–53.

Sinai, Ya. 1976. *Introduction to Ergodic Theory*. Princeton: Princeton University Press.(**)

Sklar, L. 1967. "Types of Inter-Theoretical Reduction." *British Journal for the Philosophy of Science* 18: 109–124.

Sklar, L. 1973. "Statistical Explanation and Ergodic Theory." *Philosophy of Science* 40: 194–212.

Sklar, L. 1974. *Space, Time, and Spacetime*. Berkeley: University of California Press.

Sklar, L. 1985. *Philosophy and Spacetime Physics*. Berkeley: University of California Press.

Sklar, L. 1987. "The Elusive Object of Desire: In Pursuit of the Kinetic Equation and the Second Law." In A. Fine and P. Machamer, eds., *PSA–1986*, Volume 2. East Lansing, MI: Philosophy of Science Association.

Tisza, L. 1966. *Generalized Thermodynamics*. Cambridge, MA: MIT Press.(*)

Toda, M., Kubo, R., and Saito, N. 1978. *Statistical Physics – I*. Berlin: Springer-Verlag.(*)

Tolman, R. 1934. *Relativity, Thermodynamics and Cosmology*. Oxford: Oxford University Press.(*)

Tolman, R. 1938. *The Principles of Statistical Mechanics*. Oxford: Oxford University Press.(*)

Truesdell, C. 1961. "Ergodic Theory in Classical Statistical Mechanics." In P. Caldirola, ed., *Ergodic Theories: International School of Theoretical Physics "Enrico Fermi,"* number 14.(*)

Truesdell, C. 1977. *The Concepts and Logic of Thermodynamics.* New York: Springer-Verlag.(*)

Truesdell, C. 1984. Rational Thermodynamics. New York: Springer-Verlag.(*)

van Fraassen, B. 1982. "Rational Belief and the Common Cause Principle." In R. MacLaughlin, ed., *What? Where? When? Why?.* Dordrecht: Reidel.

van Kampen, N. 1962. "Fundamental Problems in the Statistical Mechanics of Irreversible Processes." In E. Cohen, ed., *Fundamental Problems in Statistical Mechanics.* Amsterdam: North-Holland.(*)

Watanabe, S. 1969. *Knowing and Guessing.* New York: Wiley.

Wightman, A. 1985. "Regular and Chaotic Motion in Dynamical Systems." In G. Velo and A. Wightman, eds., *Regular and Chaotic Motions in Dynamical Systems.* New York: Plenum Press.

Yablo, S. 1992. "Mental Causation." *Philosophical Review* 101: 248–280.

Zubarev, D. 1974. *Nonequilibrium Statistical Thermodynamics.* New York: Plenum Press.(*)

Zwanzig, R. 1960. "Statistical Mechanics of Irreversible Processes." In *Lectures in Theoretical Physics* 3: 106–141.(**)

Zwanzig, S. 1964. "On the Identity of Three Generalized Master Equations." *Physica* 30: 1109–1123.

Index